Sensors and Microsystems

Piero Malcovati • Andrea Baschirotto
Arnaldo D'Amico • Corrado di Natale

Editors

Sensors and Microsystems

AISEM 2009 Proceedings

Springer

Editors
Piero Malcovati
Dipartimento di Ingegneria Elettrica
Università di Pavia
Via Ferrata, 1
27100 Pavia
Italy
piero.malcovati@unipv.it

Andrea Baschirotto
Dipartimento di Fisica
Università Milano-Bicocca
Piazza della Scienza, 3
20126 Milano
Italy
andrea.baschirotto@unimib.it

Arnaldo D'Amico
Dipartimento di Elettronica
Università di Roma, Tor Vergata
Via del Politecnico, 1
00133 Roma
Italy
damico@eln.uniroma2.it

Corrado di Natale
Dipartimento di Elettronica
Università di Roma, Tor Vergata
Via del Politecnico, 1
00133 Roma
Italy
dinatale@eln.uniroma2.it

ISBN 978-94-007-3136-3 e-ISBN 978-90-481-3606-3
DOI 10.1007/978-90-481-3606-3
Springer Dordrecht Heidelberg London New York

Printed on acid-free paper

Springer is part of Springer Science+Business Media (www.springer.com)

FOREWORD

This book is the collection of most of the papers presented at the 14th Italian Conference on Sensors and Microsystems, promoted by the Italian Association on Sensors and Microsystems (AISEM). This Conference edition, organized by the University of Pavia, was held from 24th to 26th February 2009 in the historical Aula Foscolo of the University of Pavia. The book includes also three tutorial papers, which address basic concepts in the area of sensors and microsystems.

The 2009 AISEM Conference edition was opened by the Rector of the University of Pavia, Prof. Angiolino Stella, who is well known for his research activity in the sensor and microsystems community. Moreover, this edition was characterized by four important invited talks from the industrial world, which demonstrated the successful exploitation of the research in the field of sensors and microsystems.

Benedetto Vigna (STMicroelectronics) illustrated the strategy, both at technical and management level, that allowed STMicroelectronics to become in 2009 the world leader in the MEMS market. Christoph Hagleitner (IBM Research Zurich) described the recent development of the "Millipede Project", that uses a nanotechnology approach to increase the performance of data storage systems. Marco Sabatini (Pirelli) explained the application of several microsensor techniques in the realization of a Cyber-Tyre under development at Pirelli. Finally, Giorgio Fagnani (Gefran Sensori) spoke about the recent developments on sensors for pressure measurements at high temperature.

More than 100 researchers participated to the event ,organized with both oral and poster sessions. At the end of the Conference, two awards were delivered for the best paper and the best posters.

The Best Paper Award entitled to Angelo Rizzo, aimed to support young researchers, sponsored by CNR-IMM and delivered by Dr. Pietro Siciliano, was assigned to Marco Grassi for the paper "A Multisensor System for High Reliability People Fall Detection in Home Environment".

The Best Poster Award, sponsored by the Conference Organization, was assigned ex-aequo to the papers "Actively Controlled Power Conversion Techniques for Piezoelectric Energy Harvesting Applications" by A. Romani, C.

Tamburini, R. P. Paganelli, A. Golfarelli, R. Codeluppi, M. Dini, E. Sangiorgi, M. Tartagni and "A Novel Based Protein Microarray for the Simultaneous Analysis of Activated Caspases" by I. Lamberti, L. Mosiello, C. Cenciarelli, A. Antoccia, C. Tanzarella.

Special thanks are given to Dr. Alessandro Cabrini of the University of Pavia for his effort and dedication in the organization of the Conference.

Andrea Baschirotto
Università di Milano-Bicocca

Piero Malcovati
Università di Pavia

Arnaldo D'Amico, Corrado Di Natale
Università di Roma "Tor Vergata"

AISEM COMMITTEES

Conference Chairman

Prof. Andrea Baschirotto
Università di Milano-Bicocca

Technical Program Chairman

Prof. Piero Malcovati
Università di Pavia

Local Organization Chairman

Dr. Alessandro Cabrini
Università di Pavia

AISEM Directive Committee

A. D'Amico
Università di Roma
"Tor Vergata"
Presidente AISEM

L. Campanella
Università di Roma
"La Sapienza"

P. Siciliano
CNR-IMM Lecce

C. Mari
Università di Milano

G. Martinelli
Università di Ferrara

U. Mastromatteo
STMicroelectronics

A. G. Mignani
CNR-IFAC Firenze

M. Prudenziati
Università di Modena

G. Sberveglieri
Università di Brescia

G. Soncini
Università di Trento

AISEM Scientific Committee

M. C. Carotta Università di Ferrara	P. Dario Scuola Superiore S. Anna Pisa	F. Davide Telecom Italia Roma
A. Diligenti Università di Pisa	C. Di Natale Università di Roma "Tor Vergata"	L. Dori CNR-IMM-LAMEL Bologna
G. Faglia Università di Brescia	C. Malvicino CR Fiat Orbassano (To)	G. Martinelli Università di Ferrara
M. Mascini Università di Firenze	N. Minnaja Polo Navacchio Cascina (PI)	B. Morten Università di Modena
G. Palleschi Università di Roma "Tor Vergata"	F. Villa STMicroelectronics	M. Zen FBK Trento

TABLE OF CONTENTS

TUTORIALS

WITH THE EYE OF THE BEHOLDER: AN INTRODUCTION TO THE OBSERVATION OF MULTIDIMENSIONAL DATA WITH THE PRINCIPAL COMPONENT ANALYSIS

C. DI NATALE

Deparment of Electronic Engineering, University of Rome "Tor Vergata". Via del Politecnico 1,00133 Roma, Italy, e-mail: dinatale@uniroma2.it

Abstract. Multivariate data analysis is the necessary tool to study complex phenomena and to analyze data of complex analytical techniques such as chromatography and spectrophotometer. One of the most useful approaches in science to experimental data interpretation is the visualization of data. This fundamental operation cannot be simply performed with multivariate data. In this paper, an introduction to principal component analysis is offered as one of the method that can provide a meaningful representation of data in a projection plane. The choice of the projection plane corresponds to the determination of an optimal point of observation where multidimensional data can display most of their meaning.

1. Data, patterns, matrices, and vector spaces

The graphical representation of experimental data is one of the most important and immediate approaches for the interpretation of scientific experiments. The process of description of data in a geometrical space is guided by some fundamental assumptions. For instance the connection of experimental data with points in a Euclidean space is the basic operation for any data representation. It is out of the scope of this paper to discuss the consequences of the introduction of non-Euclidean spaces, but it would be interesting to reflect that in the last century several demonstrations about the physical meaning of non-Euclidean spaces were provided.

Data are elementary entities that describe partial aspects of a phenomenon. When physical quantities are involved, data are the results of the application of predetermined measurement instruments to the phenomenon under study. Measurement instruments and their measured quantities are predefined entities that are used, with the same meaning, in very different context. Let us consider, for instance, the measure of length; this can be applied to measure the height of an individual or the distance traveled by a vehicle. The resulting data will be formally the same but their meaning is, of course, extremely different. In some case such a data will not be adequate to describe a phenomena, as it will be discussed later.

P. Malcovati et al. (eds.), *Sensors and Microsystems: AISEM 2009 Proceedings*, Lecture Notes in Electrical Engineering 54, DOI 10.1007/978-90-481-3606-3_1,

A great difference about data is related to their dimension. Simple data are scalar quantities complemented by a unit of measure, such as the data provided by a ruler. The treatment of scalar (or univariate) data is well known and it is a background provided by any scientific or technical training.

In this paper the description of multivariate data will be considered. Multivariate data are provided either by multidimensional instruments or collecting together a number of univariate data in order to describe complex phenomena.

Multidimensional instruments are those measurement systems that, as a result of a measure, provide a sequence of ordered numerical quantities. Examples of this are gas-chromatographers, spectrophotometers, or sensor arrays and in particular those particular classes of arrays of partially selective sensors known as electronic noses. On the other hand, the description of complex phenomena requires the use of many individual data. As an example, to define the meteorological conditions is necessary to get at the same time data about the atmospheric pressure, the magnitude and direction of wind, the temperature, the relative humidity and so on.

Multivariate data are naturally composed in ordered sequences that are called patterns. The mathematical entity suitable to represent collections of patterns is the matrix. Linear algebra is then utilized to manipulate the patterns [1]. In this context the known relationship between matrices and vector spaces is also utilized to describe pattern. In practice, each pattern is associated to a vector in a suitable Euclidean vector where each basis vector is one of the dimensions of the multivariate data.

It is important to consider that observations take place at dimension 2, and then when the dimension is larger than 2, in order to observe the data is necessary to consider the point of view of the observation. This is a common knowledge that the perception of a three-dimensional object may change according to the point of view. As well known to any person taking a photograph there is a set of optimal point of views where the interesting aspects of an object can be projected onto the camera focal plane.

As an example, let us consider for instance a set of peaches of different cultivars for which the following parameters have been measured: total antocyanins content, brix degree, chlorophyll content, details about this experiment are found in [2]. In Fig. 1 the data, plotted in a three dimensional Euclidean space, are observed from two different points of view. It is straightforward that different points of view allows for different appreciation of the characteristics of the studied samples. For instance data points cluster in three groups in one case and in two groups in the other.

In the following of this paper the observation of patterns of dimensions larger than 3 will be discussed. Even if the original space goes beyond a sensible appraisal, the method is the same described for the display of data of dimension 3 and namely the projection from a point of observation onto a plane.

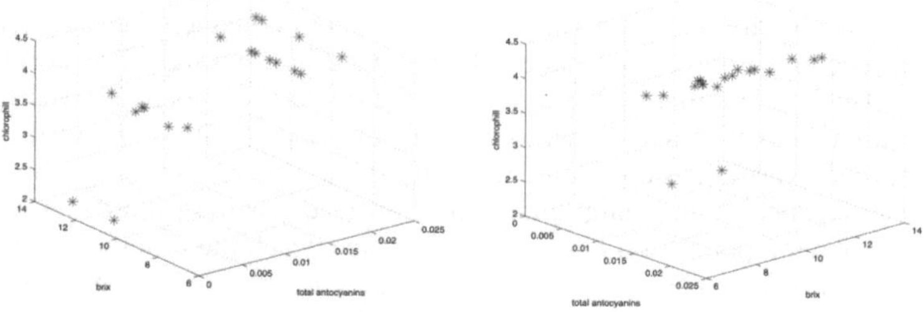

Figure 1. Observation of the same data set from different points of view reveals or conceal the relationship between the data.

2. Data correlation

In the previous section it has been mentioned that the kind of data that can be acquired from a given phenomenon is predefined and it is limited to the set of quantities that are measurable. This means that for a specific phenomenon a sort of inadequacy of the measurable data can emerge. The inadequacy of data is manifested by the fact that different quantities tend to describe the same aspect of a phenomenon. In practice, it appears that a phenomenon is described by a set of internal processes that affects, at the same time, quantities that are physically different among them. To elucidate this point let us consider the following example. Let us suppose to consider a population of individuals and to measure for each of them the height and the weight. From a physical point of view length and mass are distinct quantities, and actually they are part of the three basic quantities from which all the other mechanical units of measure are formed. If the collected data are considered to form patterns and if these patterns are plotted in a bidimensional plane it is likely to observe that the points are almost aligned along a line. This means that the two data are in some way related one each other, in particular if an individual has a large weight it probably has a large height too. The connection between height and weight is absolutely not due to the nature of the measured quantity but rather to the fact that humans are characterized by law according to which the body mass tend to be vertically distributed. Deviations from this general trend indicate the peculiar properties of each individual that tend to derogate from the common law.

Data connected each other by an internal mutual dependence are said correlated. The degree of correlation is measured, for instance, by the linear correlation coefficient. Correlation can be either positive or negative describing if at a positive growth of one variable corresponds an increment or a decrement of the other. The correlation coefficient takes value from -1 to $+1$ indicating with ± 1 positive and negative correlation and with 0 the absence of correlation.

Given a matrix of patterns, the calculus of the mutual correlation coefficients Provides evidences about the existence of internal relationships connecting variables among them. It is worth to mention that these kinds of conclusions derive exclusively from an observation of data and they have to be corroborated and explained by the science pertaining to the kind of considered phenomenon. The mutual correlation coefficients can be arranged in a matrix called correlation matrix. Of course the matrix is symmetric (the correlations between variables 1 and 2, and variables 2 and 1, are obviously the same) with 1 along the main diagonal (each variable is correlated to itself).

Let us consider as an example an extension of the pattern matrix previously introduced about a population of peaches. For each peach the following quantities have been measured: acidity, antocyanins, brix, carotene, and chlorophyll. The correlation matrix of these data is shown in Table 1.

Table 1. Correlation matrix of a number of variables measured in a population of peaches.

	Acid	Antoc.	Brix	Carot.	Chloroph.
Acid	1.00	−0.48	0.34	0.15	−0.32
Antoc.	−0.48	1.00	−0.70	−0.56	0.88
Brix	0.34	−0.70	1.00	0.30	−0.76
Carot.	0.15	−0.56	0.30	1.00	−0.25
Chloroph.	−0.32	0.88	−0.76	−0.25	1.00

The correlation coefficients suggest for instance that since the couple antocyanins and chlorophyll are largely correlated, the color of peaches is a blend of red and green. Another suggestion comes from the negative correlation of the couple chlorophyll and brix degree indicating that green peaches are not sweet.

In terms of data appearance in the Euclidean space, the existence of correlation indicates that the data point tend to be aligned along a line. Then two correlated quantities instead of uniformly filling the space tend to occupy a space volume that is more than unidimensional but not completely bidimensional. This characteristic influences greatly the determination of an optimal point of view where multidimensional data can be observed preserving the relationship between the data.

3. Principal component analysis

Principal Component Analysis (PCA) is a method to decompose a set of multivariate patterns into non-correlated variables [3]. In practice, PCA defines for a given set of multivariate data a number of novel variables that are not directly measurable but are defined as a linear combination of measurable variables. The main property of these novel variables (called principal components) is their non-correlation. They can be interpreted as a sort of virtual data each describing non-correlated properties of the phenomenon under study. The contribution to each principal component of the original measured quantities helps in understanding the processes occurring in the phenomenon as it will be clear in the example that will be discussed later.

The calculus of PCA is based on the statistical properties of the whole set of patterns. For this reason it is necessary, before to discuss PCA, to introduce the main aspects related to multivariate Probability Distribution Function (PDF) and in particular the multivariate Gaussian PDF [4].

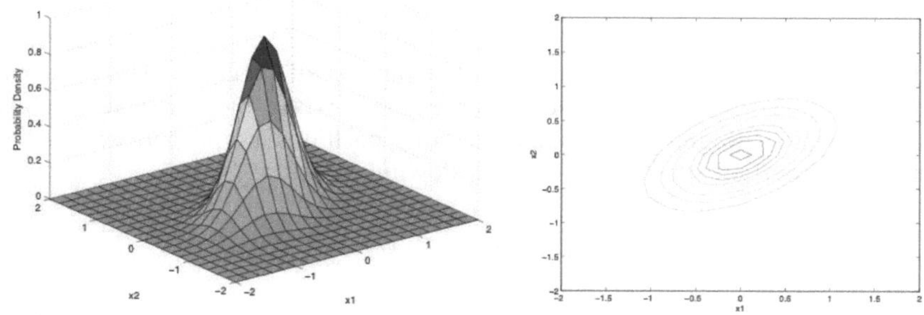

Figure 2. Example of a bivariate Gaussian PDF with mean = [0.0] and covariance matrix defined by $\sigma^2_{x1} = 0.25$, $\sigma^2_{x2} = 0.15$, $\rho\sigma_{x1}\sigma_{x2}$. Right figure shows the iso-probability curves in the variable space.

3.1. *Multivariate Gaussian PDF*

The Gaussian PDF, also known as normal PDF, of a univariate quantity x is given by the following equation:

$$f(x) = \frac{1}{\sqrt{2\pi\sigma}} \cdot \exp\left[-\frac{(x-m)^2}{2\sigma^2}\right]$$

where m and σ are the mean and the variance of the variable x. The importance of Gaussian distribution in experimental science derives from the so-called central theorem limits that claims that a measurable quantity that is due to the simultaneous occurrence of elementary phenomena differently distributed tend to be normally distributed as the number of elementary phenomena tend to infinite. Since practically all the measurable quantities are macroscopic manifestation of a large number of microscopic phenomena (e.g. the electric current) all the experimental data can be considered distributed according to the Gaussian PDF.

In order to generalize the univariate PDF to the multivariate case is necessary to define the statistical descriptors (mean and variance) that characterize the Normal function. Given a matrix of patterns, the multivariate mean is simply a vector composed by the mean of each variable. The generalization is more complex in case of variance. It is worth to recall that the variance is the expected value of the square deviation from the mean:

$$\sigma^2 = E\left[(x-m)^2\right]$$

When x is a vector the previous definition becomes:

$$\Sigma = \sigma_{ij}^2 = E\left[(x_i - m)^T \cdot (x_j - m)\right]$$

Where the indexes *i* and *j* go from 1 to the number of variables. The previous definition gives rise to a symmetric square matrix where the diagonal contains the variances of single variables and the other positions are proportional to the correlation coefficient according to the following equation:

$$\Sigma_{ij} = \rho_{ij} \cdot \sigma_i \cdot \sigma_j$$

Figure 2 shows a bivariate Gaussian PDF with evidenced the iso-probability surfaces (in this case curves). These curves are defined by the quadratic form built with the covariance matrix.

$$\begin{bmatrix} x - \bar{x} & y - \bar{y} \end{bmatrix} \cdot \begin{bmatrix} \sigma_x^2 & \rho_{xy}\sigma_x\sigma_y \\ -\rho_{xy}\sigma_x\sigma_y & \sigma_y^2 \end{bmatrix} \cdot \begin{bmatrix} x - \bar{x} \\ y - \bar{y} \end{bmatrix} = k$$

Since the covariance matrix is symmetric, according to the properties of quadratic forms, the iso-probability curves of bivariate Gaussian PDF are ellipse, in case of multivariate Gaussian PDF the iso-probability curves are ellipsoids.

3.2. *Covariance matrix and principal components*

The orientation of the curves associated to the covariance matrix provides immediate information about the correlations among the variables. Figure 3 shows two important configurations related to correlated and non-correlated variables. For sake of simplicity let us consider all the variables reduced to zero mean. The removal of mean leaves the covariance matrix as the unique meaningful descriptor of the set patterns. It is worth to remind that this is a peculiarity of Gaussian function.

It is clear that, in terms of iso-probability ellipsoids, a covariance matrix describing correlated variables is not written in canonical forms, and on the contrary when the iso-probability ellipsoid appears in canonical form it describes non-correlated variables. Let us remind that an ellipsoid is in canonical form when the principal axes of the ellipsoid coincide with the basis vectors.

The calculus of the principal components can then be reduced to the calculus of a novel basis in the patterns space where the principal axes of the iso-probability ellipsoid are the basis vectors.

It is known that the principal axis of an ellipsoid are the eigenvectors of the matrix associated to the curve, then the eigenvectors of the covariance matrix of the patterns are the principal components of the matrix of patterns.

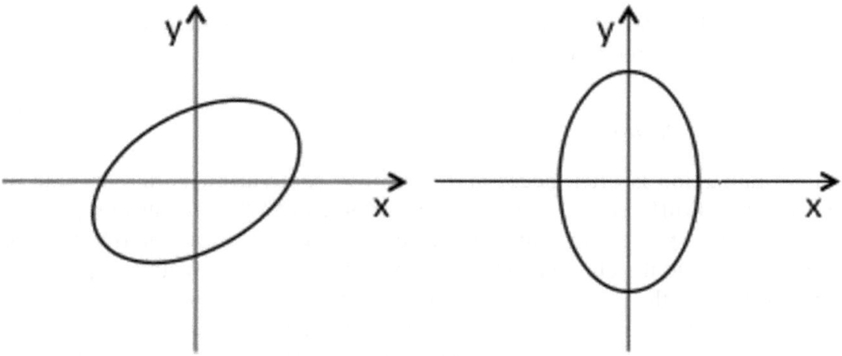

Figure 3. Examples of iso-probability ellipses occurring in case of correlated (*left*) and non-correlated (*right*) variables.

Besides the eigenvectors, the associated eigenvalues defines the importance of each principal component. Eigenvalues provide a measure of the elongation of the ellipse along the associated eigenvectors and it is a measure of the variance explained by each principal component. Indeed, since the principal components are non-correlated the total variance of the data set is given by the sum of the variances along each principal component. Therefore, the eigenvalues measure the "importance" of each associated principal components.

In practice the calculus of PCA can be reassumed in the following procedure.

Let us consider a patterns matrix **X** where each row of the matrix is a pattern describing a different sample and each column is one measurable variable. The corresponding covariance matrix is $\mathbf{Cov(X)} = \mathbf{X}^{\mathbf{T}}\mathbf{X}$. The principal components of the matrix **X** are the eigenvectors of the corresponding covariance matrix. The coefficients defining the eigenvectors in the original variables basis are called loadings. The loadings form a matrix called **P**. The coordinates of the patterns in the principal components basis are called scores, these form a matrix called **T**. The relationship between scores (**T**), loadings (**P**) and the original matrix (**X**) is the following: $\mathbf{T} = \mathbf{XP}$.

Since the eigenvalues associated to the eigenvectors provide a measure of the amount of the total variance explained along the eigenvector, the representation can be limited only to the most meaningful principal components allowing for a reduction of the dimensions and the representation of the patterns in space of reduced dimensionality. Once this is done with two principal components, they identify a projection plane where data, even coming from a high dimension space are plotted. Larger is the variance in the representation space smaller is the approximation error.

In practice, PCA defines a hierarchical list of points of view from which to observe, through a linear projection, the multidimensional patterns. These views are optimized in order to preserve, as much as possible, the variance of the data. This condition, in geometrical terms means to provide images where the data points span the largest extension of the space.

Before to illustrate through some examples the application of PCA it is important to discuss the role of data normalization.

3.3. *Data normalization*

Data normalization is often necessary to remove some artifacts that are due to mere numerical differences between the variables. Indeed, if for instance, some of the variables span a larger numerical range the whole representation will be limited to them and the other variables, that may be of great qualitative importance do not influence the data representation. For this scope, two important data normalization are commonly adopted: zero mean and autoscaling. Zero-mean has been introduced in the previous section, and it simply consists in shifting the range of variability of each variable in order to have a null mean for all the variables. After zero mean, covariance is assumed to be only meaningful data descriptor. Autoscaling is a further normalization step where not only the mean is null, but the variance of each variable is made equal to 1. In this way each variable play the same role in defining the patter disregarding its numerical level and the spanned range. It has to be remarked that while zero mean can be applied to any kind of data, autoscaling is meaningful when the pattern is composed of data coming from different measurement instruments, on the other hand auto-scaling applied, for instance, to spectra destroy the peculiar spectral characteristics removing the presence of peaks and then the meaning of individual spectra.

4. An example of PCA

In order to illustrate the application of PCA let us consider a number of peaches each of a different cultivar. For each peach the following quantities were measured: pH, sucrose, glucose, fructose, malic acid, and citric acid. Experimental details about the data set are found in [5]. The scope of the experiment was to study the relationship between acids and sugars in peaches. Figure 4 shows, in a color

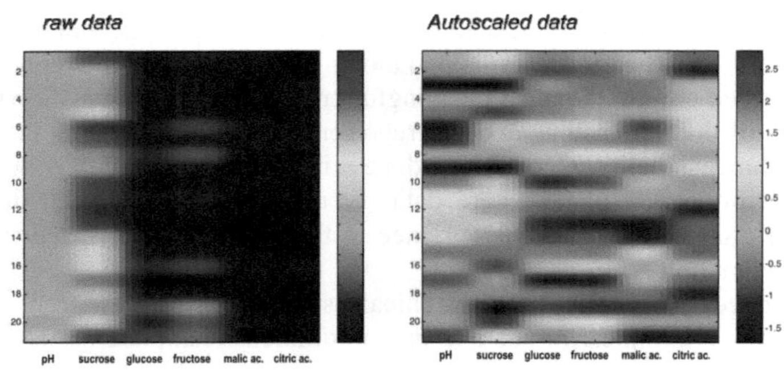

Figure 4. Peaches data matrices displayed in a color code. Original data (*left*) and autoscaled data (*right*).

code, the data matrix before and after the application of autoscaling. Original data matrix is dominated by pH and sucrose whose numerical values exceed that of the other variables. Autoscaled data matrix on the contrary displays an homogenous distribution of the variables.

Table 2. Eigenvalues and variance contribution for each principal component.

Principal component	Eigenvalue	Percentage of variance per PC	Percentage of cumulative variance
1	2.2872	38.1195	38.1195
2	2.0393	33.9881	72.1076
3	1.1284	18.8064	90.9139
4	0.4475	7.4588	98.3727
5	0.0758	1.2632	99.6359
6	0.0218	0.3641	100.0000

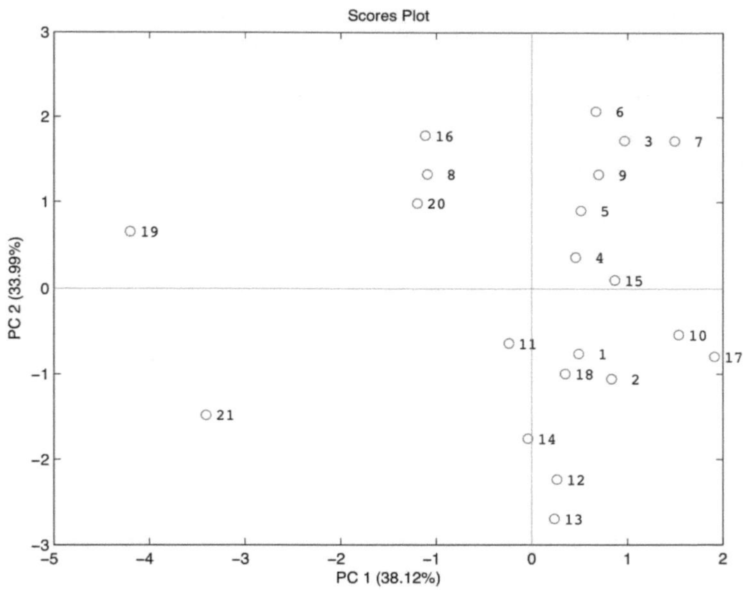

Figure 5. Score plot of the first two principal components.

The PCA is then calculated on the autoscaled data matrix. The data are formed by six variables, and then they are represented in a vectorial space of dimension 6. As it was discussed before, due to the data occupy a sub-set of the six-dimension space. Important information about the real dimension of the data set is acquired plotting the eigenvalues of each principal component. In table the eigenvalue, the explained variance and the cumulative variance for each principal component are listed. After the third principal component more than 90% of the total variance is explained, and the total variance is recovered with six principal components. It suggests that only a moderate correlation among the variables takes place.

Figure 5 shows the data points projected in the plane formed by the first two principal components. A plot of this kind is called scores plot. According to Table 2 the plot takes into account only 72% of the total variance of the data set. Each point of the plot corresponds to a peach whose original pattern can be seen on the matrices of Fig. 4.

The exiguity of the variance explained in Fig. 5 gives the opportunity to mention that PCA is a matrix decomposition according to the following equation:

$$X_{nm} = S_{nq} \cdot P_{qn}^{T}$$

Where n is the number of samples (peaches in this case), m is the number of variables (six in this case), and q is the number of considered principal components. In case $q = m$ the matrix **X** is completely recovered, but if $q > m$, such as the plot on Fig. 5 where $q = 2$, there is a residual difference between the original matrix and the scores plot representation. In Fig. 6 a third axis is added to Fig. 5 and the residual is plotted for each data point

The residuals are of the same entity for all data except data number 16. In Fig. 5 this peach formed a group with peaches 8 and 20. Figure 6 suggests that this similarity is only apparent, and that really sample 16 is different from those close to him in Fig. 5.

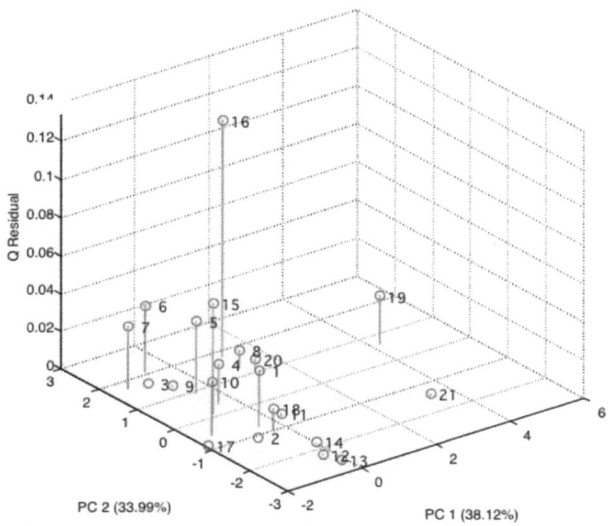

Figure 6. A third axes carrying the residual, is added to the plot in Fig. 5 to illustrate that each point is affected by a different PCA approximation error.

Furthermore, important information that can be gathered by PCA is given by the loadings, namely the coefficients of the principal components. The loadings provide information about the role played by original variables.

In Table 3 the loadings of the first two principal components are shown. In Fig. 5 peaches are mostly distributed along the second principal components in the positive and negative semi plane.

In Table 2 it is observed that the positive part of PC2 is mostly due to the acids, it has to be observed that pH being defined as the logarithm of the concentration of hydrogen ions decreases as the acids concentration increases.

Along the first principal component few peculiar peaches are found in particular in the negative semi plane that is mainly defined by glucose and fructose concentrations.

Table 3. Loadings of the first two principal components.

	pH	Sucrose	Glucose	Fructose	Malic acid	Citric acid
PC1	−0.4875	0.1244	−0.5527	−0.5816	0.2931	0.1315
PC2	−0.4262	−0.3659	0.3346	0.2922	0.4008	0.5713

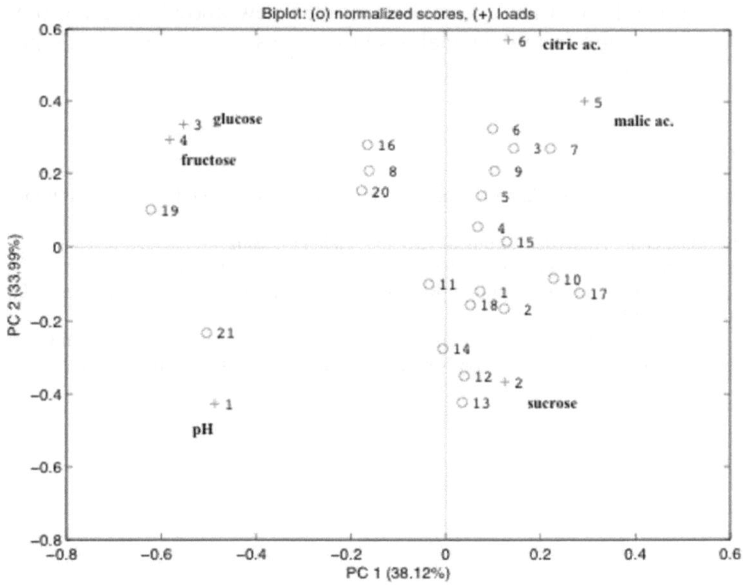

Figure 7. Biplot of the first two principal components.

A synthetic way to appraise scores and loadings, and then to study the relationship between variables and samples is obtained plotting contemporaneously scores and loadings in a plot called biplot. Loadings in this case are simply the projection of the original axis in the principal components plane. Figure 7 shows the biplot of the peaches data the same conclusions that were obtained analyzing the table of loadings and the position of peaches data can be easily achieved looking at the biplot.

5. Caveat and conclusions

PCA is a powerful algorithm allowing for the definition of an optimal point of observation of a set of multivariate data. However, it is necessary to keep always in mind that the representation is automatically obtained maximizing the variance represented. Then it is important for a correct interpretation of the scores to evaluate the residual to appraise if some of the points are misrepresented. Nonetheless, even if residuals are moderate the score plot can still conceal important relationships among the data.

On this basis, PCA provides a simple and efficient instrument to investigate the relationship among data, and to get conclusions in phenomena even without knowledge of the processes occurring in the samples under studies. For instance, in this paper some example of alleged relationships between acids and sugars in peaches have been given without considering fruit physiology issues. Finally, it has to be remarked that statistics does not do science but it only offers either confirmation of a known law or hints to develop new knowledge, and then any conclusion we can get from data analysis has always to be scientifically justified on theoretical basis.

References

1. G. Golub, C. Van Loan, Matrix Computations, J. Hopkins University press, Baltimore, MD, 1996
2. C. Di Natale et al., Analytica Chimica Acta 459, 107–117, 2002
3. T. Jolliffe, Principal Component Analysis, Springer-Verlag, New York, 1986
4. R. Johnson, D. Wichern, Applied Multivariate Statistical Analysis, Pearson Education, Prentice-Hall, 2002
5. M. Esti et al., in Artificial and Natural Perception: Proceedings of the 2nd Italian Conference on Sensors and Microsystems (C. Di Natale, A. D'Amico, F. Davide, editors), World Scientific Publ. 1998

BIOSENSOR TECHNOLOGY: A BRIEF HISTORY

I. PALCHETTI AND M. MASCINI
Dipartimento di Chimica, Università degli Studi di Firenze, Via della Lastruccia 3, 50019 Sesto Fiorentino, Italy

Abstract. Biosensors use a combination of biological receptor compounds (antibody, enzyme, nucleic acid, etc.) and the physical or physic-chemical transducer directing, in most cases, "real-time" observation of a specific biological event (e.g. antibody-antigen interaction). They allow the detection of a broad spectrum of analytes in complex sample matrices, and have shown great promise in areas such as clinical diagnostics, food analysis, bio-process and environmental monitoring. Biosensors may be divided into six basic groups, depending on the method of signal transduction: optical, mass, electrochemical, magnetic, micromechanical and thermal sensors. This paper aims to give an brief tutorial of bio-sensor technology.

1. Brief history of biosensors

The vast literature in the last 50 years related to the keyword *Biosensor* reveals without doubt that the scientific field is *attractive.* We realized at once that several researchers with different background are involved in this field of research, from chemistry to physics, to microbiologists and of course to electrical engineering, all are deeply involved in several facets of the assembly of the object "Biosensor".

Looking at the past we realize also that the concept of Biosensor has evolved.

For some authors, especially at the beginning of this research activity, i.e. about 50 years ago, Biosensor is a self contained analytical device that responds to the concentration of chemical species in *biological samples.* This is clearly wrong, but it has been very difficult to clarify this point. No mention of a biological active material involved in the device. Thus any physical (thermometer) or chemical sensor (microelectrode implanted in animal tissue) operating in biological samples could be considered a *Biosensor*. We agree that a biosensor can be defined as a device that couples a biological sensing material (we can call it a molecular biological recognition element) associated with a transducer.

Recently the concept evolved again in the tentative to replace or mimic the biological material with synthetic chemical compounds.

In 1956 Professor Leland C. Clark publishes his paper on the development of an oxygen probe and based on this research activity he expanded the range of analytes that could be measured in 1962 in a Conference at a Symposium in the New York Academy of Sciences where he described how "to make electrochemical

P. Malcovati et al. (eds.), *Sensors and Microsystems: AISEM 2009 Proceedings*,
Lecture Notes in Electrical Engineering 54, DOI 10.1007/978-90-481-3606-3_2,

sensors (pH, polarographic, potentiometric or conductometric) more intelligent" by adding "enzyme transducers as membrane enclosed sandwiches" [1]. The first example was illustrated by entrapping the enzyme Glucose Oxidase in a dialysis membrane over an oxygen probe. The addition of glucose determined the decrease of oxygen concentration in proportional relation. The first biosensor was described in the published paper coining the term "enzyme electrode" [2]. Then subsequently in 1967 Updike and Hicks use the same term "enzyme electrode" to describe a similar device where again the enzyme glucose oxidase was immobilized in a polyacrylamide gel onto a surface of an oxygen electrode for the rapid and quantitative determination of glucose [3].

Besides amperometry Guilbault and Montalvo in 1969 use glass electrodes coupled with urease to measure urea concentration by potentiometric measurement [4].

Starting from 1970, several others authors start to prove the concept of Biosensors, the coupling of an enzyme and electrochemical sensors. This was at the beginning a Biosensor, a strange research where biological elements were combined with electrochemical sensors.

In the electrochemical community at that period the research on Ion selective electrodes (ISE) was very active and the idea to extend the range of sensors to non electrochemical active compounds, and even to non ionic compounds, like glucose, has been very well accepted. We saw at that time the possibility to extend much more the research activity. The groups active in ISE development have been definitively the first to shift to the development of electroanalytical biosensors.

Professor G. Rechnitz developed of an "amygdaline" sensor based on the coupling of an Ion Selective Electrode (cyanide ISE) with betaglucosidase to give benzaldeyde and cyanide [5].

But this was just the beginning of a large activity where obtained couplings have been multiplied by changing the "biological element" and the kind of transducer. Enzyme, multiple enzyme, organelles, bacteria, specialized biological tissue, containing specific enzymes were coupled to potentiometric or amperometric devices, then optical, thermometric, piezoelectric, etc. We continue also today to enlarge the list of physical sensors with the last entry of "magnetic devices".

Enzymes (and all biological elements based on the enzymes contained in it) represent the class of what is now called "catalytic elements". The other important class is represented by the "affinity elements", namely antibodies, lectins, nucleic acids (DNA and RNA) and recently synthetic ligands.

Biomolecolar sensing can be then defined as the possibility to detect analytes of biological interest, like metabolites, but also of environmental concerns or of any other technological field where the concentration of a specific compound is important to be quantified in a complex sample.

The exploitation of the selectivity of the biological element is the "driving force" of the Biosensor.

2. The problem of amplification

Catalytic events or affinity events have not the same scheme of transduction. If the biological recognition element present in the sensing layer is an enzyme or generally a biocatalyst, a reaction takes place in the presence of the specific target analyte and an increasing amount of co-reactant or product is consumed or formed, respectively, in a short time depending on the turnover. In this scheme the amplification step is inherent and a large chemical amount can be obtained from the sensing layer.

In contrast the use of the antibodies for the detection of antigens has not an amplification stage involved and then the “affinity” reaction should be amplified in order to have a clear transduction. We have two possibilities, one is the use of a bioconjugate involving a bound enzyme, like in the classical ELISA test; the second is the inherent amplification given by the mass of the biological element involved, a piezoelectric device (sensitive to mass) can detect minute amount of large proteins (like antibodies) if they are attracted on the surface of the sensor.

With the same scheme surface plasmon resonance can be sensitive to minute amount of large molecule reacting at the surface of the electrode.

3. The biological system

The main problem of the biological system, catalytic or affinity, is the associated fragility and the operational activity. Most proteins have an optimal pH range in which their activity is maximal; this pH range should be compatible with transducer. Moreover the most of the biological systems have a very narrow range of temperature (15–40°C).

The most important problem and main drawback for industrial exploitation is the short lifetime associated with the biological elements. Many research efforts were specifically dedicated to this point. Lifetime or at least shelf lifetime of months or few years are the prerequisite for a suitable market and the fragility of the assembled systems has always limited the diffusion of biosensors in the market.

4. Immobilization of the biological system

The technique of the immobilization of the biological elements has changed according with the different events, catalytic or affinity. The simplest way to retain enzymes on the tip of a transducer is to trap them behind a perm-selective membrane. This method has been mainly used in addition to embedding procedures in polyacrilammide gels. Then, mainly in the 80th, the trend shifted to use disposable membranes with bound bioactive material. Several companies put on the market pre-activated membranes suitable for the immediate preparation of any bioactive membrane and this appeared as a real improvement at least for the easy use of enzyme sensors.

The removal of interference has been also the other important aspect for the wide use of biosensors for industrial processes. The two problems have been solved by using multilayer membranes, with the enzyme sandwiched between a special cellulose acetate membrane and a polycarbonate membrane. The main role of the membrane is to prevent proteins and other macromolecules from passing into the bioactive layer. Cellulose acetate membrane allows only molecule of the size of hydrogen peroxide to cross and contact the platinum anode, thus preventing interference from ascorbic acid or uric acid, for example, at the fixed potential. Such configuration has been used by several researchers in their biosensor assembly. But at the same time several recipes of immobilization of enzymes were published and several laboratories developed their own procedure for immobilizing the biological element, sometimes also patented.

One approach was also the development of disposable sensor, based on combination of screen printed electrochemical sensors with enzyme adsorbed on the electrode surface (in this case mainly carbon). The use of the sensor just for one measurement limited the use of complicated immobilization procedures to simplest as possible, like only based on adsorption on carbon surface. This electrode surface acted as a sponge, and the large protein was easily immobilized even if the bond was weak. This approach was useful only for a quick and rapid measurement.

The immobilization of antibodies soon revealed that random immobilization of proteins was not effective and a new research in this direction started. Several researchers start to think how to immobilize proteins using an exact deposition. Technology, like self-assembling, based on gold surface and thiol groups prove to have a high potential. Proteins immobilized on the surface of the transducers were now aligned and their ligands group were directed toward the exterior ready to fix the metabolite.

This technique become important in the antigen-antibody reactions and even more in immobilization of nucleic acids such as 20–30 bases' oligonucleotides, which single strand conformations were looking for their complementary sequence of bases in the sample material.

5. Important steps in the biosensor research

This report concentrate on three points which are considered very relevant in the Biosensors development in the last 30 years:

5.1. *The case of glucose pen*

In 1984 Cass et al. publishes a scientific paper where the team prove the use of ferrocene and its derivatives as mediators for amperometric biosensors [6]. Few years later the Medisense Exatech Glucose Meter was launched in the market and become the world's best selling biosensor product. The initial product was a pen-shaped meter with a disposable screen printed electrodes.

There were several advances in this product; first of all, the miniaturized instrument, just a pen with a small screen, where the current, transferred in mg/dL of glucose was directly displayed. Then the concept of disposable screen printed electrodes, which allowed discarding the sensor after the use, and more important allowed the elimination of the calibration step. This was an incredible step in the Sensor community. All sensors known, from pH glass electrodes to all kind of ISE, etc. should pass the “calibration step”, where the sensor must be calibrated every day and sometime before every measurement. The disposable screen printed electrodes do not need the calibration which simplifies enormously the use of it. The sensors became simple and user-friendly objects and diabetic people started to use it for individual monitoring and a large market was created. The performance and design of several strip analyzers based on different electrochemical principles and meters has been published [7]. The book described around ten different home glucose meters but it is not the final number because the new instruments continue to appear on the market.

5.2. *The appearance of BIAcore on the market*

In 1982 researchers from Pharmacia started to work jointly with physics and biochemistry professors at Linkoping University in order to develop a new bioanalytical instrument able to monitor the interactions between biomolecules. In 1984 a new company Pharmacia Biosensor was created. The company introduced a new instrument, BIAcore, in 1990. The instrument had a very high impact on the Biosensor community. The price of the instrument was more than 100 times higher than any other electrochemical or optical apparatus. The instrument based on SPR technology was a fully automated instrument which monitored the biomolecular interactions and included a sample handling equipment. The instrument performed the immobilization of the biomolecules, the SPR analysis and the regeneration of the sensor surface automatically by a microprocessor and this was a great advantage over the more traditional sensor technologies.

The autosampler able to handle up to 192 samples without operator assistance was also included in order to increase the reproducibility of the analysis and provide a large sample capacity. The instrument provided also the stabilization of the temperature at 0.1°C and allowed to study biomolecular kinetics. The BIAcore instrument also eliminates the moving components that are generally associated with prism-based instrumentation. However the real advancement of the instrument was the unique sensor chip technology to simplify the immobilization of biomolecules to the sensor surface.

Figure 1 is a scheme of the Biacore sensor chip. The chip consisted of a glass slide embedded in a plastic support; the glass surface is 1 cm^2 and has approximately 50 nm of gold coated on one side of the glass. The gold layer was then covered with dextran acting as a linking layer in order to facilitate the binding of biomolecules. Dextran, an old Pharmacia product, acted as a support for biomolecules but also protected the gold layer from unspecific binding which is the main problem of this kind of apparatus. Tipical protein concentrations required for immobilization are in the range 10–100 μg/ml. Typically, the chips can be used for more than 50 measurements without a notable loss in sensitivity and reproducibility. Moreover the company provided the sensor with alternative linking and binding layers (http://biacore.com). The flow injection system has been designed with miniaturized sample loops, valves and conduits reducing drastically sample and reagent volumes with help of silicone layers. This is, of course, very important when dealing with valuable biological reagents.

The instrument was sold initially mainly to the pharmaceutical companies looking for monoclonal antibodies constants (antibodies were ranked and selected for specific conditions). The instrument was found very powerful in its automatic performance which significantly cut the time for the evaluation of the binding constant between antibodies and antigens. Consequently it was also applied to study several other affinity reactions and became a leader instrument in the research of several research laboratories.

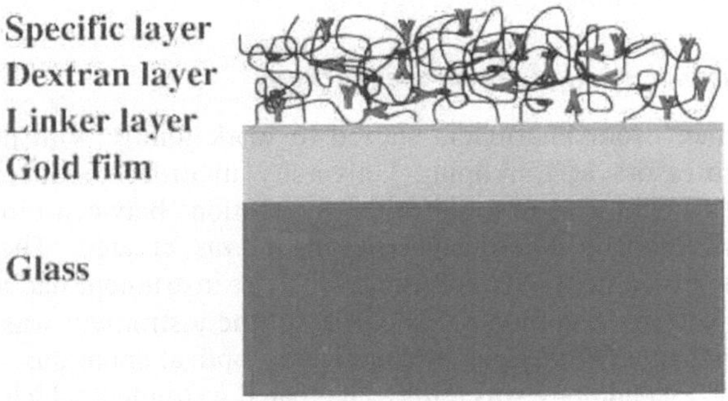

Figure 1. Biancore sensor chip.

5.3. *Nucleic acid based biosensor*

A nucleic acid (NA) is a macromolecule composed of nucleotide chains. The most common nucleic acids are deoxyribonucleic acid (DNA) and ribonucleic acid (RNA). Nucleic acids are universal in living things, as they are found in all cells and they are also found in viruses. Traditionally, NA molecules were thought to have the sole function of carrying and passing genetic information from one generation to another. Since the early 1990s, many NA molecules – known as aptamers – have been isolated that are able to bind a broad range of molecules with high affinity and specificity [7]. The molecules that can be recognized by aptamers range from small organic molecules to proteins, cells and even intact viral particles. A further advance in the development of function DNAs was made in 1994, when DNA was shown to act as a catalyst for the first time. Herein, these catalytic DNA molecules are called DNAzymes (also described as DNA enzymes, deoxyribozymes or catalytic DNA elsewhere). Moreover, DNAzymes and aptamers have been combined to form allosteric DNAzymes or aptazymes. These DNAzymes, aptamers and aptazymes are collectively called functional DNAs, whose functions extend beyond the Watson-Crick base pair recognition of complementary strands [7].

Naturally occurring NA as well as DNA-templated materials, that are responsive to chemical stimuli, have been used to assemble biosensor.

A NA biosensor is defined as an analytical device incorporating an oligonucleotide, even a modified one, with a known sequence of bases, or a complex structure of NA (like DNA from calf thymus) either integrated within or intimately associated with a signal transducer. NA biosensors can be used to detect DNA/RNA fragments or either biological or chemical species. Most nucleic acid biosensors are based on the highly specific hybridization of complementary strands of DNA or RNA molecules; this kind of biosensor is also called genosensor. In other applications (Aptasensor) selected NAs play the role of highly specific receptor of biologic and/or chemical species, such as target proteins, pollutants or drugs. In addition, the interaction of chemical compounds with DNA molecules have been exploited for toxicity screening assays. Finally catalytic RNA and DNA can be used to develop functional NA biosensor based (Fig. 2).

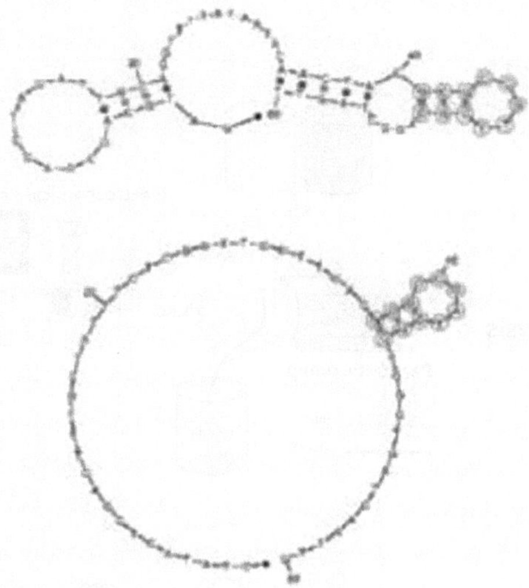

Figure 2. Examples of Aptameric structures.

6. New trends and conclusions

Micro Total Analysis Systems (μTAS) or Lab-on-a-chip (LOC) (term identifying for devices that integrate (multiple) laboratory functions on a single chip of only millimeters to a few square centimeters in size and that are capable of handling extremely small fluid volumes down to less than pico liters) are the future to reduce reaction time, amount of reagents, labor and cost (Fig. 3).

The integration of nanotechnology, microfluidics, and bioanalytical systems clearly represent one of the future directions of biosensor research. Possibilities seem endless when combining these different areas and hopes are high to solve many current problems ranging from sample preparation to real portability, from single molecule detection to true analytical speed, to reliability. Many obstacles will have to be overcome in order to fulfill all of these hopes; however, some exciting examples, typically of subsystems and not complete bioanalytical sensors, can already be found in the current literature.

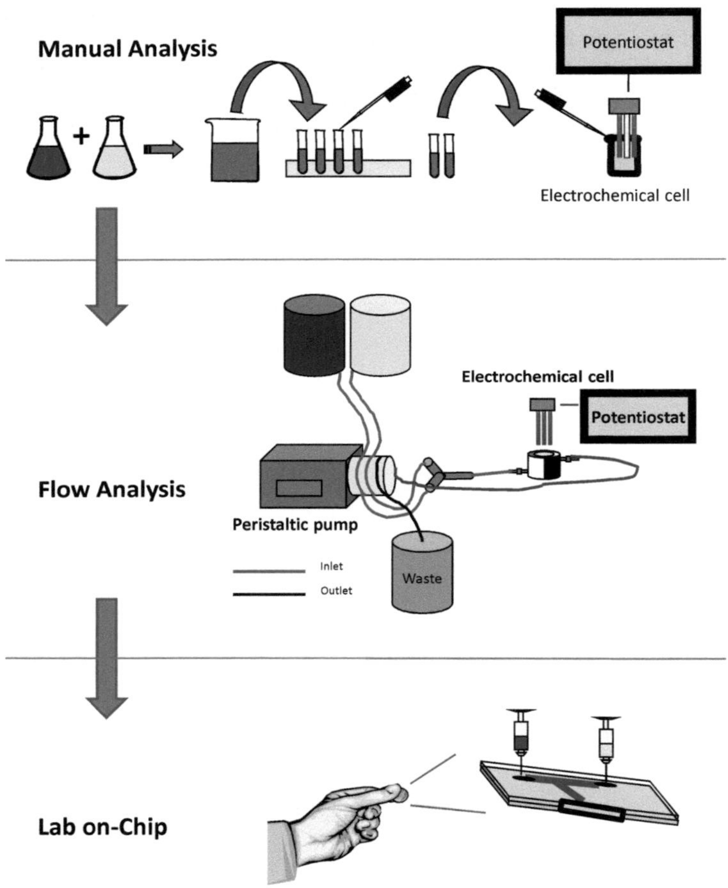

Figure 3. Lab on a chip technology.

A variety of microstructures have been proposed for on-chip amplification and analysis of nucleic acid [7]. This undoubtedly will benefit area such as biosensors assays.

In the near future, biosensor technology itself, will undoubtedly benefit from the integration of biosensors into micro- or nanodevices. Nanomaterials are claimed to improve transducer sensitivity. Thus, Biosensor will greatly benefit from nanotechnology and μTAS technology.

Acknowledgment ASI in the framework of MoMa project for fundings.

References

1. Clark LC. Monitor and control of blood and tissue oxygenation. Trans. Am. Soc. Artif. Intern. Organs; 2:41–48, 1956
2. Clark LC, Lyons C. Electrode systems for continuous monitoring cardiovascular surgery. Ann. N. Y. Acad. Sci; 102:29–45, 1962
3. Updike SJ, Hicks GP. The enzyme electrode. Nature; 214:986–988, 1967
4. Guilbault GG, Montalvo J. A Urea Specific Enzyme Electrode. JACS; 91:2164–2169, 1969
5. Rechnitz GA, Llenado R. Improved enzyme electrode for amygdalin. Anal. Chem.; 43:1457–1461, 1971
6. Cass AEG, Francis DG, Hill HAO, Aston WJ, Higgins IJ, Plotkin EV, Scott LDL, Turner APF. Ferrocene-mediated enzyme electrode for amperometric determination of glucose. Anal. Chem.; 56:667–671, 1984
7. Palchetti I, Mascini M,. Nucleic acid biosensors for environmental pollution monitoring. Analyst; 133:846–854, 2008

FUNDAMENTAL LIMITATIONS IN RESISTIVE WIDE-RANGE GAS-SENSOR INTERFACE CIRCUITS DESIGN

M. GRASSI[1], P. MALCOVATI[1] AND A. BASCHIROTTO[2]

[1]*Department of Electrical Engineering, University of Pavia, 27100, Pavia, Italy, e-mail: [marco.grassi, piero.malcovati]@unipv*

[2]*Department of Physics, University of Milano-Bicocca, 20126, Milano, Italy, e-mail: andrea.baschirotto@unimib.it*

Abstract. Due to their physical operating principle, resistive gas sensors demand for very wide-range interface circuits. As a consequence, the design of read-out devices, with particular emphasis on integrated circuit solutions, has to face off several limitations, like noise contribution and limited voltage supply. The combination of these constrains would lead, in a traditional linear design approach with digital output, to the exploitation of conditioning network circuits and A/D converters with unreliable specifications. This situation is described in the following to define the fundamental limitations. Moreover a design optimization technique for a state-of-the art solution is reported.

1 Introduction

Portable, low-cost gas-sensing systems demand is constantly growing [1] due to the concurrent increasing interest on several application fields, like for safety and security issues to improve ambient pollution and air quality monitoring. Moreover high performance and portable gas-sensing devices may be also employed in order to avoid or reduce human exposure to dangerous gasses in industrial plants or in other possible exposed environments, providing them, for instance, in defense and firemen equipment or clothes.

In order to satisfy portability, integrated technology approach is mandatory for both the electronics and the sensor, and this is nowadays reliable thanks to newer advances in micromachining solutions [2, 3]. For instance, a resistive metal-oxide (MOX) gas sensor operating at the temperature of about 250°C exhibits a reasonable sensitivity in few tens of milliseconds [4] after power up, consuming few milliwatts of power, thanks to the reduced mass of the device and to the good insulation of the active MOX membrane with respect to substrate. For this reason, such devices, if connected to suitable low-power, small size, integrated interface circuits are excellent candidates for realizing high-efficiency and high-reliability portable gas-sensing systems. Figure 1 shows the block diagram of a complete modern gas sensing system including an array of gas sensors, a closed loop temperature control circuit, an interface

P. Malcovati et al. (eds.), *Sensors and Microsystems: AISEM 2009 Proceedings*, Lecture Notes in Electrical Engineering 54, DOI 10.1007/978-90-481-3606-3_3,

circuit to read-out the sensors and a digital processor to perform pattern recognition operation on acquired data. In this way a self consistent gas sensing system may be developed without any bulky and expensive laboratory instruments.

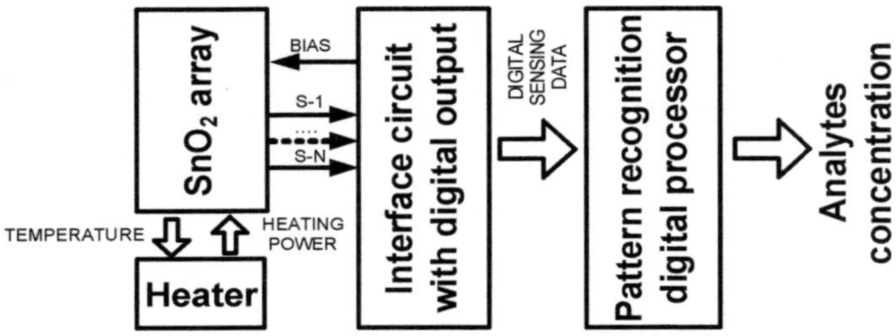

Figure 1. Simplified block diagram of a modern portable gas-sensing system.

The use of an array of MOX gas sensors allows performing the required selectivity, in particular when the desired target is a set of gasses instead of a single one. The use of different sensors types, optimized for the detection of different gasses, and the use of pattern recognition algorithms may be applied to gathered data in order to perform different analytes concentration extraction, as it is usually requested in air quality monitoring.

In Fig. 2 a simplified example of a single chip approach system (a) and a channel of a sensor read-out exploiting multichip approach (b) are shown respectively.

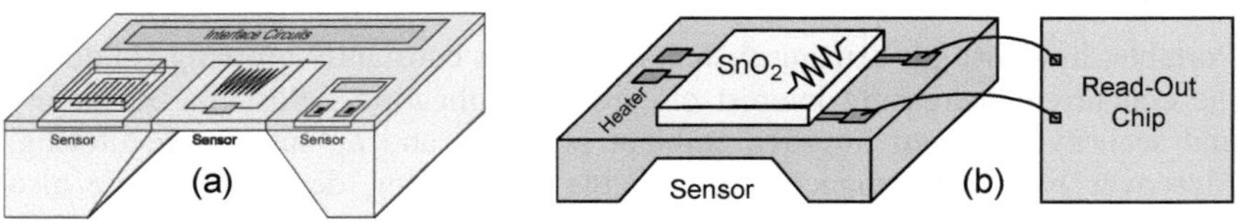

Figure 2. (**a**) A single chip approach system. (**b**) A sensor read-out multichip approach channel.

This chapter is outlined as following: section 1 is a brief introduction to portable gas-sensing systems. The development of integrated solutions both for sensing device and interface circuit is considered mandatory for such instruments. Feasibility is then justified by advantages given by recent improvements in research and development in micromachining technology. Section 2 reports main specifications, with particular emphasis on in-air electrical conductivity order of magnitude and on its variation with gas concentration exposure, of latest MOX gas-sensors. In fact these specifications are the starting point of most of the design considerations and strategies reported in this work. Section 3 describes the main limitations in the design of traditional gas-sensors linear wide-range interface circuits based on pre-amplifier conditioning network followed by an A/D converter (basically noise contribution and limited supply swing). Section 4 starts from the outlined limitations and illustrates a design strategy able to face-off such issues in order to get the best performance from the developed device by splitting the huge required dynamic range for gas sensing over different sub-ranges while correcting boundary non linearity exploiting a digital assisted solution. Section 5 finally draws few conclusions about faced design issues and how they have been overridden.

2 General specifications of resistive MOX gas-sensors

To define the specifications for a gas-sensing electronic read-out circuit, it is useful to recall the sensor response behavior and the input data format needed by the subsequent digital pattern analyzer. This has to be done in terms of rate and resolution for every gas-sensing-system equivalent-channel, actually consisting of a given sensor from the array with associated heater and the interface circuit with embedded A/D converter, as shown in Fig. 2b. MOX gas sensors are based on direct analyte adsorption and charge transfer processes between the gas molecules and the MOX surface. Since this causes an electrical resistance variation of the gas sensing elements, they are characterized by an effective resistive electrical behaviour [2]. This relationship between gas concentration and resistance value, even if it is not strictly linear and the processing by the subsequent DSP is almost mandatory, is one of the main reasons for the popularity of such devices. More in detail, considering the state-of-the-art in sensor manufacturing [3], the sensor resistance value may vary across several decades, being the effect of three variable components:

- The baseline ($R_{baseline}$), or resistance in air, which typically ranges, depending on process doping concentration, for metal oxide devices, from a moderated value, like 100 kΩ, to a much higher one, like 10 MΩ.
- The deviation, even up to 1 decade, from the baseline (ΔR), due to technology spread and sensor age as well as temperature.
- The resistance variation (R_{gas}), due to gas concentration, negative for oxidizing gas types and positive for reducing ones, which may be as large as a couple of decades from baseline [4–7].

The overall sensor resistance can then be written as:

$$R_{sens} = R_{baseline} + \Delta R + R_{gas} \tag{1}$$

In addition, the main requirements of the cascaded pattern analyzer DSP have to be considered. In fact, there are several techniques to process the data produced by the sensor interface to increase specific performance, like, for instance, selectivity. Recently developed dynamic pattern recognition techniques for gas-sensing are able to extract important information not only from the static value of the resistance of the sensor, but also from its transient behavior while varying gas concentration or temperature operation, through a morphological descriptor of the sensor trajectory in a phase space [4, 6, 8]. As a consequence, this technique forces a specific requirement to the electronics read-out, i.e. that the electronic interface circuit should be sufficiently fast to preserve the transient data, e.g. able to process all the information coming from the sensor array with an adequate throughput rate. It has been verified that an output rate of 10 Hz is sufficient to satisfy major dynamic pattern recognition requirements, even if a bandwidth up to 50–100 Hz is recommended in few particular contexts [4, 6, 8]. Furthermore, for most chemical targets, in order to detect every kind of gas of interest with enough accuracy, usually in the order of parts-per-million (ppm) after pattern analysis like for CO, NO_2, CH_4 or % vol units for alcohols, it is important to measure the sensor resistance value, R_{sens}, with a precision that could range from 10% to 0.1% depending on the specific application. For instance, for rough detection, 10% accuracy is considered sufficient, while around 1% and 0.1% accuracy are recommended for indoor and outdoor air quality monitoring respectively. This leads, as an example, to a Signal-to-Noise-and-Distortion-Ratio (SNDR) requirement for the interface circuit up to 60 dB for outdoor applications. Combining the three above reported resistance value range contributions in (1) together with the required precision for R_{sens}, a dynamic range (DR) of about 160 dB is required for outdoor modules, while a DR of about 140 dB could be sufficient for mentioned indoor applications [4–7], where the DR is the product of the relative precision of the electrical value to be measured (Accuracy) and the accommodated input range width (SensRange).

$$DR_{dB} = Accuracy_{dB} + SensRange_{dB} \tag{2}$$

The interface circuit has then to be designed in order to satisfy the above reported requirements. Calibration operation or additional operating modes may be exploited in order to raise the dynamic range performance even few over 160 dB, paying in terms of additional complexity, or in order to distribute more efficiently the required DR between SNDR and sensor resistance range.

It will be demonstrated that traditional front-end circuits based on linear pre-amplification followed by A/D conversion are not suitable to accommodate such a large dynamic range and alternative design solutions have to be investigated. This intrinsic limit of traditional front-end circuits could be bypassed, for instance, by exploiting different kinds of analog compression of the input electrical value. This technique could be based on devices with exponential (PN junctions or bipolar transistors) or quadratic (MOS transistors) behaviour, leading to a logarithmic or square root actual dynamic swing reduction respectively for a given input signal range (SensRange) specification. In Fig. 3 an example of current based logarithmic compressor [9] using current mirrors and diodes is shown. The sensor current drown by R_{sens}, biased at constant voltage V_{REF} by means of a buffer, is mirrored into one diode, while a reference constant current I_{REF} is drown by another matched diode. Thus the output V_0, seen as the difference between the two diodes voltage drops is:

$$V_0 = -V_T \cdot \ln\left(R_{sens} \cdot \frac{I_{REF}}{V_{REF}} \right) \tag{3}$$

where V_T is the thermal voltage of the matched diodes at ambient temperature.

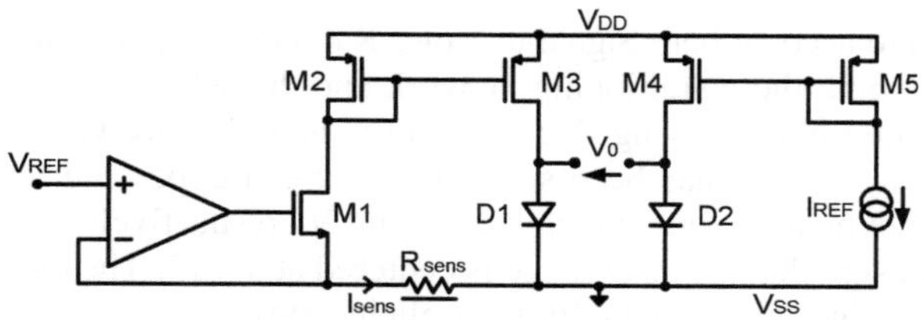

Figure 3. Example of logarithmic compression technique to read-out R_{sens} over a wide range.

The main results of the actual implementations based on this approach [9] can satisfy a DR requirement up to 115 dB given by a 4 decade input range value and a precision better than 2%. This limitation mainly comes from the mismatch behaviour between diodes and current mirror linearity over the wide interval of currents to be processed.

3 Main limitations in wide-range gas-sensors interface circuits design

A traditional solution for wide-range gas-sensors interface circuit to read-out resistive sensors is given by the cascade of an analog conditioning network followed by an A/D converter, as depicted in Fig. 4.

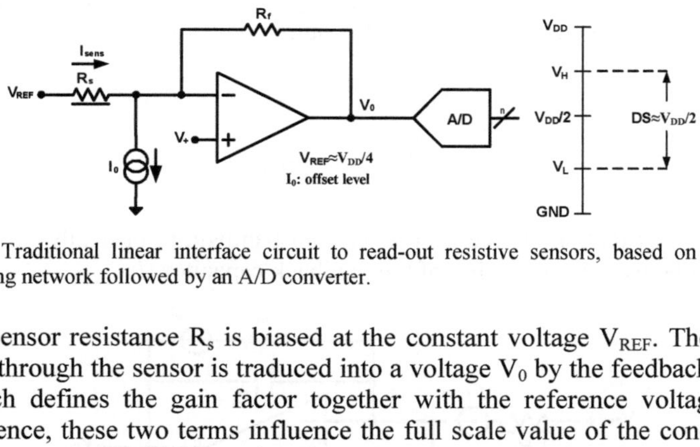

Figure 4. Traditional linear interface circuit to read-out resistive sensors, based on an analog conditioning network followed by an A/D converter.

The sensor resistance R_s is biased at the constant voltage V_{REF}. The current flowing through the sensor is traduced into a voltage V_0 by the feedback resistor R_f, which defines the gain factor together with the reference voltage. As a consequence, these two terms influence the full scale value of the conditioning network, as written in (4). For simplicity all voltages are referred to V_+ value.

$$V_0 = -R_f \cdot \left(\frac{V_{REF}}{R_S} - I_0 \right) \tag{4}$$

The current I_0 extracted from virtual ground allows adjusting the DC operating point of the circuit, thus permitting offset correction and dynamic range optimization. In order to accommodate the full range of the input possible value for R_s (SensRange), design parameters V_{REF} and R_f must be set to an adequate value taking into account both Signal-to-Noise Ratio (SNR) performance and that, unfortunately, both the conditioning network and the subsequent A/D converter will have limited output swing due to finite voltage supply value, V_{DD}. As an example, in Fig. 4, V_{REF} has been set to $V_{DD}/4$. Such upper and lower limits in analog output dynamic swing are called V_H and V_L respectively, while it may be assumed by design that the A/D range is centered at $V_{DD}/2$. The available analog output signal swing will be called for simplicity $SSw = V_H - V_L$. Please note that I_0 could be used to cancel $R_{baseline}$ and ΔR contributions in (1) by adequately shifting the DC operating point of the Analog Front-End (AFE).

3.1. *Analog front-end signal-to-noise ratio limitations*

Considering the analog conditioning network of Fig. 4 operating within its optimal output signal swing $[V_L, V_H]$, in first approximation non idealities are dominated by noise rather than distortion in the allowed Signal Swing (SSw), so that SNDR≈SNR. Main noise sources in the traditional AFE for resistive sensors depicted in Fig. 5 are: the thermal noise of the sensor itself and of the feedback resistor, the reference voltage V_{REF} noise, the op-amp noise contribution, and the I_0 programmable generator noise. Noise coming from V_{DD} may be included in op-amp contribution and neglected if the employed op-amp features a reasonable power supply rejection ratio. From Fig. 5, considering the resistive thermal noise for sensor and feedback resistor respectively (k is the Boltzmann constant and T the absolute temperature) the integrated output power of noise is:

$$< V_{n_s}^2 >= 4kTR_sB \tag{5}$$

$$< V_{n_f}^2 >= 4kTR_fB \tag{6}$$

where $B = (2\pi \cdot R_1 \cdot C_f)^{-1}$ is the AFE operating bandwidth, basically limited to a desired value by a proper sizing of the feedback capacitor C_f.

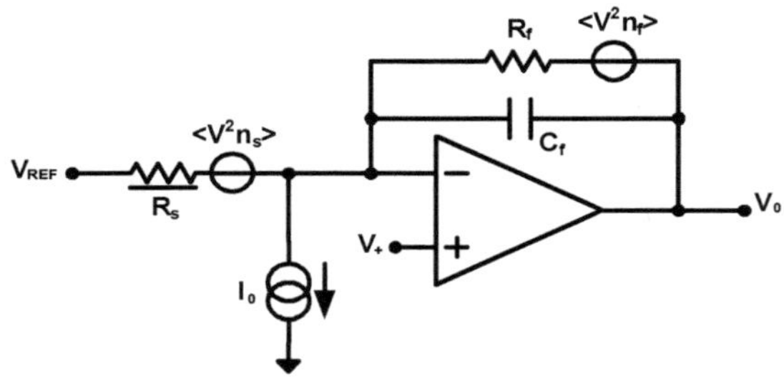

Figure 5. Thermal noise model for a traditional analog front-end for resistive sensors.

According to (4–6) the total integrated thermal noise power referred at the AFE output due to the sensor and to the feedback resistor is:

$$< N_0^2 >= 4kTR_sB\left(\frac{R_f}{R_s}\right)^2 \tag{7}$$

while, considering that the output signal power due to sensor response is:

$$< S_0^2 >= \left(\frac{R_f}{R_s} \cdot V_{REF}\right)^2 \tag{8}$$

by combining (7) and (8) the AFE Signal-to-Noise Ratio is derived:

$$SNR = \sqrt{\frac{< S_0^2 >}{< N_0^2 >}} = \sqrt{\frac{R_f V_{REF}^2}{4kTBR_s(R_s + R_f)}} \tag{9}$$

Since SNR rises as sensor resistance drops (9), a target SNR is achieved by setting $R_f > R_s$ for every possible sensor resistance value in the SensRange of interest. This means that R_f could be considered as the nominal ceil value for the resistive sensor to be read-out even if, when needed, few additional over-range may be added in order to accommodate possible gain error or offset of the circuit. In this way the in-range minimum value for SNR with $R_s = R_f$ is:

$$SNR_{min} = \sqrt{\frac{< S_0^2 >}{< N_0^2 >}}\Bigg|_{R_s=R_f} = \frac{V_{REF}}{\sqrt{8kTBR_f}} \tag{10}$$

For a target SNR_{min}, the reference voltage V_{REF} is defined and corresponds (in first approximation, i.e. neglecting distortion contribution) to the precision required in sensor resistance read-out. For instance, this means that a 0.1% precision requirement corresponds to a SNR_{min} of 60 dB. At higher SNR, also V_{REF} and op-amp noise contribute. This condition must be guaranteed by adequate low noise op-amp design and V_{REF} generator low-pass filtering. The reported relationships can be used to verify which integrated circuit technologies, according to their voltage supply, are suitable to meet a given AFE precision specification or vice versa, once given an IC technology, to calculate its SNR performance in standard AFE application for resistive sensors read-out. In a typical situation, assuming $V_{REF} = V_{DD}/4$, full-scale value $R_{s,\,max} = R_f = 10\ M\Omega$ and bandwidth B = 100 Hz, any given available integrated technology is suitable to guarantee, for instance, a precision of 0.1% (ten bits) or vice-versa that any desirable precision up to 17 bits (0.001%) may be met having access, for instance, to a CMOS 0.35 μm technology with 3.3 V supply. This verification is also explained in Fig. 6 and 7.

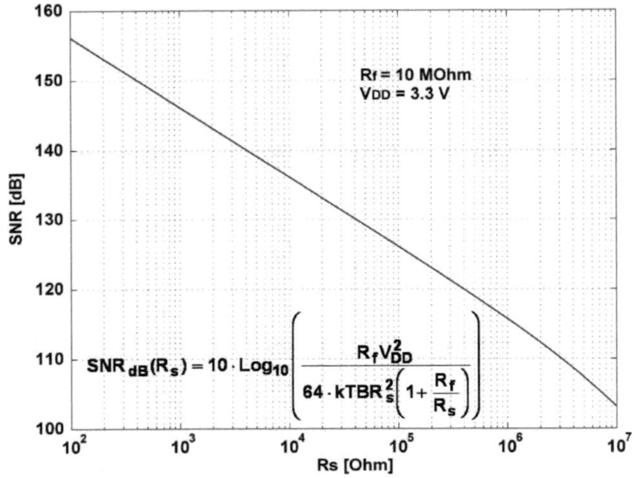

Figure 6. Analog front-end SNR performance over input sensor sweep value R_s = [100 Ω–10 MΩ] given V_{DD} = 3.3 V and a full-scale value R_f = 10 MΩ.

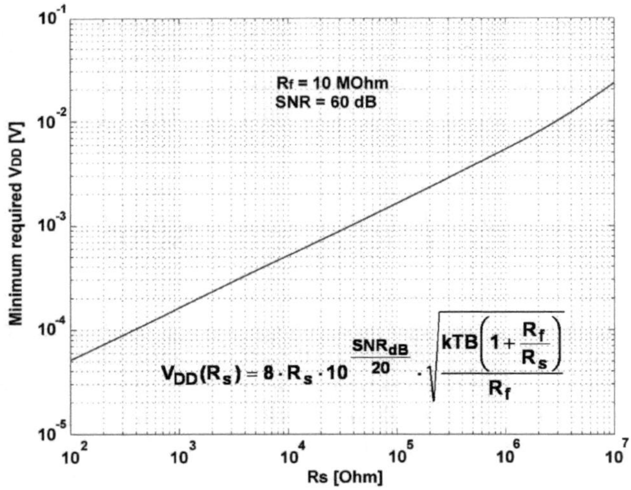

Figure 7. Analog front-end minimum voltage supply requirement over input sensor sweep value R_s = [100 Ω–10 MΩ] given a SNR needed performance of 60 dB and a full scale value R_f = 10 MΩ.

3.2. *Analog front-end signal swing limitations*

Typically, bare SNR requirements are relaxed in AFE design. On the other hand, challenging points arise taking into account also signal swing limitation, which is basically a voltage in traditional operational amplifier AFE architecture represented in Figure 4. In fact, for the typical resistance value range of MOX sensors, an interval up to 5 decades for the input signal is not suitable to be processed by a linear voltage or trans-resistance amplifier. Several solutions are available in state-of-the-art literature, like continuous, as in Fig. 3 [9] or local [5] compression of the input value. The hypothesis of considering V_{REF} as a V_{DD} ratio, making the voltage reference and thus voltage supply as large as possible according to the available technology in order to maximize SNR figure, is not anymore applicable when considering signal swing limitation. In fact V_{REF}, which may be thought uncorrelated to V_{DD}, should be as small as possible leading to a trade-off on reference voltage value that will be shown to be almost impossible to practice for input sensors range width (SensRange) larger than a couple of decades. On the other hand, according to swing limitations, maximum required AFE output voltage interval (hereinafter SSw) could be considered for simplicity a ratio of the minimum required V_{DD} value. If the maximum applicable sensor resistance value is $R_{s,\ max} = R_f$, and the minimum is $R_{s,\ min} = R_f/\text{SensRange}$, then the minimum required V_{DD} could be, for instance, twice SSw. Taking into account that SensRange is expected to be very wide (decades), the output swing results to be:

$$|V_0| = V_{REF}\frac{R_f}{R_s} \tag{11}$$

$$SSw = |V_0|_{MAX} = V_{REF}|1 - \text{SensRange}| \approx V_{REF} \cdot \text{SensRange} \tag{12}$$

$$V_{DD,min} = 2 \cdot SSw_{min} \approx 2 \cdot V_{REF,min} \cdot \text{SensRange} \tag{13}$$

where $V_{REF,min}$ is the minimum value for reference voltage that fulfills also SNR requirement for every input sensor resistance value in the interval of interest $[R_f/\text{SensRange};\ R_f]$ for the AFE circuit as illustrated in previous paragraph. Since it has been verified the SNR worst case to be $R_s = R_f$, the minimum required output voltage swing for the AFE can be calculated. To do this the V_{REF} expression from (12) is considered, i.e. $V_{REF} \approx SSw/\text{SensRange}$, and after substituting it in (10) to get (14), the equation for SSw (15) can be solved.

$$SNR_{min} = \sqrt{\frac{< S_0^2 >}{< N_0^2 >}}\Bigg|_{R_s = R_f} = \frac{SSw}{\text{SensRange}} \cdot \frac{1}{\sqrt{8kTBR_f}} \tag{14}$$

$$SSw = \frac{\text{SensRange}}{\sqrt{8kTBR_f}} \tag{15}$$

The SSw can be expressed in terms of the system dynamic range DR according to (2), being SNR_{min} in (14) the precision specification of the front-end circuit:

$$SSw = \text{SensRange} \cdot SNR_{min} \cdot \sqrt{8kTBR_f} = DR \cdot \sqrt{8kTBR_f} \tag{16}$$

According to the hypothesis that SSw can be seen as a ratio of the minimum required voltage supply, e.g. its half part, V_{DD} is extracted from (15) or (16). In the realistic system specifications for a sensor front-end circuit for outdoor environmental monitoring, given by a local resistance value accuracy of 0.1% (SNR_{min} = 60 dB), an input sensor value range of 5 decades (SensRange = 10^5), a bandwidth (B) of 100 Hz, and a maximum input value for sensor resistance ($R_s = R_f$) of 10 MΩ, according to (15) and (16) an analog front-end with such features is really far away from being feasible with any available technology, being the theoretical required voltage supply of the order of thousand volts.

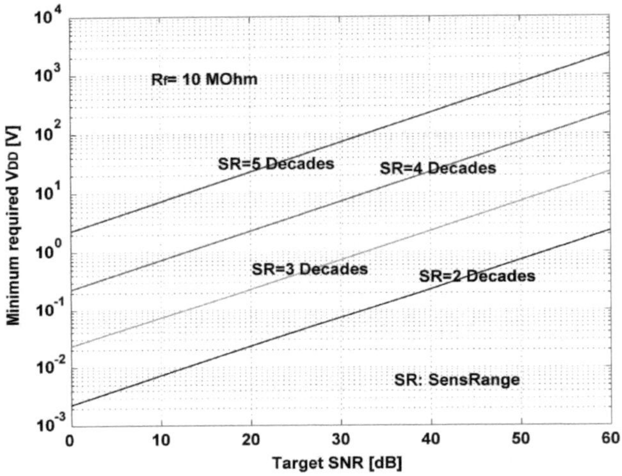

Figure 8. Analog front-end minimum voltage supply requirement in function of SNR specification (i.e. read-out precision) and desired input sensor range width, given a full-scale value R_f = 10 MΩ.

According to the plot of Fig. 8, reducing the desired SensRange to a couple of decades, the front-end circuit can be implemented using a low voltage CMOS technology, like 0.18 μm, being the needed V_{DD} less than 2 V, while for instance, in order to get a SensRange of 3 decades more that 10 V supply is needed, forcing the implementation to move to PCB solutions with off the shelf components if following a pure analog front-end circuit design solution. Relaxing instead SNR_{min} specification to 40 dB, i.e. reducing the precision of the read-out to 1%, a 0.18 μm CMOS implementation could reach 3 decades of SensRange, which, as mentioned in first paragraph, is sometimes enough for some indoor gas concentration measurements and surely satisfactory for safety threshold based gas detection.

3.3. *Analog-to-digital converter resolution limitations*

The analog preamplifier output feeds the ADC, as shown in Fig. 4. This means that the ADC also has to be able to process the entire DR given by the combination of the desired input sensor range width and the required precision on the resistance value to be read-out. Basically the A/D converter, which is supposed to be linear in its trans-characteristic function, must discriminate even the minimum output signal over the whole input desired SensRange always with the required precision.

3.3.1. *Maximum least significant bit amplitude according to AFE SNR_{min}*

The value of the ADC Least Significant Bit (LSB) may be calculated, according to SNR considerations, by applying the full-scale input sensor resistance value ($R_s = R_f$), which leads to the minimum signal and SNR feature, according to (10) and (17), once dimensioned V_{REF} by design to the minimum value which grants the SNR performance. In this way the maximum allowed LSB according to SNR issues for the ADC is given by the expression in (18).

$$S_{0,min} = S_0\big|_{R_s=R_f} = V_{REF} \tag{17}$$

$$LSB_{MAX}(V_{REF,min}) = \frac{V_{REF}}{SNR} = \sqrt{8kTBR_f} \tag{18}$$

With the same design specifications, related to outdoor portable air quality environment device, indicated in the example at the end of sub-paragraph 3.2 and used to calculate the required V_{DD}, the absolute maximum value for the LSB can be calculated thanks to (18), which is thus of the order of 6 μV. Such LSB, having access for instance to a technology with a supply voltage of 3.3 V and considering a SSw equal to half supply, could be satisfied with an ADC having a resolution, in bit, equal to $N_{ADC} = Log_2(SSw/LSB) \approx 18$, that, considering the usual almost narrow band of 100 Hz, would be feasible with the available technology exploiting for instance a high order sigma delta architecture for high resolution sensors applications [10, 11].

3.3.2. *Effective resolution requirement for the ADC according to real DR*

The ADC effective required LSB specification is typically dominated by the actual DR specification rather than bare SNR combined with voltage supply limitation issues. In fact, the necessary front-end overall dynamic range given by the specifications in bit would be:

$$N_{ADC} = Log_2(DR) \tag{19}$$

leading, for instance, taking account of the usual 160 dB for outdoor environmental monitoring, to a resolution of about 27 bit, which is unreliable.

4 Analog front-end design solutions and guidelines

Taking account of the main limitations in wide-range interface read-out circuits illustrated in previous paragraph, it is clear that a traditional solution, based on a single wide range scale voltage or trans-resistance amplifier followed by a high resolution A/D converter, as in Fig. 4, is not reliable for an effective dynamic range (DR) higher than 100 dB. The read-out solution exploiting compression depicted in Fig. 3, instead, is able to reach a DR near to 120 dB, but such performance may not be easily split optimizing its two components by design: SensRange and Accuracy, according to required specifications, being actually the second term strongly limited by devices mismatch factor [9]. There are at least two possible solutions to go beyond the 100 dB dynamic range limit without exploiting current-voltage device based compression [5, 7] and both of them may be assisted by a digital logic in order to get from interface circuit the best dynamic range possible performance. In [5], for instance, the huge required SensRange specification for the analog front end is relaxed by splitting it into different sub-ranges and an additional digital calibration logic is employed to cancel boundary non linearity leading to a feasible outdoor portable air monitoring system with a DR of about 160 dB. In [7], instead, the large SensRange issue is faced off first avoiding to traduce the resistance of the read-out sensor into a voltage by measuring directly the current through the sensor and then accommodating the large measured current range along time exploiting an integrator and a set of digital counters, also assisted by a suitable logic. This second solution has the strong advantage of not requiring any kind of calibration, leading to an air monitoring system which is cheaper and does not require periodic maintenance nor ageing related calibration, but due to unavoidable non linearity in very large current range measurement, as also happens for the scheme of Fig. 3, the maximum achievable dynamic range is of the order of 140 dB, making the solution suitable basically for indoor fine detection applications. Paying in terms of complexity, i.e. adding calibration or in terms of speed (renouncing to the possibility of exploiting dynamic pattern recognition), i.e. assigning a larger time integration window it is possible to extend the DR to a value comparable to the first solution.

4.1. *Multi-scale analog front-end design solution*

It has been verified that, in traditional AFE, if the minimum required SNR performance is achieved at full-scale, there are basically no limitations regarding minimum required supply voltage, while a reasonable voltage signal swing (SSw) will be, with any technology, insufficient to fulfill SensRange requirement. On the other side, if the design target is a given signal swing, a reference voltage V_{REF} orders of magnitude lower than available SSw would be needed, leading to unavoidable noise issues. Nevertheless, the resolution of the A/D converter would be unreliable.

In order to accommodate the huge SensRange over a relatively small SSw fulfilling noise requirement, an effective solution is [5] to split the overall range to be read-out in multiple sub-ranges. This could be done, for instance, by selecting the resistor R_f of Fig. 4 among a set of NS different components featuring different resistance value $[R_{f,i}]$, while, in order to optimize the split effectiveness, such provided resistance value portfolio should be stepped logarithmically, i.e the ratio between consecutive elements must be constant.

4.1.1. *Noise limitations in multi-scale AFE solution*

All given sub-ranges i = [1, 2, ..., NS], associated to the feedback resistor $R_{f,\,i}$ will have to fulfill SNR requirement in (10), reminding that the core interval of the sub-scale, whose related input sensor resistance value range width is SR_s<< SensRange, will be [$R_{f,i}/SR_s$, $R_{f,i}$], where of course also SR_s is a ratiometric constant term (e.g. a SR_s of 100 means 2 decades of possible input signal width). Furthermore, overall noise, including op-amp and voltage references, must be SNR_{min} times lower than minimum signal to discriminate. Such noise does not depend on the selected $R_{f,i}$ and, named N^{+}_{RMS}, is reported in (20) referring its RMS value to the positive input of the op-amp:

$$N^{+}_{RMS} = \frac{1}{SNR_{min}} \cdot \left. \sqrt{\frac{\left(V_{REF} \cdot \frac{R_{f,i}}{R_s}\right)^2}{\left(1 + \frac{R_{f,i}}{R_s}\right)^2}} \right|_{R_s = R_{f,i}} = \frac{V_{REF}}{2 \cdot SNR_{min}} \quad (20)$$

Noise limitations reported in (10) and (20) usually do not represent an actual issue in multi-scale design approach.

4.1.2. *Signal swing limitations in multi-scale AFE solution*

For a given analog front-end circuit, non idealities associated to linearity issues may be neglected if the circuit works within its optimal output voltage boundary values, i.e. within its SSw. In order to guarantee this hypothesis, the overall SensRange has been divided, and, in order to face off unavoidable inter-range gain and DC offset mismatch non-idealities, adjacent scales must be partially overlapped, as depicted in Fig. 9.

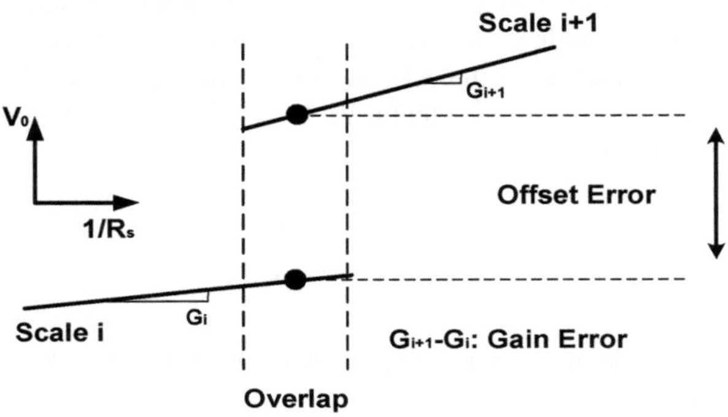

Figure 9. Offset and gain mismatch in adjacent sub-ranges in AFE multi-scale approach.

Such non idealities will be corrected thanks to the assistance of an additional digital circuit, which optimizes the overall trans-characteristic by correcting inter-range gain errors trimming, for each scale, the reference voltage V_{REF} by means of a fine tuning DAC. Such logic will also heal in minimizing overall offset error among successive scales acting on the programmable current generator of Fig. 4, selecting the best value for I_0 in every available subrange. In order to cover also the overlapping interval, hereinafter scale overrange, the required dynamic range for every scale will be larger that the one obtained simply dividing the system specification by NS, i.e. the number of available feedback resistors $R_{f,i}$. Finally, according to the described architecture, the output signal swing SSw for all AFE scales will be typically identical.

4.1.3. *ADC limitations in multi-scale AFE solution*

In order to accommodate the augmented required dynamic range of a single scale with respect to its theoretical value due to the introduction of over-ranges, given the basic input signal range width SR_s, an A/D converter with resolution, or SNDR, larger than bare system desired accuracy for sensor resistance read-out is demanded. This means for instance that, with half output signal swing for the bare over-range to perform non idealities digital assisted correction, one additional bit of resolution for the ADC is required. This means that, in order to meet 0.1% accuracy (60 dB) over one decade SR_s input signal (20 dB), 14 bits (86 dB) are needed instead of 13 bits (80 dB) to allocate upper and lower scale overlaps. Thus the minimum required resolution for the ADC will be:

$$N_{ADC} = \mathrm{Log}_2\left(\frac{SR_S \cdot SNDR_{min}}{ASOR}\right) \tag{21}$$

where $SNDR_{min}$ is the overall system accuracy, which coincides with SNR_{min} if neglecting distortion contribution, while ASOR is the adjacent scales overlapping ratio with respect to nominal sub-range input signal range width (ASOR = 0.5 requires one additional bit for the ADC). Finally, being always SSw the output swing for the AFE, using an ADC featuring N_{ADC} bits, the least significant bit of the converter will be given by:

$$LSB_{ADC} = \frac{SSw}{2^{N_{ADC}}} \tag{22}$$

while the single scale input sensor resistance range width, SR_s, may also be written by combining (21) and (22):

$$SR_S = \frac{2^{N_{ADC}} \cdot ASOR}{SNDR_{min}} = \frac{SSw \cdot ASOR}{SNDR_{min} \cdot LSB_{ADC}} \tag{23}$$

4.1.4. *Minimum number of sub-ranges in multi-scale AFE solution*

Once known the resolution of the ADC (N_{ADC}) to be employed, which also depends on the sub-range analog front-end achievable performance, and the adjacent scales overlapping ratio (ASOR), the minimum number of scales for the system may be calculated by finding out how many (NS) times SR_s is required to build the overall system SensRange:

$$NS > Log_{SR_s}(SensRange) = \frac{Log_2(SensRange)}{Log_2(SR_s)} \quad (24)$$

The second expression in (24) is obtained by changing the logarithm base from SR_S to binary. Finally considering the overall system range in decades (DecRange) and that the required resolution for the ADC is the product of SR_S and Accuracy (i.e. system $SNDR_{min}$ or SNR_{min} neglecting distortion), plus the additional contribution due to scale overlap, as depicted in (21), it results:

$$NS > Log_{SR_s}(SensRange) \approx \frac{3.32 \cdot DecRange}{N_{ADC} - SA_{BIT} + Log_2(ASOR)} \quad (25)$$

where SA_{BIT} indicates the system accuracy required in sensor resistance read-out expressed in bits, i.e. $SA_{BIT}=Log_2(Accuracy)$.

For the usual example of outdoor air monitoring recommended specifications for MOX resistive gas sensors read-out, 5 decades range and 0.1% accuracy are considered. Setting by design an ASOR of ½ and considering a technology related maximum SSw of 2 V, a 13-bit ADC can be used, leading to a reasonable LSB of about 250 μV, which does not introduce noise or other implementation issues, and to a scale range width SR_s of about 8 (few less than a decade) not including over-range. Finally, being the required system accuracy for sensor resistance read-out about ten bits (0.1%), the minimum number of required scales results to be:

$$NS > \frac{3.32 \cdot 5}{13 - 10 - 1} \approx 8.3 \quad (26)$$

From (26) such enumerated specifications for MOX sensor resistance read-out require a multi-scale AFE with a minimum number of nine sub-ranges.

5 Conclusions

In this chapter fundamental limitations in resistive wide-range gas-sensors read-out circuits design have been analyzed. Starting from the description of basic characteristics and working principle of MOX gas sensors, the main required design specifications for the interfaced electronic read-out circuit have been defined. Considering a traditional linear implementation for resistance sensing, consisting of a trans-impedance amplifier, acting as conditioning network, followed by an A/D converter, it has been demonstrated that, in order to fulfill defined gas-sensing specifications, both the amplification stage analog front-end and the ADC would require unreliable performance. This issue, once optimized the architecture to meet SNR specification, is mainly due to limited available voltage swing, an unavoidable constrain in analog integrated circuits. Such limitation may be for instance bypassed by splitting the overall wide input sensor resistance value range into multiple scales and by exploiting a digital logic to correct boundary non idealities. For this approach, design optimization theory and guidelines have been discussed, outlining potential results.

Acknowledgments The study reported in this chapter has been inspired by the research work carried out within Italian Government funded projects PRIN 2003091427 and PRIN 2005092937.

References

1. C. L. Dai, M. C. Liu, "Nanoparticle SnO_2 gas sensor with circuit and micro heater on chip fabricated using CMOS-MEMS technique", Proceedings of Nano-Micro Engineered and Molecular Systems Conference, pp. 959–963, IEEE, 2007.
2. S. Capone, P. Siciliano, "Gas sensors from nanostructered metal oxides", Encyclopedia of Nanoscience and Nanotechnology, vol. 3, pp. 769–804, American Scientific Publishers, 2004.
3. G. Sberveglieri, "Recent developments in semiconducting thin-film gas sensors", Sensors and Actuators B, Chemical, vol. 23, pp. 103–109, Elsevier, 1995.
4. A. Lombardi, L. Bruno, M. Grassi, P. Malcovati, S. Capone, L. Francioso, P. Siciliano, A. Baschirotto, "Integrated read-out and temperature control interface with digital I/O for a gas-sensing system based on a SnO_2 microhotplate thin film gas sensor", Proceedings of International Conference on Sensors, pp. 596–599, IEEE, 2008.
5. M. Grassi, P. Malcovati, A. Baschirotto, "A 160dB equivalent dynamic range auto-scaling interface for resistive gas sensors arrays", Journal of Solid-State Circuits, vol. 42, pp. 518–528, IEEE, 2007.
6. A. Baschirotto, S. Capone, A. D'Amico, C. Di Natale, V. Ferragina, G. Ferri, L. Francioso, M. Grassi, N. Guerrini, P. Malcovati, E. Martinelli, P. Siciliano, "A portable integrated wide-range gas sensing system with smart A/D front-end", Sensors and Actuators B, vol. 130, pp. 164–174, Elsevier, 2008.
7. M. Grassi, P. Malcovati, A. Baschirotto, "A 141dB dynamic range CMOS gas-sensor interface circuit without calibration with 16-bit digital output word", Journal of Solid-State Circuits, vol. 42, pp. 1543–1554, IEEE, 2007.
8. A. Burresi, A. Fort, S. Rocchi, M. B. Serrano Santos, N. Ulivieri, V. Vignoli, "Temperature profile investigation of SnO_2 sensors for CO detection enhancement", Transactions on Instrumentation and Measurement, vol. 54, pp. 79–86, IEEE, 2005.
9. D. Barrettino, M. Graf, S. Hafizovic, S. Taschini, C. Hagleitner, A. Hierlemann, H. Baltes, "A single-chip CMOS micro-hotplate array for hazardous-gas detection and material characterization", Proceedings of International Solid State Circuits Conference, pp. 312–313, IEEE, 2004.
10. F. Maloberti, "Data Converters", Springer, 2007.
11. A. Baschirotto, P. Malcovati, "Technology-driven alternatives for smart sensor interfaces", in H. Baltes, G. Fedder, J. Korvink, ed., "Sensors Update", vol. 13, pp. 45–81, Wiley-VCH, Weinhein, 2003.

MATERIALS AND PROCESSES

ADVANCES IN SILICON PERIODIC MICROSTRUCTURES WITH PHOTONIC BAND GAPS IN THE NEAR INFRARED REGION

G. BARILLARO[1], A. DILIGENTI[1], L. M. STRAMBINI[1], V. ANNOVAZZI-LODI[2], M. BENEDETTI[2], S. MERLO[2] AND S. RICCARDI[2]

[1]*Dipartimento di Ingegneria dell'Informazione: Elettronica, Informatica, Telecomunicazioni, Università di Pisa, Via G. Caruso 16, 56122 Pisa, Italy*
[2]*Dipartimento di Elettronica, Università degli Studi di Pavia, Via Ferrata 1, 27100 Pavia, Italy*

Abstract. In this work we present experimental and numerical results relative to reflectivity spectra of micromachined one-dimensional photonic crystals (1D-PhCs), evaluated in the near infrared region at normal incidence with an improved experimental setup, based on single-mode optical fibers with lensed terminations. A Gaussian stochastic process was exploited in the numerical calculation to evaluate the effects of fabrication-induced non-idealities of the devices and limitation of the optical setup, on the reflectivity spectra.

1. Introduction

One-dimensional photonic crystal (1D-PhC) structures have attracted a lot of attention for different applications in optical communications [1] and biomedicine [2]. A 1D-PhC can be described as a structure with a periodic variation of the refractive index, realized through a multilayer of two dielectrics with fairly different refractive indexes. Periodicity gives rise to Photonic Band Gaps (PBGs), i.e. wavelength intervals over which the propagation of the electromagnetic radiation is forbidden. In our previous works [3, 4], we demonstrated the feasibility of large-area vertical devices, with high aspect-ratio, by silicon electrochemical micromachining (ECM). In particular, reflectors based on the so-called hybrid quarter-wavelength structure with a band gap centered at 1.55 μm were designed, fabricated and tested. These reflectors consisted of arrays of 1–2-μm-thick silicon walls with period of 3–4 μm and etching depth of 100 μm, fabricated by ECM. A SEM cross-section of a typical 1D-PhC is reported in Fig. 1. PhCs used as reflectors for optical communications typically exploit low-order PBGs, and, in turn, small period structure, to obtain a reasonably large PBG width at 1.55 μm. In order to use such kind of 1D-PhCs for the fast-growing field of sensing applications, a more complicated line-shape of the reflectivity spectrum is desirable, in order to enhance optical changes induced by suitable reactions occurring within the structure. Increasing the spatial period of the structure allows the fabrication of 1D-PhCs with better mechanical stability, especially in

P. Malcovati et al. (eds.), *Sensors and Microsystems: AISEM 2009 Proceedings*, Lecture Notes in Electrical Engineering 54, DOI 10.1007/978-90-481-3606-3_4,

the presence of liquids (typical of biosensing applications) and, at the same time, with high reflectivity order at the selected wavelengths. For this purpose, hybrid structures with a high-order (up to 25th) transmission gap at 1.55 μm were fabricated by Silicon ECM.

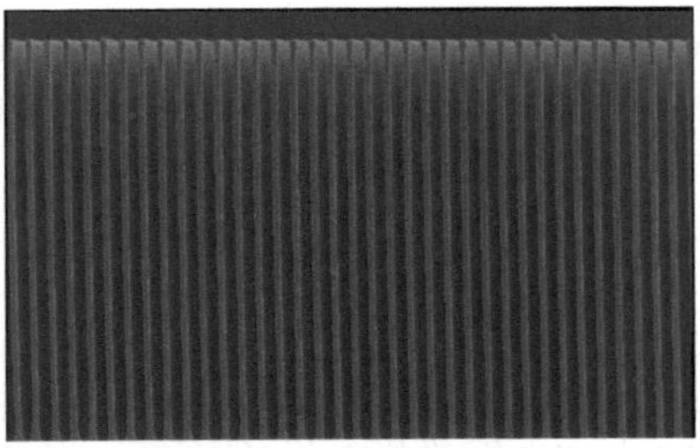

Figure 1. SEM cross-section of a Si/air 1D-PhC microstructure fabricated by silicon ECM.

2. Experimental results

Measurements of spectral reflectivity (in the range of 1.0–1.7 μm at orthogonal incidence) on samples of fabricated 1D-PhCs with period of 4 and 8 μm were performed with an improved optical setup. In particular, we used standard telecommunication single-mode fibers with a lensed termination in front of the silicon structure, for back-reflection measurements. The instrumental configuration is reported in Fig. 2.

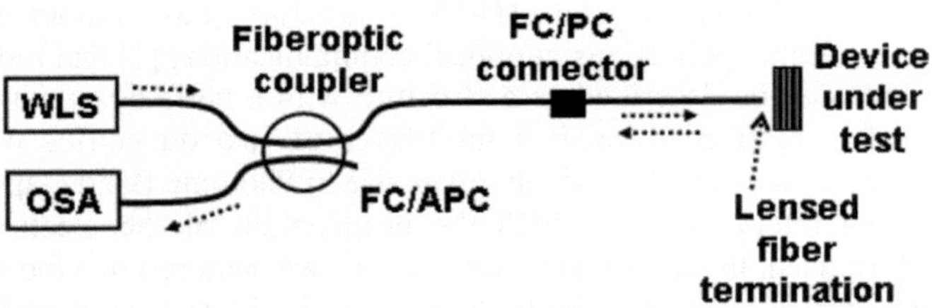

Figure 2. Optical setup for spectral reflectivity measurements. WLS: White Light Source. OSA: Optical Spectrum Analyzer.

The acquired power spectra were normalized to the power spectrum expected from an ideal reflector placed at the output port of the coupler, in order to obtain a good estimation of the absolute spectral reflectivity of the devices.

Figure 3 shows an experimental reflectivity spectrum of a 4-μm-period PhC, obtained with the setup described above, and using a focusing lensed fiber with a 5-μm-spot diameter at a working distance of 26 μm. Figure 4 shows the experimental spectrum, measured with the same optical setup, for a 8-μm-period PhC. It can be clearly seen that, at the same wavelengths, PhCs with a longer spatial period show narrower reflectivity peaks, since they correspond to PBGs of a much higher order. High absolute reflectivity was observed (−0.8 dB at 1,525 nm): since coupling efficiency between fiber tip and device cannot be quantified, we assume that intrinsic losses of the fabricated 1D-PhC are even lower, indicating a very high optical quality of the silicon surfaces, as also verified with a profilometer and an interferometric microscope.

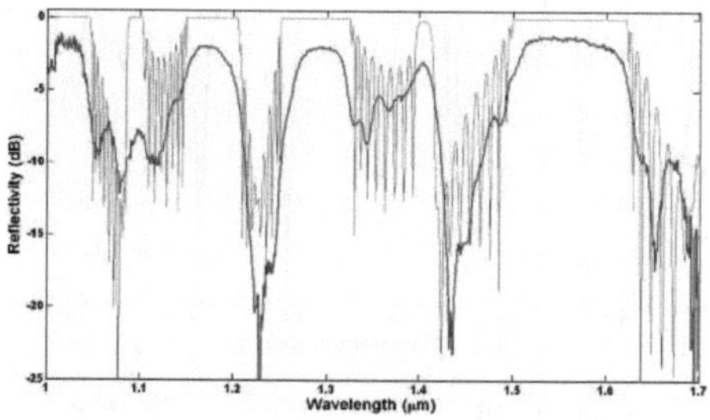

Figure 3. Black trace: experimental data of measured reflectivity for a 4 μm period PhC (RB = 10 nm). Gray trace: calculated ideal reflectivity for the same sample.

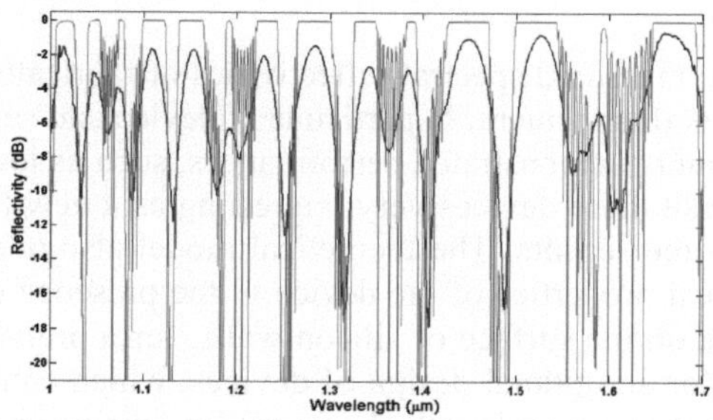

Figure 4. Black trace: experimental data of measured reflectivity for a 8 μm period PhC (RB = 10 nm). Gray trace: calculated ideal reflectivity for the same sample.

3. Theoretical analysis

Theoretical reflectivity spectra of 1D-PhCs shown in Figs. 3 and 4 were calculated using the characteristic matrix method and supposing an ideal device and measuring setup. In those figures, a good agreement between experimental and calculated reflectivity spectra, with regard to the positions of the PBGs, is observed. However, to improve the model for simulating the PhCs behaviour, surface roughness of the samples was taken into account introducing a Gaussian stochastic variable for the porosity (the ratio between air layer thickness and spatial period). Note that if roughness amplitude is small compared to the wavelength, it doesn't contribute significantly to scattering losses. Under these assumptions, we can calculate reflectivity as the statistical mean value of different realizations of the stochastic process [5]. Figure 5 shows the comparison between the theoretical reflectivity spectrum, calculated keeping into account roughness at the surface of silicon walls composing the structure and Resolution Bandwidth (RB) of the OSA, and the experimental data, for a 8 μm period PhC, demonstrating the great improvement in the modelling.

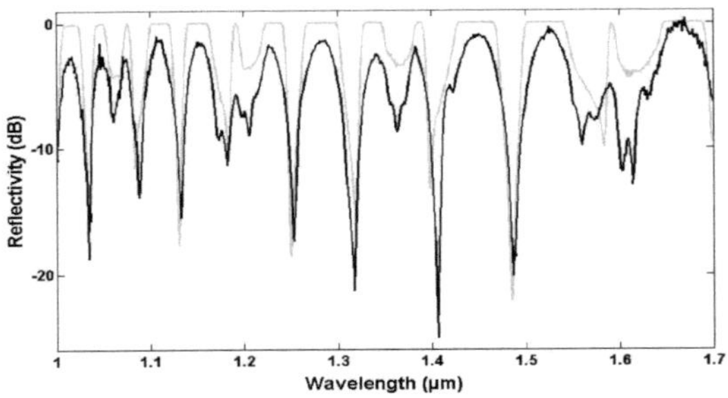

Figure 5. Black trace: experimental reflectivity for a 8 μm period PhC (RB = 5 nm). Gray trace: calculated reflectivity for the same sample, taking into account surface roughness and RB.

4. Conclusions

The theoretical and measured spectral reflectivities of fabricated high-order 1D-PhCs are in very good agreement, in particular if device and setup non-idealities are taken into account. Demonstrated performances, such as low losses and high dynamic range, make these devices very interesting as a new three-dimensional platform for label-free sensors. The theoretical model also allows us to predict effects on the optical properties of the device in the presence of additional bio-layers immobilized on the surface of silicon walls, for a preliminary evaluation of sensitivity and for an optimal design of devices, which enhances reflectivity variations at a convenient wavelength (1.55 μm) for sensing applications.

Acknowledgments This work was supported in part by MUR under a PRIN-COFIN 2007 and by the CARIPLO Foundation.

References

1. A. Lipson, E. M. Yeatman, "A 1-D Photonic Band Gap Tunable Optical Filter in (110) Silicon," Journal of Microelectromechanical Systems, vol. 16(3), 521–527, 2007.
2. C.J. Choi, B. T. Cunningham, "Single-step fabrication and characterization of photonic crystal biosensors with polymer microfluidic channels," Lab on a Chip, vol. 6, 1373–1380, 2006.
3. G. Barillaro, V. Annovazzi-Lodi, M. Benedetti, S. Merlo, "Reflection properties of hybrid quarter-wavelength silicon microstructures," Applied Physics Letters, vol. 90, 121110, 2007.
4. G. Barillaro, S. Merlo, L. M. Strambini, "Band gap tuning of silicon micromachined 1D photonic crystals by thermal oxidation," IEEE Journal of Selected Topics in Quantum Electronics, vol. 14(4), 1074–1081, 2008.
5. G .Barillaro, L. M. Strambini, V. Annovazzi-Lodi, S. Merlo, "Optical characterization of High-Order 1D Silicon Photonic Crystals," IEEE Journal of Selected Topics in Quantum Electronics, vol. 15(5), 2009.

INVESTIGATION OF THE SWELLING PROPERTIES OF PHEMA AND PHEMA/CB FOR SENSING APPLICATION

V. LA FERRARA, E. MASSERA, M. L. MIGLIETTA, T. POLICHETTI, G. RAMETTA AND G. DI FRANCIA

CR ENEA Portici, piazzale E.Fermi, 1, I-80055 Portici (Napoli), Italy

Abstract. In the present study, we report on the swelling properties of poly(2-hydroxyethyl methacrylate) (PHEMA) and PHEMA/carbon black (CB) nanocomposite layers, used for volatile organic compounds (VOCs) detection. In particular, by means of a simple optical method based on thin film interferometry, the swelling, is studied and measured in ethanol vapors (max 1%). The method is very sensitive and the swelling in the range of only few nanometers is measured. Interestingly we find that the nanocomposite undergoes a more pronounced swelling process with respect to the pristine polymer. Ethanol diffusion coefficient in the nanocomposite was evaluated too.

1. Introduction

Polymer nanocomposite-based chemical sensors work on the basis of a change of their electrical properties when they come into contact with a specific substance. Due to the solvent absorption, the volume of the polymer matrix increases (swelling) and, at the same time, the volumetric fraction of carbon particles decreases [1–5]. Therefore, the change in film volume results in a modification of the conduction paths throughout the film that is measured as a change in the electrical resistance. Swelling measurement is thus very important to study the sensing layer operating mechanism and to optimize the sensor device performances.

Optical methods, such as white light interferometry are commonly used in commercial measuring system [6] to estimate transparent film thickness on reflective substrate. A method based on white light interferometry for the evaluation of the swelling of non-scattering materials, such as (PHEMA) has been already proposed [2].

Here we show that white light interferometry can be profitably used for the swelling evaluation of scattering materials such as polymer/carbon black nanocomposites.

2. Experimental

The PHEMA series of samples were prepared by dissolving the polymer (4 wt%) in hexafluor-2-propanol (HFiP). PHEMA/CB layers were obtained by

P. Malcovati et al. (eds.), *Sensors and Microsystems: AISEM 2009 Proceedings*, Lecture Notes in Electrical Engineering 54, DOI 10.1007/978-90-481-3606-3_5,

dispersing carbon black Pearls® 2000 (CABOT Corporation) (5.5 wt% with respect to polymer) in the PHEMA (1 wt%)/HFiP solution. After sonication, the solutions were spin-coated on SiO_2 coated crystalline Si wafers. For this study films with thickness between 100 and 300 nm have been deposited. Reflectance VIS spectrum fringes were acquired using a S2000 Oceanoptics spectrometer. Thickness was evaluated using an Ocean Optics Nanocalc software [6].

The sensible layers were mounted in the test chamber and analyzed upon VOCs exposure. The experimental setup consists of a small chamber based on a modified multiple tube-fitting connector (Fig. 1a, b). Sample is placed on the top of an adjustable stage (Fig. 1a-1). Fiber optic reflection probes are connected to a VIS-NIR light source through SMA connector and aligned with the sample (Fig. 1a-3). The reflected light beam is collected by the fiber and directed to the spectrophotometer.

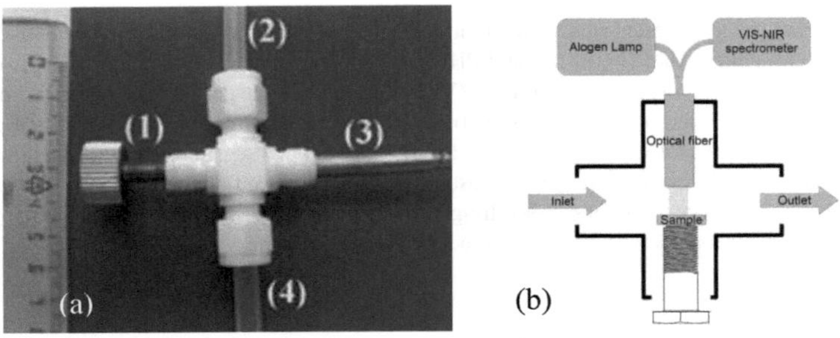

Figure 1. (**a**) photograph of the chamber for optical characterization in controlled environment: (1) sample holder, (2) gas inlet, (3) fiber optic holder, (4) gas outlet; (**b**) a scheme of the chamber for optical characterization.

Measurements in controlled environment were performed using a Gas Sensor Characterization System better described elsewhere [7] connecting the optical chamber (Fig. 1a-2) to the gas output of the electrical test chamber. Ethanol at concentration of 3,000 ppm and 12,500 ppm in N_2 was used as testing analyte.

3. Results and discussion

In Fig. 2a the effect of 30 nm polymer swelling on the optical spectrum is shown. As expected the film expansion produces a fringe shift. It can be noted that the curve slope near the inflection points quite rigidly translates. The result of a thickness variation may then be particularly marked: at a fixed wavelength a less than 4% variation of the total thickness yield a reflectance variation larger than 50%.

This effect is more clearly visible in Fig. 2b where the theoretical variation, at a fixed wavelength, of the reflectance intensity as a function of the layer expansion is reported. If we fix the wavelength at an inflection point, a ±15 nm thickness variation results in a ±40% linear reflectance change, so that extremely small swelling values can be evaluated.

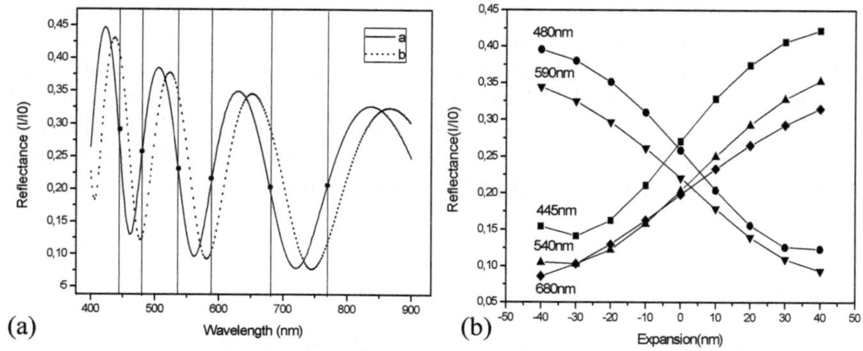

Figure 2. (**a**) Reflectance spectrum of a 866 nm thick PHEMA layer (dot line) and fringe shift for a swelling of 30 nm (solid line). Black dots indicate inflection point. (**b**) Simulated reflectance vs swelling for a 866 nm thick layer taken at the five inflection point wavelength: negative inflection point (IP) at 445 nm; positive IP at 480 nm; negative IP at 540 nm; positive IP at 590 nm.

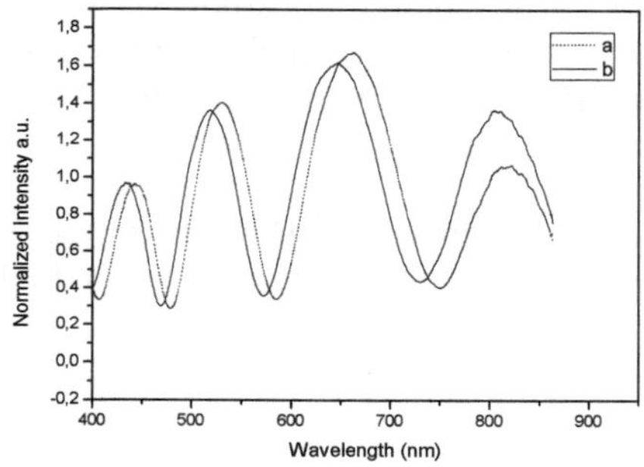

Figure 3. Reflected Light intensity spectra of PHEMA sample under nitrogen and exposed to 12,500 ppm of ethanol vapors.

Figure 3 shows the typical interference fringe shift due to film expansion; the picture relates to PHEMA sample exposed to 12,500 ppm of ethanol vapors; a clearly observable effect of the volume change is evident. Using the simulated reflectance variation vs swelling, a layer expansion of 19 nm with a confidence of 10% is calculated.

The swelling evaluation on PHEMA/CB gives a film expansion of 10 nm for the sample exposed to 3,000 ppm while a swelling of 31 nm is found when the same sample is exposed to 12,500 ppm of ethanol. As expected, the film expansion amount depends on the concentration of the analyte. Nanocomposite layers exhibit a slightly larger swelling as compared to pure PHEMA layers. A similar result has been previously reported for nanostructured fluoropolymer (CF_x) and metal-CF_x composite films [8], where a possible active role of the metal filler has been suggested. In our case, preliminary BET investigations have highlighted the occurrence of porosity in the nanocomposite that could then account for the observed swelling.

The swelling kinetics of PHEMA/CB sample at two different analyte concentrations exhibits the typical $(t)^{1/2}$ law at least for the first 600 s, so a diffusive mechanism can be then assumed. Using the Fick equation, the diffusion coefficient of PHEMA exposed to 12,500 ppm of ethanol is equal to 1×10^{-14} cm^2/s, while the diffusion coefficients for PHEMA/CB exposed to 3,000 ppm and 12,500 ppm of ethanol are 1×10^{-16} cm^2/s and 0.5×10^{-14} cm^2/s respectively. Data for pure PHEMA are comparable with those reported by Goustouridis et al. [2].

4. Conclusions

In this work we developed a simple optical method based on thin film white light interferometry to measure the swelling sensitivity of polymer and nanocomposite layers. The method allows to easily correlate polymer nanocomposite film expansion to light intensity variation at a fixed wavelength, as long as the swelling is small with respect to the whole layer thickness. It was successfully applied to PHEMA and to nanocomposite carbon black/PHEMA layers which are a typical choice in chemical sensor technology. Interestingly, we have found that the nanocomposite undergoes a more pronounced swelling process with respect to pure polymers.

Acknowledgment The authors gratefully acknowledge the Ministry of Universities and Research for financial support as part of Public-Private Laboratory TRIPODE.

References

1. S. Chatzandroulis, D. Goustouridis, and I. Raptis, "Characterization of polymer films for use in bimorph chemical sensors", Journal of Physics: Conference Series [10], pp. 297–300, 2005.
2. D. Goustouridis, K. Manoli, S. Chatzandroulis, M. Sanopoulou, and I. Raptis, "Characterization of polymer layers for silicon micromachined bilayer chemical sensors using white light interferometry", Sensors and Actuators B [111–112], pp. 549–554, 2005.
3. A. Convertino, G.Leo, M. Tamborra, C. Sciancalepore, M. Striccoli, M.L. Curri, and A. Agostiano., "TiO_2 colloidal nanocrystals functionalization of PMMA: A tailoring of optical properties and chemical adsorption", Sensors and Actuators B [126], pp.138–143, 2007.
4. A. Carrillo, I. R. Martín-Domínguez, and A. Márquez-Lucero, "Modeling and experimental testing of the effect of solvent absorption on the electric properties of styrene butadiene rubber/carbon black chemical sensors", Sensors and Actuators B [113], pp. 477–486, 2006.
5. M.A. Ryan, M. L. Homer, H. Zhou, K. Manatt, and A. Manfreda, "Toward a second generation electronic nose at JPL: Sensing film optimization studies", Society of Automotive Engineers, 2001-01-2308.
6. http://www.oceanoptics.com/products/nanocalc.asp.
7. L. Quercia F. Cerullo, V. La Ferrara, G. Di Francia, C. Baratto, and G. Faglia, "Fabrication and characterization of a sensing device based on porous silicon", Phys Status Solidi A [182], pp. 473–477, 2000.
8. N. Cioffi, L. Torsi, I. Farella, D. Altamura, A. Valentini, M. Quinto, L. Sabbatini, and P.G. Zambonin, "The swelling of vapor-sensitive fluoropolymers modified with metal nanoparticles: interpretation of the material-vapor interaction mechanism", Sensors and Actuators B [100], pp. 9–16, 2004.

OPTICAL SENSING PROPERTIES TOWARDS ETHANOL VAPORS OF AU-POLYIMIDE NANOCOMPOSITE FILMS SYNTHESIZED BY DIFFERENT CHEMICAL ROUTES

S. CARTURAN[1], A. ANTONACI[1], G. MAGGIONI[1], A. QUARANTA[2], M. TONEZZER[2], R. MILAN[2], G. MATTEI[3] AND P. MAZZOLDI[3]

[1]*I.N.F.N. – Laboratori Nazionali di Legnaro, Legnaro (PD)*
[2]*Department of Materials Engineering and Industrial Technologies, University of Trento, Trento, Italy*
[3]*Department of Physics, University of Padova, Via Marzolo 8, 35131 Padova, Italy*

Abstract. Nanocomposite films containing Au nanoparticles embedded in a fluorinated polyimide (BPDA-3F) were prepared by two different chemical routes and the synthesis was followed by the estimation of the optical sensing capability of the samples in ethanol vapors. This work arises from the well-known ascertainment that optochemical sensing of volatile organic compounds (VOCs) using MNP (metal nanoparticles) characteristic surface plasmon resonance (SPR) absorption peak as the sensing probe is an attractive technique for its simplicity, high stability and sensitivity [1, 2].

1. Introduction

Nanocomposite films obtained through the dispersion of metal nanoparticles in a matrix of insulating materials such as polymers, ceramics or glasses have proved themselves to be of great interest as advanced technological materials for their unique physical properties. Such nanocomposites can in fact be used as materials for novel functional applications in optics, electronics, magnetic, etc. When this kind of hybrid materials are exposed to light, they exhibit a strong absorption maximum at a particular wavelength due to the photon induced oscillation of the electron charge density: this oscillation of the surface electrons is called surface plasmon resonance (SPR) in the case of thin films. Optochemical sensors of volatile organic compounds (VOC's), exploiting the characteristic SPR absorption peak of the metal nanoparticles as the sensing probe, are only one of the possible application of this particular class of materials. In such sensors, the intensity and the position of the plasmon absorption bands, located in the visible region for metals such as silver, gold, copper and nickel, are variable with the dielectric constant of the nanocluster surrounding medium, according to the Mye theory. Nanocomposite films formed by metal nanoparticles dispersed in different hosts are currently studied as suitable systems for gas sensors [3]. In general, polyimides are particularly attractive polymers as the

P. Malcovati et al. (eds.), *Sensors and Microsystems: AISEM 2009 Proceedings*,
Lecture Notes in Electrical Engineering 54, DOI 10.1007/978-90-481-3606-3_6,

dielectric matrix in composites for their outstanding physico-chemical properties, such as chemical inertness, radiation resistance and high softening temperatures (ranging from 300°C to 450°C) [4, 5]. In particular, fluorinated polyimides are transparent in the visible part of the spectrum allowing the detection of the SPRP of the embedded metal nanoparticles. For all these reasons, polyimides are suitable hosts for the entrapment of transition metal nanoparticles for the production of optical gas sensors.

2. Material synthesis

This paper deals with the synthesis of Au-polyimide nanocomposite films and the starting polyimide was the 1, 1-bis(4-aminophenyl)-1-phenyl-2,2,2-trifluoroethane (BPDA-3F). For the synthesis of the nanocomposite films two different chemical routes, described in Fig. 1, were followed: mixing chloroauric acid with the polyamic acid, which is the precursor of the final polyimide, or with a solution in strongly polar solvents containing a fully cured fluorinated polyimide. The solution was added with a certain amount of porogen azodicarbonamide (ADC) and the final resin was spun on different substrates in order to obtain composite thin films. After thermal treatments, porous polyimide thin films containing Au nanoclusters of about 10 nm of diameter were obtained.

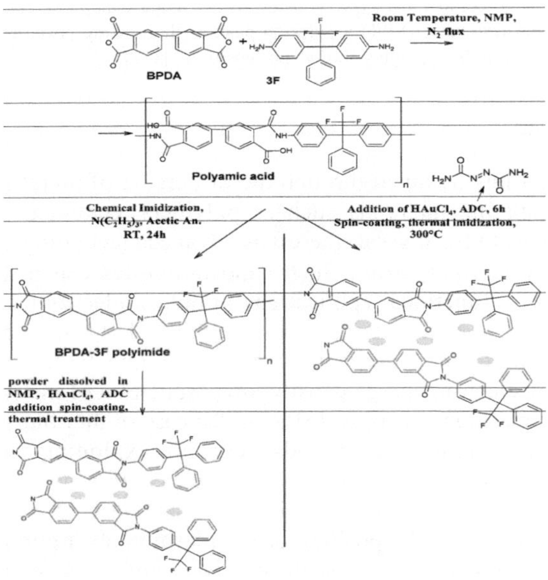

Figure 1. Scheme of the synthetic procedure.

3. Experimental results

The optical sensing capability of the samples was tested by monitoring the changes of the optical absorbance in ethanol vapors. Optical absorption spectroscopy measurements evidenced the SPR peak of gold nanoclusters arising at about 540 nm, as can be observed in the spectra of Fig. 2.

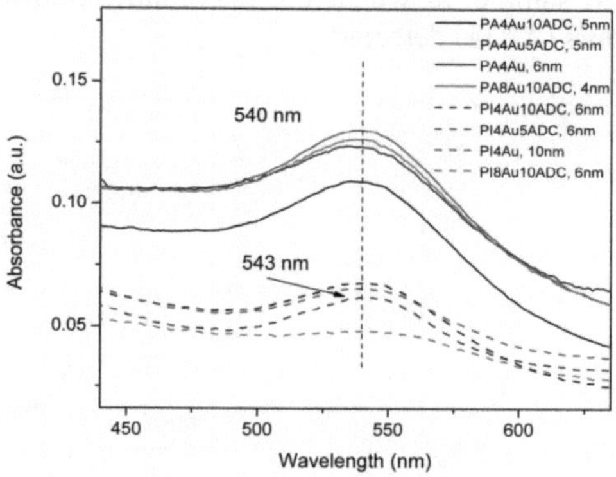

Figure 2. Optical absorption spectra of the thin composite films, collected in air. The SPR absorption maximum is 540 nm for all the samples, except for the sample PI4Au. The nanocrystals size in the legend was derived from XRD measurements.

In Fig. 3 the absorption spectra of the sample PA4Au10ADC, collected upon exposure to 29,500 ppm of ethanol vapors, are reported. As in the other samples (spectra not reported), the absorption features are almost completely recovered by exposing the sample to pure nitrogen, thus proving that the sensing process is fully reversible.

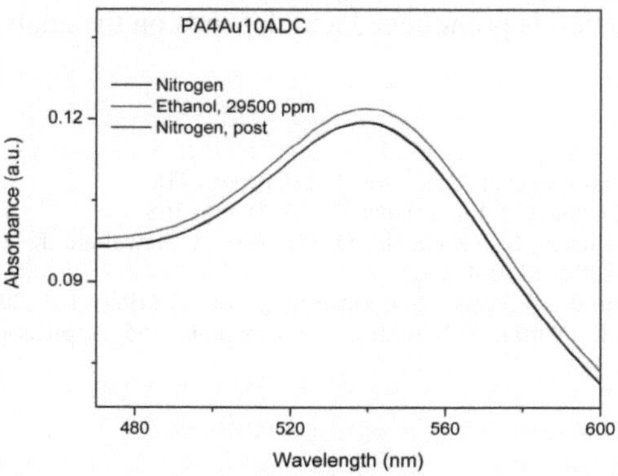

Figure 3. Spectral changes in absorption for the sample PA4Au10ADC upon exposure to ethanol vapors (29,500 ppm).

The swift response and recovery of the samples and reproducibility has been demonstrated by dynamic measurements, as shown in Fig. 4a for the sample PA4Au reported as an example. The kinetic response curve in Fig. 4b has been collected by exposing the sample to alternated pulses of pure nitrogen and vapors of ethanol in nitrogen at different volume concentrations. The absorption signal as a function of time has been collected at the fixed wavelength of 566 nm, for the selected sample, at which the maximum intensity variation upon exposure to the vapors (ΔA) is detected.

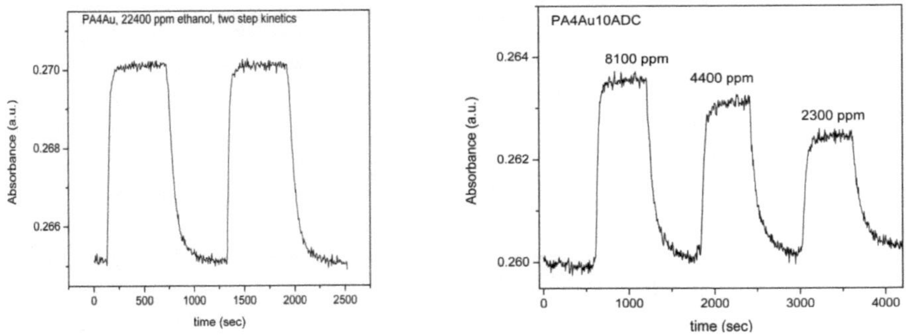

Figure 4. (**a**) Two step kinetics for the sample PA4Au, upon exposure to ethanol vapors (22,400 ppm). (**b**) Optical response profiles of PA4Au10ADC upon exposure to different concentrations of ethanol vapors.

4. Concluding remarks

The experimental results show that the synthesized nanocomposite films are sensible to the exposure to ethanol vapors in a fully reversible way and that the optical signal variation is pronounced and depends on the analyte concentration.

References

1. A. Quaranta, S. Carturan et al., Sens. Act. B 118 (2006) 418.
2. C.S. Cheng, Y.Q. Chen, C.J. Lu, Talanta 73 (2007) 358–365.
3. J. Matsui, K. Akamatsu, S. Nishiguchi, D. Miyoshi, H. Nawafune, K. Tamaki, N. Sugimoto, Anal. Chem. 76 (2004) 1310–1315.
4. C.E. Sroog, Polyimides, J. Polym. Sci. Macromol. Rev. 11 (1976) 161–208.
5. M.K. Ghosh, K.L. Mittal, Polyimides: Fundamentals and Applications; Marcel Dekker: New York, 1996.

OPTICAL VAPORS SENSING CAPABILITIES OF POLYMERS OF INTRINSIC MICROPOROSITY

S. CARTURAN[1], A. ANTONACI[1], G. MAGGIONI[1], A. QUARANTA[2], M. TONEZZER[2], R. MILAN[2] AND G. DELLA MEA[2]

[1]*I.N.F.N. – Laboratori Nazionali di Legnaro, Legnaro (PD)*
[2]*Department of Materials Engineering and Industrial Technologies, University of Trento, Mesiano (TN)*

Abstract. Polymers of Intrinsic Microporosity (PIMs) represent a novel class of polymer-based organic microporous materials that offer unrivaled scope for membrane separations, heterogeneous catalysis and hydrogen storage. In this work, we have explored for the first time the vapors sensing capability of PIMs. The solubility of the polymer in aprotic polar solvents allowed the incorporation of a strongly fluorescent dye (Nile Red). The materials, synthesized by a chemical route, were analyzed by absorption and emission spectroscopy, FT-IR spectroscopy, nitrogen physisorption measurements and X-ray diffraction. In particular the sensing capabilities of the hybrids toward ethanol vapors were tested and the kinetic profiles highlighted the optimal behavior of the films as related to sensitivity, response and recovery speed and reproducibility.

1. Introduction

For definition microporous materials are solids that contain interconnected pores of less than 2 nm in size and this structural characteristic let them possess large and accessible surface areas, typically in the range of 300–2,000 $m^2\ g^{-1}$ [1]. In particular the new Polymers of Intrinsic Microporosity (PIMs), discovered by Budd and co-workers, show that a network structure is not strictly necessary for the generation and maintenance of microporosity within an organic material. On the contrary microporosity can be simply due to the highly rigid and contorted molecular structure so that space-efficient packing in the solid state is not allowed. In general optochemical sensing of volatile organic compounds (VOCs) using dye molecules embedded in polymers is an attractive technique for its simplicity, high stability and sensitivity [2]. In such sensors, the intensity and the position of the absorption or emission bands is variable with the polarity and H-bonding features of the dye molecule surrounding medium. However, great concern derives from the low free volume of polymers, derived from the close packing of macromolecules, which hinder efficient vapor permeation, thus leading to delayed sensor response. In the case of PIMs, a high free volume, and consequently high vapors permeation rate, is expected owing to the high specific surface area (>300 m^2/g) and the presence of interconnected microcavities (size

P. Malcovati et al. (eds.), *Sensors and Microsystems: AISEM 2009 Proceedings*, Lecture Notes in Electrical Engineering 54, DOI 10.1007/978-90-481-3606-3_7,

0.7–1.1 nm). The aim of this work is the valuation of the optical vapors sensing capabilities of the hybrids toward ethanol vapors and for this purpose a strongly fluorescent dye (Nile Red) has been incorporated to the polymer, thanks to the good solubility of the polymeric matrix in aprotic polar solvents: the fluorescence features of Nile Red undergo visible changes upon exposure to alcohol vapors, owing to the strong solvatochromic effect induced by the change in polarity and H-bonding capability of the dye microenvironment.

2. Material synthesis

Powders of PIMs were prepared by the aromatic nucleophilic substitution reaction of cathecol with perfluorobiphenyl in the presence of potassium carbonate, as detailed in Fig. 1. In a 50-mL round-bottom flask equipped with a condenser 3.34 g (0.01 mol) of perfluorobiphenyl, 1.702 g (0.005 mol) of cathecol and 2.76 g (0.02 mol) of anhydrous potassium carbonate were charged with 25 mL of *N,N*-dimethylformamide (DMF). The reaction was carried out at 120°C for about 5 h. The precipitated white solid was collected by filtration, dried under vacuum, dissolved again in tetrahydrofuran (THF) and precipitated in methanol. IR (KBr): 1,100–990 cm^{-1} (C–F stretching), 1,317 cm^{-1} (C–O–C stretching), 1,490 cm^{-1} (C=C stretching). Powders dissolution in THF and addition of Nile Red, followed by spin-coating, afforded thin films of hybrid materials.

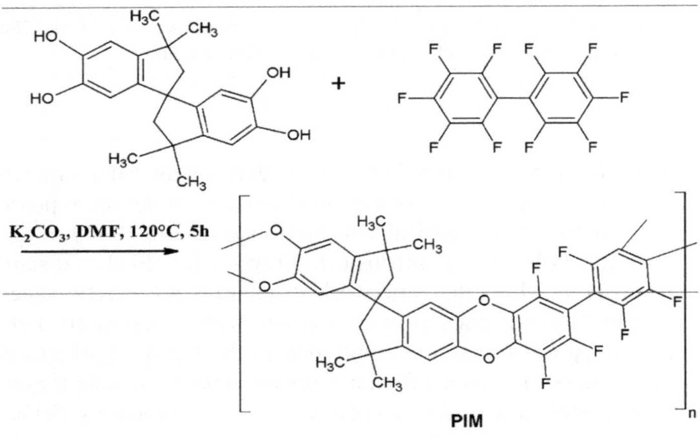

Figure 1. Scheme of the PIM synthetic procedure.

3. Experimental results

3.1. *BET surface area analysis and XRD*

The actual microporosity was verified throughout measurements of the particle surface area and of the pore size, using an instrument (ASAP 2020, Micromeritics) able to measure the adsorption and desorption isotherms of nitrogen at 77 K.

After a very strong degassing (2.10^{-3} m bar), the surface area was determined by the BET method: the presence of the micropores was proved by the high gas adsorption at low pressures (Fig. 2) and a surface area of 308 m^2/g was calculated.

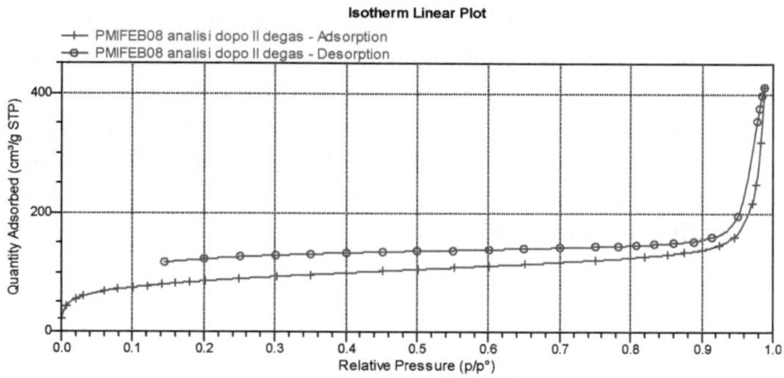

Figure 2. Adsorption and desorption isotherms of nitrogen at 77 K.

The powders were also analyzed by XRD, in order to derive the *d*-spacing of the macromolecular network, through the Bragg formula. The results, reported in Fig. 3, pointed out a bimodal distribution of macromolecules spacing at 1.4 nm and 0.7 nm, in agreement with the micropore size distribution derived from physisorption measurements.

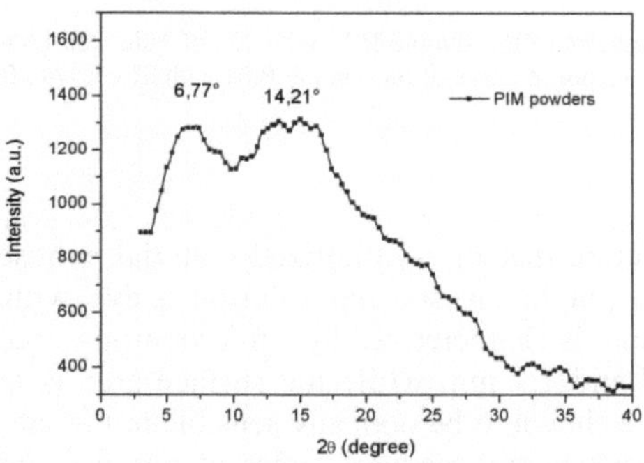

Figure 3. XRD spectra of the PIM powders: the peaks position, used for the calculation of d-spacings through the Bragg rule, is reported.

3.2. *Optical characterization of PIM*

The spin coated thin films of the hybrid materials were subjected to an optical characterization in order to value the use of them as optical sensors of organic vapors. The emission spectra, recorded in static air, fluxing nitrogen and saturated ethanol vapors are reported in Fig. 4a. Both a strong red-shift of the emission band and a hyperchromic effect were observed in presence of the analyte. The optical features display a rapid and almost complete recovery upon exposure to nitrogen flux. The change of intensity and position of the peak of fluorescence was correlated to the solvatochromic effects and to specific effects, like the formation of H-bonding between the dye and the analyte. The kinetic response curve in Fig. 4b was collected by exposing the sample to alternated pulses of pure nitrogen and vapors of ethanol in nitrogen at constant volume concentrations. The emission signal as a function of time was collected at the fixed wavelength of 620 nm, at which the maximum intensity variation upon exposure to the vapors (ΔI) is detected; the optical signal variation is pronounced and reproducible and drift phenomena in the signal intensity in presence of both nitrogen and analyte are negligible.

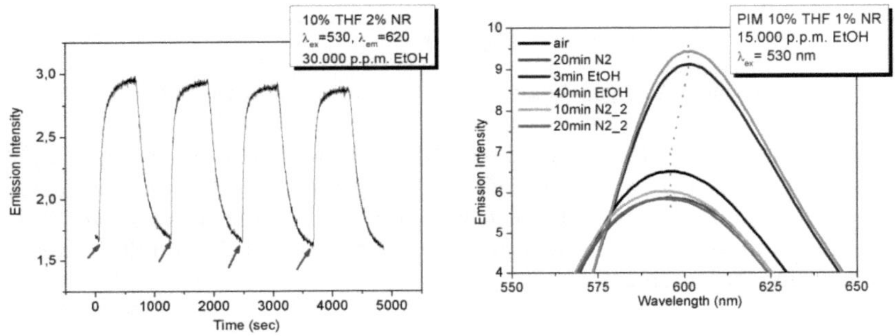

Figure 4. (**a**) Emission spectra of the sample PIM with 1% of Nile Red upon exposure to different atmospheres. (**b**) Kinetic response curve of the sample PIM with 2% of Nile Red.

4. Conclusions

The results demonstrate that the synthesized material, whose very good transparency allowed to put in the starting solution a dye with an high yield of fluorescence emission, is characterized by a microporous structure: the pore size varies between 0.7 and 1.1 nm, while the surface area is about of 300 m^2/g. Moreover PIMs have shown to be optically sensible to the introduction of vapors of ethanol, with response and recovery times of few minutes and with a good sensibility also at low concentrations.

References

1. N. B. Mckeown and P. M. Budd, “Polymers of intrinsic microporosity (PIMs): organic materials for membrane separations, heterogeneous catalysis and hydrogen storage”, Chemical Society Reviews, 17 March 2006.
2. D. Li, C. A. Mills, and J. M. Cooper, Sens. Act. B 92 (2003) 73.

FOCUSED ION BEAM AND DIELECTROPHORESIS AS GROW-IN-PLACE ARCHITECTURE FOR CHEMICAL SENSOR

V. LA FERRARA[1], B. ALFANO[2], E. MASSERA[1] AND G. DI FRANCIA[1]

[1]*ENEA Research Center, P.le E. Fermi, 1 80055 Portici (NA), Italy*
[2]*University "Federico II", Piazzale Tecchio 80, 80125 Napoli, Italy*

Abstract. In this work, we present a novel architecture for nanodevice manufacturing, combining an electrochemical process, such as dielectrophoresis, with focused ion beam apparatus. This approach simplifies the growth-pattern fabrication and simultaneously demonstrate the possibility to produce single nanowires of different materials with fully controlled dimensions, position, alignment and electrical contacting. Focused Ion Beam is used to deposit platinum microelectrodes on silicon/silicon nitride substrate. Dielectrophoresis is employed for assembling the single nanowire and precisely positioning it between the nanocontacts. Single nanowire based devices are tested as chemical sensors confirming the reliability of this innovative technology.

1. Introduction

Recently efforts to utilize single nanowire in the realization of field effect transistors [1–3], chemical sensors [4–6] and photodetector [7] have been reported. The techniques used to fabricate these devices are random dispersions of suspended nanowires in solution and subsequent selection of single nanowire onto a substrate with prepatterned electrodes [1, 2, 4, 7], electron and ion beam induced deposition [3, 5] and electrochemical process [6]. Although these methods are adequate for demonstrating the operational characteristics of individual devices, they have some drawbacks. Random nanowires dispersions based devices deteriorate single nanowire through manipulation and show a limited controllability for nanowire spatial location [1, 2, 4, 7]. Single nanowire growth by means of electron and ion beam induced deposition takes place in highly localized sites but the nanowire can be fabricated only in platinum or in tungsten depending on the metal-organic gas precursor used [3, 5, 8]. As concern electrochemical process, it results a good technique both for growing single nanowire and for having a variety of nanowires from nanoparticles suspended in aqueous solution [6, 9]. Such devices can be improved, for example, in the growth-pattern procedure.

In this work, we present an architecture for nanodevice manufacturing, combining an electrochemical process, such as dielectrophoresis (DEP), with focused

P. Malcovati et al. (eds.), *Sensors and Microsystems: AISEM 2009 Proceedings*,
Lecture Notes in Electrical Engineering 54, DOI 10.1007/978-90-481-3606-3_8,

ion beam (FIB) apparatus. This approach simplifies the growth-pattern fabrication and simultaneously demonstrate the possibility to produce single nanowires of different materials with fully controlled dimensions, position, alignment and electrical contacting.

FIB is used to deposit platinum microelectrodes on silicon/silicon nitride substrate. DEP is employed for assembling the single nanowire and precisely positioning it between the nanocontacts. Single nanowire based devices are tested as chemical sensors confirming the reliability of this innovative technology.

2. Experimental

A crystalline silicon wafer coated with 100 nm of Si3N4 has been used as starting substrate. FIB column, integrated in a FEI Quanta 200 3D dual beam system, has been used for platinum nanoelectrodes patterning. Highly focused gallium ion beam interacts with the platinum organometallic precursor gas. Metal-organic compounds are dissociated and the non volatile part of the compound is deposited onto the sample surface to form electrodes. Using 30 kV as accelerating voltage and 10 pA as emission current for ion beam emission, platinum nanoelectrodes about 8 μm spaced are deposited onto the substrate. The feed solution for DEP processing has been prepared dissolving crystalline Pd(acetate)2 in 10 mM HEPES buffer solution. A function generator and an oscilloscope have been used as the alternating current (AC) source and for monitoring the applied electrical signal, respectively. 2 μl of saturated solution have been deposited by casting onto the substrate, previously washed in isopropyl alcohol, deionised water and dried in a nitrogen flow.

Palladium nanowire has been grown between the electrodes by applying a 10 Vpp (Vpeak-peak) sinusoidal electric field at 300 kHz. The process stops when the opposite electrodes are short-circuited by the forming nanowire.

The single palladium nanowire based device works as a hydrogen sensor. The device has been first characterized in direct current (DC) condition at room temperature and in ambient air, showing an ohmic current-voltage (I–V) characteristic between –1 V ÷ 1 V, with a typical electrical resistance of about 40 kΩ.

A volt-amperometric technique, at constant bias, has been then employed for sensor DC electrical characterization in a controlled gas-flow environment, premixed with dry carrier in the desired percentage by mass flow meters and continuously controlled by means of an on-line Fourier transform infrared spectrometer [10]. All the tested devices have been biased at 5 mV. Total gas flow has been set at 500 sccm.

3. Results and discussion

In Fig. 1 an ionic image of the device, tested as hydrogen sensor, is shown. The single palladium nanowire is branched, thin, and its diameter is about 100 nm. Nanowire composition has been investigated by energy dispersive spectrometry (EDS) analysis confirming that the wire is only composed by palladium [11].

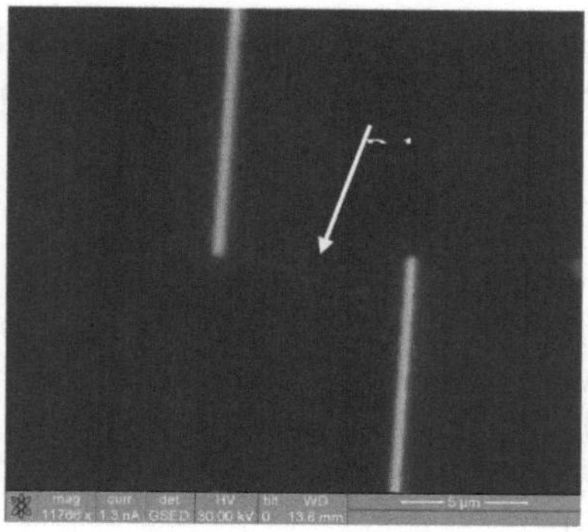

Figure 1. Single palladium nanowire growth by DEP onto two FIB patterned electrodes.

A typical electrical response, when the device is under hydrogen gas, is shown in Fig. 2. Device based on single palladium nanowire has been maintained, for 30 min, in a dry air flow for the zero current (baseline) monitoring. Hydrogen concentrations ranging from 0.1% to 1% have been then introduced in the test chamber and the device electrical response has been recorded.

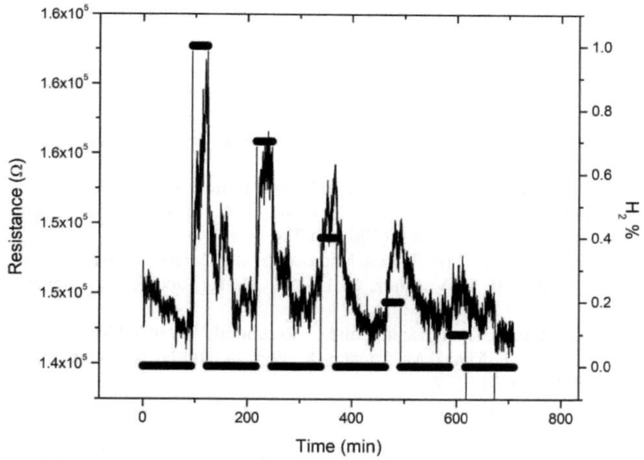

Figure 2. Sensor resistance (left axis) vs time, recorded at room temperature ranging from 1% to 0.1% H_2 concentration in dry air (right axis).

Upon exposure to hydrogen, palladium reacts to form the more resistive palladium hydride (PdHx where x is the atomic ratio H/Pd), as a consequence electric current decreases.

The permeation of hydrogen through palladium nanowire is a reversible dissociative chemisorption process: when dry air is introduced in test chamber, device shows a reversible behavior [12]. Device response time is 2 min ± 0.1 min at 0.5% hydrogen exposure.

Device resistance linearly increases between 0.1% and 1% and the sensitivity of sensor has been calculated to be 0.48 Ω/ppm ± 0.02 Ω/ppm.

4. Conclusions

We have shown how a device based on a single palladium nanowire can be fabricated in well controllable conditions using FIB and DEP combination. DEP is versatile and nanowire length, diameter and position can be controlled if nanopatterned electrodes are realized. A sensor device has been fabricated and it operates as a hydrogen sensor at room temperature. Within the testing range, the sensor is sensible and quite reversible. This result shows that the combined use of FIB and DEP techniques is an innovative and powerful strategy for realizing nanodevices based on single nanowire.

Acknowledgment The authors gratefully acknowledge the Ministry of Universities and Research (MUR) for financial support as part of Public-Private Laboratory TRIPODE (Technologies and Research for the application of polymers in electronics devices).

References

1. C. Liu, L. Dai, R. M. Ma, W. Q. Yang, and G.G. Qin, Journal of Applied Physics 104, (2008) 034302
2. K. Kim, H. Kang, H. Kim, J. S. Lee, S. Kim, W. Kang, and G.-T. Kim, Applied Physics A, (2008) DOI 10.1007/s00339-008-4787-5
3. L. Rotkina, J. -F. Lin, and J. P. Bird, Applied Physics Letters 83, (2003) 4426
4. L. Liao, Z. Zheng, B. Yan, J. X. Zhang, H. Gong, J. C. Li, C. Liu, Z. X. Shen, and T. Yu, Journal of Physical Chemistry C 112, (2008) 10784
5. T. -Y. Choi, B. Kang, and D. Poulikakos, Microscopy and Microanalysis 13, (2007) 1498
6. M. Yun, N. V. Myung, R. P. Vasquez, C. Lee, E. Menke, and R. M. Penner, Nanoletters 4, (2004) 419
7. G. A. O'Brien, A. J. Quinn, D. A. Tanner, and G. Redmond, Advanced Materials 18, (2006) 2379
8. K. Mitsuishi, T. Noda, T. Mano, M. Tanaka, K. Furuya, and N. Koguchi, Japanese Journal of Applied Physics 46, (2007) 6277
9. K. Ramanathan, M. A. Bangar, M. Yun, W. Chen, A. Mulchandani, and N. V. Myung, Electroanalysis 19, (2007) 793
10. L. Quercia, F. Cerullo, V. La Ferrara, G. Di Francia, C. Baratto, and G. Faglia, Physica Status Solidi 182, (2000) 473
11. V. La Ferrara, B. Alfano, E. Massera, and G. Di Francia, Journal of Nanoscience and Nanotechnology 9, (2009) 2931
12. T. B. Flanagan and W. A. Oates, Annual Review of Materials Science 21, (1991) 269

A NOVEL APPROACH FOR THE PREPARATION OF METAL OXIDE/CNTs COMPOSITES FOR SENSING APPLICATIONS

G. NERI[1*], A. BONAVITA[1], G. MICALI[1], G. RIZZO[1], M.-G. WILLINGER[2], E. RAUWEL[2] AND N. PINNA[1]

[1]*Department of Industrial Chemistry and Materials Engineering, University of Messina, C/da di Dio, Vill. S. Agata, 98166 Messina, Italy*

[2]*Department of Chemistry, CICECO, University of Aveiro, 3810–193 Aveiro, Portugal*
**E-mail: neri@ingegneria.unime.it*

Abstract. In this report we describe the preparation of metal oxide coated–carbon nanotubes (MeOx/CNTs) with a controlled thickness of the inorganic coating by a new atomic layer deposition (ALD) process based on non-aqueous sol-gel reactions. The as deposited films are amorphous and the coatings on the inside and outside the nanotube present a high conformality. The electrical and sensing properties of thin films of VOx-coated carbon nanotubes prepared by ALD have been investigated. The developed VOx/CNTs-based sensor devices show enhanced performance for NO_2 monitoring compared to uncoated CNTs-based sensors.

1. Introduction

The growth of thin metal oxide films on the walls of carbon nanotubes (CNTs) is of outmost importance in many practical applications. However, a common feature of conventional deposition techniques is that the metal oxide coating does not always cover the entire nanotube surface. Moreover, the thickness reproducibility of the coating is difficult to obtain.

To overcome these drawbacks, we applied a novel synthesis procedure based on the atomic layer deposition (ALD) technique. ALD is based on a reaction between reactants that are separated into successive surface reactions [1]. In this manner, the precursors are kept separated until the adsorbed species react at the surface in a self-limiting process, i.e. without the presence of a gas phase reaction. However, in the conventional ALD process, an annealing treatment is often necessary to accomplish the final conversion to desired metal oxide.

Recently, we introduced a new non-aqueous sol-gel approach for the ALD of metal oxides [2]. By using carboxylic acids as oxygen source instead of the more traditionally used ones (e.g. water, ozone, oxygen, etc.) it was shown that several metal oxides such as hafnia and titania can be grown from the respective metal alkoxide precursors even at low temperatures. The surface reaction leading to the M–O–M bond formation is self-limited and takes place via an

P. Malcovati et al. (eds.), *Sensors and Microsystems: AISEM 2009 Proceedings*,
Lecture Notes in Electrical Engineering 54, DOI 10.1007/978-90-481-3606-3_9,

ester elimination condensation step from the reaction between a metal alkoxide and surface carboxylate species. As a result, the metal oxide growth is self-limited, does not needs of any annealing treatment and high quality and purity films can be obtained. In this report we describe the general preparation of metal oxide coated–carbon nanotubes (MeOx/CNTs) with a controlled thickness of the inorganic coating. The electrical and sensing characterization of VOx-CNTs films are also reported.

2. ALD process

The coating of the CNTs (supplied by Applied Science PR-24-PS and Nanocyl NC3101) was done in an ALD reactor working in exposure mode from metal alkoxide precursors (hafnium tert-butoxide, titanium isopropoxide and vanadium n-propoxide) and acetic acid. Pure nitrogen gas flow, varied from 5 to 50 sccm, was used as carrier and purging gas. The CNTs were treated with concentrated HNO_3 at 100°C for 2 h before deposition. The substrate temperature was set to 200°C. Metal precursor and acetic acid vapors were introduced into the reactor through an ALD valve. In a typical experiment the valves were opened for 0.02 s for carboxylic acid and 1 s for the metal alkoxides. The residence and purging periods were 20 and 15 s, respectively. The growth per cycle under these conditions was around 0.05 (hafnium), 0.06 (titanium) and 0.04 (vanadium) nm/cycle.

The coating of the CNTs walls occurs through surface reactions leading to the M–O–M bond formation via two steps: (i) formation of surface carboxylate species; (ii) ester elimination condensation step from the reaction between the metal alkoxide and surface carboxylate species (see Scheme 1).

$$\equiv\text{M–OR'} + \text{RCOOH} \rightarrow \equiv\text{M–OOCR} + \text{R'OH} \qquad (i)$$

$$\equiv\text{M–OOCR} + \text{M-OR'} \rightarrow \equiv\text{M–O–M}\equiv + \text{RCOOR'} \qquad (ii)$$

Scheme 1. Surface reactions occurring during ALD process. (i) Formation of surface carboxylate species; (ii) ester elimination condensation step from the reaction between the metal alkoxide and surface carboxylate species leading to M–O–M bond formation. R and R' indicate alkyl groups.

It is should note that the M–O–M bonds are formed without a hydrolysis step. In this case the oxygen is provided by the metal alkoxide. As a result, the metal oxide growth is self-limited, does not needs of any annealing treatment and high quality and purity films can be obtained. As such, atomic scale control over the film thickness during the whole deposition process is inherent to the process.

The morphology of the obtained hybrid materials is nicely revealed by transmission electron microscope (TEM), as shown in the images recorded for the VOx-CNTs sample (Fig.1a–c). In the TEM, the contrast rich, darker regions on the outer and inner walls of the CNTs correspond to the metal oxide layers deposited by the ALD process (Fig.1a). The elemental maps (Fig.1b) clearly prove that the tubes are coated not only on the outer surface but also on the inside. A high-resolution TEM image showing the metal oxide thickness of the VOx-CNTs sample is reported in Fig.1c. The coating shows the features of an amorphous film, only a few nanometers thick (about 2 nm), and the thickness evaluated along the whole surface of the tubes is uniform.

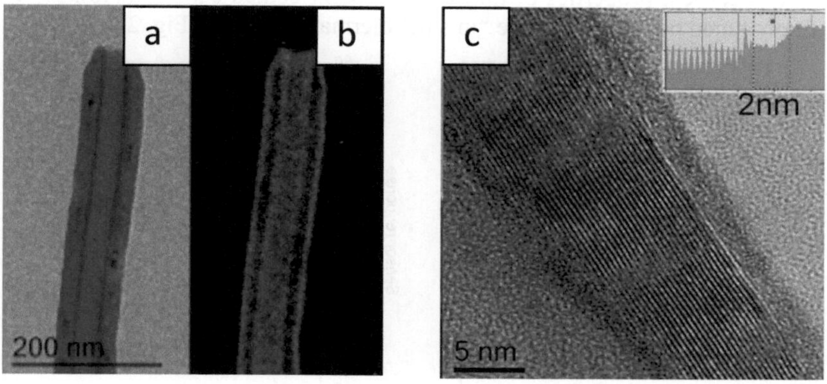

Figure 1. (**a**) TEM image; (**b**) elemental map of a VOx-CNT showing the coating film on the outside and inside of the tube walls; (**c**) high-resolution TEM image showing the metal oxide thickness.

3. Sensing tests

The combination of metal oxides such as TiO_2 or SnO_2 with CNTs has attracted much attention in view of applications as resistive gas sensors. In the aim of demonstrating the novelty and applicability of the above demonstrated ALD deposition process, we investigated the electric and sensing properties of VOx-coated carbon nanotubes.

To fabricate the sensor devices, VOx-CNTs films were deposited on alumina substrates (3 × 6 mm) supplied with interdigitated Pt electrodes and heating elements. Then, in order to stabilize the microstructure of the sensing film, a thermal treatment at 400°C for 2 h was carried out. The sensing properties of the above described V_2O_4-CNTs hybrids were evaluated via monitoring their response to NO_2 in air, under a dry gas total stream of 200 sccm,. Measurements were performed collecting the sensors resistance by an Agilent 34970A multimeter data acquisition unit and the sensor heater by DC power supply Agilent E3631A.

As a reference, the properties of the uncoated CNT-based sensor have been evaluated in order to first clarify the electric behavior and response of the CNTs support to different NO_2 concentrations. The tests were carried out by exposing the pure CNTs sensor to 6.5 ppm of NO_2 diluted in dry air at an operating temperature of 150°C. It can be observed that the addition of 6.5 ppm of NO_2 causes a decrease of the resistance (Fig. 2a), typical of a p-type behavior.

VOx-CNTs based devices behave in a different way compared to CNTs-based sensor. Indeed, the resistance increases upon exposure to NO_2 gas indicating that VOx-CNTs hybrid materials have a n-type behavior. Moreover, they show a higher response compared to the uncoated CNTs-based sensor. The thermal treatment of the sensing layer also plays a key role. The sensitivity to NO_2 increase treating the sensor in air at temperature around 200°C, suggesting that the electrical properties of thin films of VOx-coated carbon nanotubes can be finely tuned by an appropriate "in situ" thermal treatment (Fig. 2b).

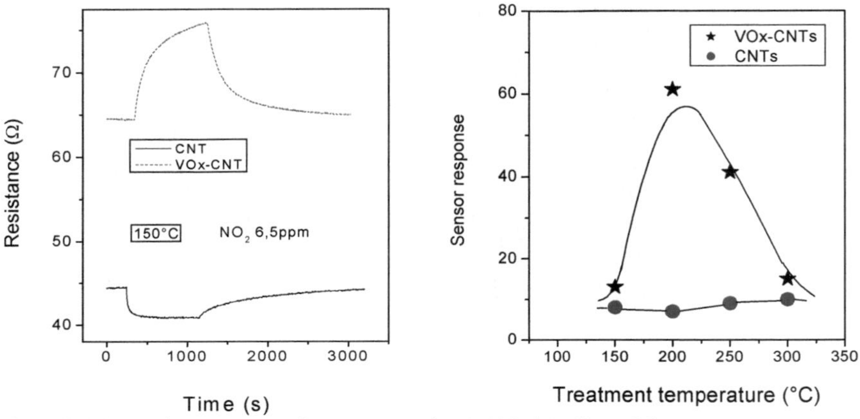

Figure 2. (**a**) Transient response of sensors treated at 150°C; (**b**) effect of the treatment temperature on the response of the VOx-CNTs and the pure CNTs sensors.

Results above reported suggest that the contact between VOx layer and CNTs contribute strongly to the improvement of sensing performance. An explanation of this synergic interaction can be given based on the formation of a p–n hetero-junction between the CNTs and the VOx coating [3, 4]. Indeed, multi-walled CNTs show a p-semiconducting behavior, whereas the VOx-CNTs nanocomposite hybrids behave as n-type semiconductors. Then, the adsorption of NO_2 on the metal oxide modifies the depletion layer at the surface of the film and, as an effect of this, also modifies the depletion layers at the n-metal oxide/p-CNTs interface. This combined effect modulates the conductivity properties of the VOx-CNTs hybrid nanocomposite and can be considered to be responsible for the improved response of the sensor.

References

1. M. Ritala and M. Leskela, In Handbook of Thin Film Materials (Ed. H. S. Nalwa), Vol. 1, Academic Press, San Diego, CA, 103–159 (2002).
2. E. Rauwel, G. Clavel, M. -G. Willinger, P. Rauwel and N. Pinna, Angew. Chem., Int. Ed., 47, 3592–3595 (2008).
3. M. -G. Willinger, G. Neri, E. Rauwel, A. Bonavita, G. Micali and N. Pinna, Nano Lett., 8, 4201–4204 (2008).
4. M. G. Willinger, G. Neri, A. Bonavita, G. Micali, E. Rauwel, T. Herntrich and N. Pinna, Phys. Chem. Chem. Phys. (in press).

THE IMPACT OF NANOPARTICLE AGGREGATION IN LIQUID SOLUTION FOR TOXICOLOGICAL AND ECOTOXICOLOGICAL STUDIES

M. L. MIGLIETTA[1], G. RAMETTA[1], G. DI FRANCIA[1], A. BRUNO[2], C. DE LISIO[2], G. LETER[3], M. MANCUSO[3], F. PACCHIEROTTI[3], S. BUONO[4] AND S. MANZO[4]

[1]*Dipartimento Tecnologie Fisiche e Nuovi Materiali Sezione Nanomateriali e tecnologie per dispositivi (FIM MAT-NANO), ENEA Centro Ricerche Portici, Piazzale E. Fermi, 80055 Portici (NA), Italy*

[2]*CNR-INFM and Dipartimento Scienze Fisiche , Università Federico II, via Cintia Complesso MSA, 80126 Napoli*

[3]*Dipartimento Biotecnologie, Protezione della Salute e degli Ecosistemi (BIOTEC) ENEA Centro Ricerche Casaccia, via Anguillarese 301, 00123 Santa Maria di Galeria, Roma*

[4]*Dipartimento Ambiente, Cambiamenti Globali e Sviluppo sostenibile (ACS), ENEA Centro Ricerche Portici, Piazzale E. Fermi, 80055 Portici (NA), Italy*

Abstract. Given the widespread applications of nanotechnology products, the evaluation of the potential hazards related to human exposures and environmental release of nanoparticles has become an urgent need. The development of a safety database for nanoscale particles is evolving as new particles, materials and exposure scenarios are being developed.

Several studies have been focused on the importance of developing adequate physicochemical characterization of nanomaterials, in the relevant media being utilized, prior to undertaking experiments for *in vitro* and *in vivo* toxicity assessment.

In this work we utilize two optical techniques: Dynamic Light Scattering (DLS) and Time Resolved Fluorescence Polarization Anisotropy (TRFPA) for measuring particle size, size distribution and aggregation kinetics of nanoparticles in aqueous solution for toxicological and ecotoxicological studies. The application of these two optical methods gives a careful description of the aggregation status of nanoparticles and its evolution on different time-scale.

1. Introduction

Given the widespread applications of nanotechnology products, the evaluation of the potential hazards related to exposures to nanoscale particles has become an urgent need. Recently, several studies have been focused on the importance of developing adequate physicochemical characterization of nanomaterials prior to undertaking experiments for *in vitro* and *in vivo* toxicity assessment [1]. Particle size, size distribution, morphology, surface area are all important properties that

P. Malcovati et al. (eds.), *Sensors and Microsystems: AISEM 2009 Proceedings*, Lecture Notes in Electrical Engineering 54, DOI 10.1007/978-90-481-3606-3_10,

must be accurately characterized in the relevant media being utilized for toxicity assessment. In fact, potential physicochemical changes can occur to nanoparticles properties while in solution and these changes may have a significant impact on the observed toxicological responses [2].

Currently, there are no well-defined techniques for the characterization of wet nanomaterials in aqueous or biological solutions [2, 3]. The use of multiple techniques should be attempted wherever possible to develop a more complete understanding of the system. Among the wide variety of available techniques for size measurements only few of them are suitable for physiological environments. Optical methods are the most desirable to this purpose due to their accuracy, flexibility and being not intrusive.

This study describes the setup and preliminary investigations devoted to carry out two nanotoxicological experiments.

2. Nanoparticles characterization

Surface area and pore size distribution of nanoparticle powders have been firstly characterized with BET technique [4]. Dynamic Light Scattering (DLS) and Time Resolved Fluorescence Polarization Anisotropy (TRFPA) have been employed for measuring particle size, size distribution and dispersion state of nanostructured materials in physiological media.

DLS measures Brownian motion by illuminating the particles with a laser and analyzing the intensity fluctuations of the scattered light. These temporal fluctuations are then related to the size of the particles.

TRFPA is a short-pulse laser technique that allows to resolve the fast rotational dynamics (in the picoseconds scale) of nanoparticles in water suspensions and, thus, to obtain nanoparticles' size by exploiting their self fluorescence.

2.1. *Evaluation of cancer effects on mice*

Mouse models of skin cancer provide an ideal *in vivo* model to identify skin carcinogens and to study the mechanisms involved on tumorigenesis. In particular, Car-S mice, phenotypically selected for susceptibility to two-stage skin carcinogenesis, are elective for studies of papilloma and squamous cell carcinoma (SCC) pathogenesis (Fig. 1).

The defense mechanisms of the organism against tumorigenesis is usually mediated by immunological responses or non-specific inflammatory reactions [5, 6]. Hereafter, the set up of a pilot study on the early inflammatory response potentially induced by multiwalled carbon nanotubes on the skin of Car-S mice is described. Mouse skin has been treated two times per day with Acetone/MWNT suspension for 5 days. A weak increment of epidermal cells inflammation (boxes) between sham ad treated groups is observed.

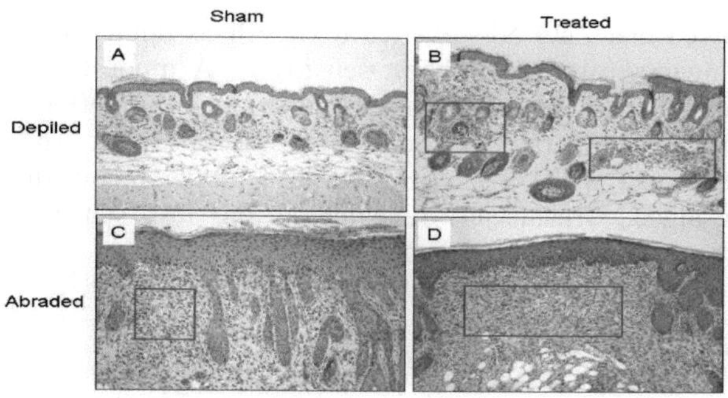

Figure 1. Figure show a different inflammatory degree in depilated or abraded (rows) Car-S mouse skin after MWNTs treatment compared with sham group (colons).

2.2. *Ecotoxicological effects*

Environmental impact of ZnO nanoparticles on aquatic systems has been deeply investigated [7], while little is known about its potential toxicity on soil and seawater. In this work, preliminary studies on potential toxicity of ZnO nanoparticles in soil and in seawater are reported. Toxicity test batteries with organisms belonging to different trophic levels and biological complexity (Table 1) have been executed. To assess if toxicity is attributable to ZnO nanoparticles or to soluble Zn^{2+} the same test batteries were carried out both on ZnO and $ZnCl_2$ contaminated soil and seawater.

Table 1. Ecotoxicity test batteries.

Soil test battery (standard method)	Ref	Seawater test battery (standard methods)	Ref
Plant germination and root inhibition test (*L. sativum*)	[8]	Bioluminescence acute test (*V. fischeri*)	[12]
Algae growth inhibition test (*S. capricornutum*)	[9]	Crustaceans acute test (*A. salina*)	[14]
Ostracod survival and growth inhibition test (*H. incongruens*)	[10, 11]	Algae growth inhibition test (*D. tertiolecta*)	[15]
Bioluminescence acute test (*V. fischeri*) .	[12]		
Genotoxicity micronuclei test (*V. faba*)	[13]		

OECD Standard soil samples have been spiked with ZnO and $ZnCl_2$ at 0.029% w/w. Analogously, toxicity tests on seawater have been performed on ASTM Standard seawater samples spiked with ZnO and $ZnCl_2$ ranging from 0.03 to 9 mg/l. All the abovementioned concentrations are referred to the zinc content.

Results showed that ZnO NPs exerted toxic effects upon all test organisms. In particular, algae showed the highest sensitivity. A modest genotoxic effect was also observed. Soluble Zn^{2+} produced comparable effects except for *L. sativum* where a moderate biostimulation was observed. The highest NP toxicity was obtained with *V. fischeri*, while a moderate biostimulation was observed for algae. Crustaceans revealed as the less sensitive organisms. Soluble Zn^{2+} seems to be responsible of toxic effects.

3. Conclusions

DLS and TRFPA have been applied to the characterization of nanoparticles dispersions in two toxicity and ecotoxicity pilot studies. The application of these optical methods gives a careful description of the aggregation status of nanoparticles and should allow the preparation of suitable nanoparticles dispersions for experimental trials.

References

1. D.B. Warheit, How meaningful are the results of nanotoxicity studies in the absence of adequate material characterization? Toxicol. Sci. 101(2), 183–185, 2008
2. K. Powers, V. Krishna, S. Wasdo, B. Moudgil, S. Roberts, Characterization of nanoscale particles for toxicological evaluation. Toxicol. Sci. 90, 296–303, 2006
3. R. Murdock, L. Braydich-Stolle, A. Schrand, J. Schlager, S. Hussain, Characterization of nanomaterial dispersion in solution prior to *in vitro* exposure using Dynamic Light Scattering technique. Toxicol. Sci. 101(2), 239–253, 2008
4. S. Brunauer, P.H. Emmett, E. Teller, Adsorption of gases in multimolecular layers. J. Am. Chem. Soc. 60, 309–319, 1938
5. A. Saran, T. Neveu, V. Covelli, D. Mouton, S. Pazzaglia, S. Rebessi, G. Doria, G. Biozzi, Genetics of chemical carcinogenesis: Analysis of bidirectional selective breeding inducing maximal resistance or maximal susceptibility to 2-stage skin tumorigenesis in the mouse, Int. J. Cancer 88, 424–431, 2000
6. M. Mancuso, S. Pazzaglia, M. Tanori, H. Hahn, P. Merola, S. Rebessi, M. J. Atkinson, V. Di Majo, V. Covelli, A. Saran, Basal cell carcinoma and its development. Cancer Res. 64, 934–941, 2004
7. A. Kahru et al., Biotests and biosensors for ecotoxicology of metal oxide nanoparticles. Sensors 8, 5153–5170, 2008
8. APAT, Guida tecnica su metodi di analisi per il suolo e i siti contaminati – Utilizzo di indicatori ecotossicologici e biologici. RTI CTN_SSC 2/2002
9. A. Baun, K.B. Justesen, N. Nyholm, Algal test with soil suspensions and elutriates: A comparative evaluation for PAH-contaminated soil. Chemosphere 46, 251–258, 2002
10. B. Chial, G. Persoone, Cyst-based toxicity tests XII – development of a short chronic sediment toxicity test with the Ostracod Crustacean Heterocypris incongruens: Selection of test parameters. Environ. Toxicol. 17, 520–527, 2002a
11. B. Chial, G. Persoone, Cyst-based toxicity tests XIII – development of a short chronic sediment toxicity test with the Ostracod Crustacean Heterocypris incongruens: Methodology and precision. Environ. Toxicol. 17, 528–532, 2002b
12. Microbics Corp. MicrotoxW users manual. Microbics Corporation, Carlsbad, CA, 1992
13. N. Kanaya et al., Vicia faba chromosomal aberration assay. Mutat. Res-Fund. Mol. M. 310(2), 231–247, 1994
14. U.S. EPA 600-4-85-013, 1985
15. IRSA, Quaderni IRSA-CNR 39:1–116, 1978

PREPARATION AND ELECTRICAL-FUNCTIONAL CHARACTERIZATION OF GAS SENSORS BASED ON TIO_2 NANOMETRIC STRIPS USING IMPEDANCE SPECTROSCOPY

C. DE PASCALI, L. FRANCIOSO, S. CAPONE AND P. SICILIANO
CNR-IMM Institute for Microelectronics and Microsystems, via Monteroni, 7310 Lecce -Italy

Abstract. TiO_2 nanometric aligned polycrystalline wires ("strips") have been prepared by top down patterning of a TiO_2 thin film. An electrical-functional characterization of the prepared TiO_2 strips based sample was carried out in comparison with the corresponding TiO_2 thin film sample. Electrical Impedance Spectroscopy (IS) was used as electrical characterization technique and a dedicated experimental set-up was arranged in order to carry out EIS measurements in controlled environment.

1. Introduction

During last years, the interest in nanostructures developed by a bottom up approach, like metal oxide single crystal nanowire (NW) assemblies, or by a top down approach, like nanopatterned thin films, experienced a fast growing, especially addressed to gas detection. Indeed, both type of nanostructures are expected to show significantly enhanced performances as gas-sensing materials, due to an high surface/volume ratio that increases the active area exposed to gas molecules, a nano-lateral dimension that modulates the electrical transport properties up to a complete depletion of charge carriers of the nanoelement. Although single-crystal nanowires grown by bottom-up technologies ensure an higher stability as compared to polycrystalline ones [1, 2], many limitations, linked to difficulties in controlling the growth process, in obtaining an ordered assembly, and in integrating single or multiple NWs in MEMS based-gas sensing devices, remain in spite of a lot of efforts devoted to find simple technological solutions.

Otherwise a top-down approach gives the advantage of an easy integration of nanostructured materials in a gas microsensor device preserving the attracting properties of a nano-dimensional structure. We chose a top down approach for the development of a nanometric structure consisting of an array of ordered TiO_2 nanowires. We demonstrated that, applying a cheap and custom nanopatterning process based on UV lithography and dry plasma etching, it's possible to obtain a controlled and reproducible integration of sol-gel nanostructured titania NWs from a solid thin film onto a silicon oxide mesa. In the following the developed

P. Malcovati et al. (eds.), *Sensors and Microsystems: AISEM 2009 Proceedings*,
Lecture Notes in Electrical Engineering 54, DOI 10.1007/978-90-481-3606-3_11,

TiO_2 nanowires will be named “strips” to stress the peculiarity of their native links to the substrate and their straightforward integration in the device. In particular, we developed a nanometric TiO_2 strips-based chemoresistive gas sensor, whose working principle is based on chemiadsorption and charge transfer processes between the analyte molecules and the surface of the sensing material, which cause a conductance variation of the semiconducting oxide. Extensive research has to be carried out in order to understand the electrical conduction properties of low-dimensional structures and to know how the nanolateral dimension can influence the final sensing behavior of the device.

In this work, electrical characterization of the TiO_2 strips array by AC Impedance Spectroscopy (IS) in controlled environment is proposed in order to investigate the influence of the nanomachining process on the electrical properties of the gas sensor in comparison to the solid sensitive film without patterning. Oppositely to a direct current measurement, which gives general and global information about the sensor electrical behaviour, an alternating current measurement method permits to discriminate and extract the contribution of different regions of the device, according to the nature of conduction processes and the mechanisms of gas/solid interactions.

2. Experimental

The fabrication process starts with the synthesis of a suitable sol-gel TiO_2 solution by a wet-chemical method, and with its deposition on a thermally oxidized silicon wafer by spin coating, followed by the necessary thermal annealings to obtain a fully crystallized (anatase phase) TiO_2 thin film with a polycrystalline morphology, as described in a previous work [4].

Figure 1. SEM image of a TiO_2 strip-based gas sensor bounded onto a TO-39 socket (on the *left*), and a magnification of the active area with the array of aligned TiO_2 strips perpendicular to the fingers interdigitated electrodes (on the *right*).

Next, a top down nanopatterning of the TiO_2 film was realized. A suitable resist mask, reproducing the desired geometry of aligned strips with low lateral dimension, was defined by a standard 365 nm ultraviolet (UV) photolithographic process, and transferred on the TiO_2 film by a next etching process in high pressure SF_6 plasma. Such top down micromachining process allowed to obtain an array of over 1,500 well-aligned TiO_2 strips; each single strip has a width ranging between 90 and 180 nm, it's 40 nm thick and about 1,400 μm long. Details of the experimental process are reported elsewhere [3]. Finally, in a front side approach interdigitated platinum electrodes and heater/thermometer were deposited in a single step onto the sensor surface. Figure 1 reports the image of the realized device.

The impedance measurements were carried out evaluating the sensor electrical response at operative temperatures ranging between 200°C and 550°C, under dry air and different concentrations of air-ethanol mixture (50, 100, 150 ppm in dry air) by a gas-mixing apparatus (Fig. 2). Complex impedance measurements were acquired using an HP 4284A precision LCR meter. The impedance measurements were performed by applying a sinusoidal potential of 30 mV (0 V of DC bias) in the frequency range 20 Hz–1 MHz.

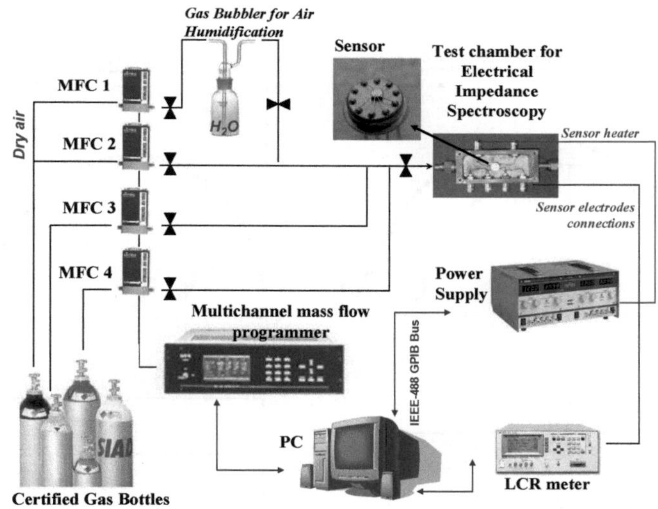

Figure 2. Scheme of the experimental apparatus for Electrical Impedance Spectroscopy measurements in controlled environment.

3. Results and discussion

Electrical-functional characterization by Electrical Impedance Spectroscopy in controlled environment were carried out in two different kinds of gas sensors, i.e. a TiO_2 strips based gas sensor and a second one based on a continuous TiO_2 film that didn't undergo the top down patterning. The aim was to investigate the effect of the sensing layer nanomachining on gas detection mechanism. Impedance spectroscopic analysis is conducted by applying a small amplitude AC voltage of variable frequency to the device and measuring the resulting current. The impedance, determined as ratio between voltage and current phasors, is a complex quantity having real component and imaginary component

(Z = Z’−jZ’’), with j imaginary unity. The frequency dispersion of the impedance data can be represented graphically through the Nyquist plot, in which the negative imaginary component of the impedance is plotted against the real impedance component at each excitation frequency.

However, analysis and interpretation of experimental results are complicated and arduous, since it’s necessary to build an equivalent circuit representing the investigated system and determine the parameter values of all the elements of the circuit. A Complex Non-linear Least Square (CNLS) fitting procedure was applied by using LEVM software [5, 6].

As expected, high resistivity of titania, whose conduction processes are thermally activated, requires relatively high sensor working temperatures; TiO_2 strip-based gas sensor, resulted more resistive than the thin film-based one, hence only at 500°C and 550°C the impedance spectra of the samples could be compared.

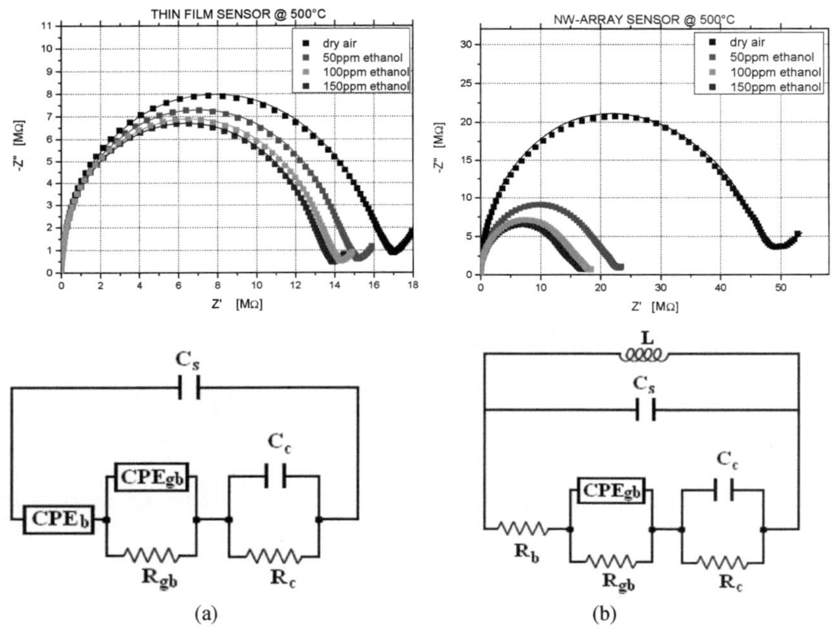

Figure 3. Fitting of impedance data associated to electrical responses of a TiO_2 thin film (**a**), and a TiO_2-strip based sample at 500°C exposed to different concentration of ethanol, with the corresponding equivalent circuit.

The experimental data were fitted using CNLS procedure by two distinct equivalent circuits, that model, even if in simplified form, the electrical behavior of two gas sensors. The complex impedance of both sensors under dry air and ethanol-air mixture at two temperatures of interest were interpolated according to the simplified equivalent circuits and Nyquist plots of experimental data (points) and fitting results (line) are reported in Fig. 3 with the corresponding equivalent circuits.

The polycrystalline structure of the TiO_2 sensitive layer were considered in building the equivalent circuits of the systems under investigation. The sensitive layer of the sensor consists of interconnected oxide grains. The bulk contribution can be represented by a resistor R_b for the TiO_2 thin film and by a constant phase element CPEg, close at an ideal resistor Rg, fot the TiO_2 strip sample. Since the ionosorption of oxygen at the grain surface produces a potential barrier and corresponding depletion layers at the intergranular contacts, the grain-grain contact can be modeled by a parallel $R_{gb}CPE_{bg}$ combination, where CPE_{gb} is close to an ideal capacitor C_{gb}: the resistor represents the high resistive depletion layers, the capacitor describes the double-layer capacitance at the interface between adjacent grains. The electrode contact can also be represented by a R_cC_c element.

The presence of an inductance in the equivalent circuit of strip array-based sensor is the aspect that countersigns the two circuital models. Its presence confirms the deep modification of the electrical behavior of the gas sensor after the patterning process. This result shows the existence of a self-induced inductive coupling between the aligned NWs of titania, probably due to their reciprocal micrometric distance. No direct dependence of fitted values of the such circuital component to the ambient gas atmosphere has been observed.

4. Conclusions

A study of electrical impedance spectroscopy (IS) of gas sensors based on an array of TiO_2 nanowires ("strips") obtained by a top down nanopatterning process of TiO_2 polycrystalline TiO_2 film prepared by sol-gel method. The IS measurements were carried out both in dry air and in presence of ethanol, and the data compared with those of a not-patterned TiO_2 thin film. Equivalent circuit models were developed and a very good fit for the impedance spectra were obtained. The presence of an inductance in the equivalent circuit of strip array-based sensor was the most relevant results. However, a lot of efforts have to be devoted to a better understanding of the gas sensing mechanisms and electrical transport by IS data analysis.

References

1. Y. Xia, P. Yang, Y. Sun, Y. Wu, B. Mayers, B. Gates, Y. Yin, F. Kim, H. Yan, Adv. Mater. 15 (2003) 353–389.
2. X. Fang, L. Zhang, J. Mat. Sc. Tech. 22 (2006) 1–18.
3. L. Francioso, A.M. Taurino, A. Forleo, P. Siciliano, Sens. Actuat. B 130 (2008) 70–76.
4. L. Francioso, D.S. Presicce, M. Epifani, P. Siciliano, A. Ficarella, Sens. Actuat. B 111–112 (2005) 52–57.
5. J.R. Macdonald, CNLS (Complex Nonlinear Least Square) Immittance Fitting Program – LEVM Manual. Version 8.0, Issue data: March 2003.
6. D.W. Marquardt, SIAM J. Appl. Math. 11 (1963) 431.

PIEZOELECTRIC LOW-CURING-TEMPERATURE INK FOR SENSORS AND POWER HARVESTING

M. FERRARI, V. FERRARI, M. GUIZZETTI AND D. MARIOLI
Dipartimento di Elettronica per l'Automazione, Università degli Studi di Brescia
Via Branze, 38–25123 Brescia

Abstract. In this paper a piezoelectric ink with curing temperature as low as 150°C is presented. It is intended to be used in MEMS, sensors, actuators and power harvesting field. Ink composition and fabrication steps are described. Electrical characterization of a ink film sample realized with screen printing technology is reported, showing a dielectric constant $\varepsilon_r = 100 \pm 10\%$. The ferroelectric hysteresis loop of the piezoelectric ink has been measured by means of a custom PC-based measurement system. A remanent polarization of 0.8 $\mu C/cm^2$ was measured with a maximum applied electric field of about 13 MV/m. In addition, the possibility of depositing the ink with a drop-on-demand technology was demonstrated, by using a commercial piezoelectric microdispensing device with a 80-μm diameter orifice. Dissolving the PZT ink in ethylene glycol diacetate, wt ratio 1:3, single drops with a diameter of 175 μm have been deposited.

1. Introduction

In the fabrication of sensored systems and microelectromechanical systems (MEMS), piezoelectric sensors, actuators and mechano-electrical energy converters are widely used [1–4]. They are often realized with thin-film or thick-film processes and materials like piezoceramics such as lead zirconate titanate (PZT). Piezoceramic materials with inorganic binders must be typically fired at temperatures above 900°C to establish suitable mechanical and piezoelectric properties. In MEMS field, the use of such high temperatures implies the possible problem of silicon contamination due to lead diffusion from the PZT into the substrate. Moreover, the use of screen printing can mechanically damage the MEMS device. In order to overcome these limitations this paper is focused on the study of a piezoelectric low-curing-temperature ink based on PZT, that can be deposited onto various types of standard (silicon, metal, alumina) and non-standard (plastics, FR4, Kapton) substrates by either screen-printing or drop-on-demand (DOD) technologies [5]. The resultant devices are intended to be used as sensors or energy converters for power harvesting.

P. Malcovati et al. (eds.), *Sensors and Microsystems: AISEM 2009 Proceedings*,
Lecture Notes in Electrical Engineering 54, DOI 10.1007/978-90-481-3606-3_12,

2. Sample realization

The prepared piezoelectric ink is a compound of PZT milled powder (ε_{33} = 4,100, $d_{31} = -260{\cdot}10^{-12}$ C/m^2, $\rho = 7.5$ g/cm^3) and low temperature polymeric vehicle-binder mixed in ratio of 2:1 wt. The obtained compound was homogenized using an ultrasonic bath to limit ink heating and consequent binder evaporation. The PZT ink was first screen printed on alumina substrate using a 325 mesh screen, dried for 1 h at room temperature and then cured at 150°C for 10 min. By repeating these three print-dry-fire steps, a film sample with a final thickness of about 125 μm was realized. The ink density was measured by weighting a known volume of ink both wet and cured, obtaining values of ρ_{wet} = 1 g/cm^3 and ρ_{cured} = 4 g/cm^3 respectively. Bottom and a top electrodes were also screen printed and fired at low temperature using a conductive Ag based polymeric ink. Figure 1 shows a typical film sample. The electrical resistance R and capacitance C of the film sample were measured with an impedance analyzer (HP4194A) at a frequency of 100 Hz obtaining the values R = 2.8 MΩ and C = 4 nF respectively. The relative dielectric permittivity ε_r was calculated from the measured capacitance C and the physical dimensions obtaining a value of ε_r = 100 ± 10%.

Dimensions:
- W = 18 mm
- Thickness = 125 μm
- L = 31 mm

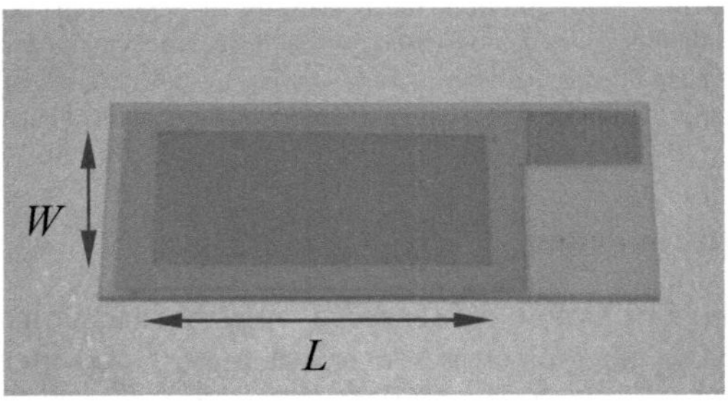

Figure 1. The manufactured piezoelectric film sample.

3. Ferroelectric hysteresis loop measurement

A tailored PC-based measurement system, based on a modified Sawyer-Tower circuit and whose block diagram is shown in Fig. 2, was designed and realized in order to measure the ferroelectric hysteresis loop of the samples.

This was obtained by applying to the sample an amplitude-modulated sinusoidal signal V_x, at a frequency of 500 Hz, and measuring the output signal V_y of the charge amplifier. Applied electric field E and induced dielectric displacement D are respectively related to V_x and V_y through sample physical dimensions. The typical obtained results are shown in Fig. 3, where in the top-left inset the applied excitation electric field is shown. Maximum applied electric field E had a peak amplitude of about 13 MV/m with a consequent remanent polarization of 0.8 μC/cm^2. A linear approximation of the D–E relationship, using the value of ε_r previously determined under small signal excitation, was plotted and compared with experimental data. Concordance between linear approximation slope and the slope of the inner hysteresis loop resulted, as shown in Fig. 3.

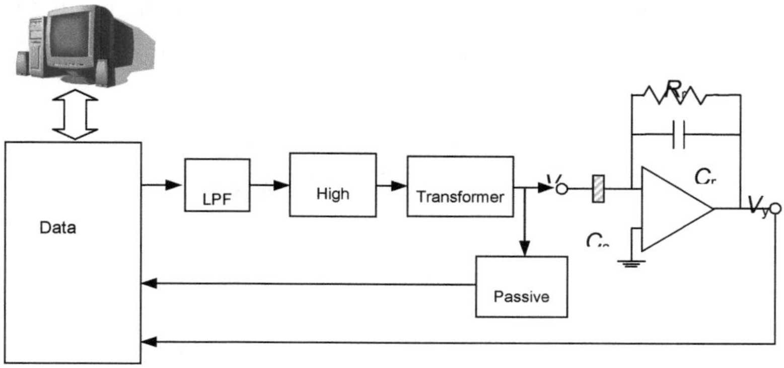

Figure 2. Block diagram of the hysteresis loop measurement system.

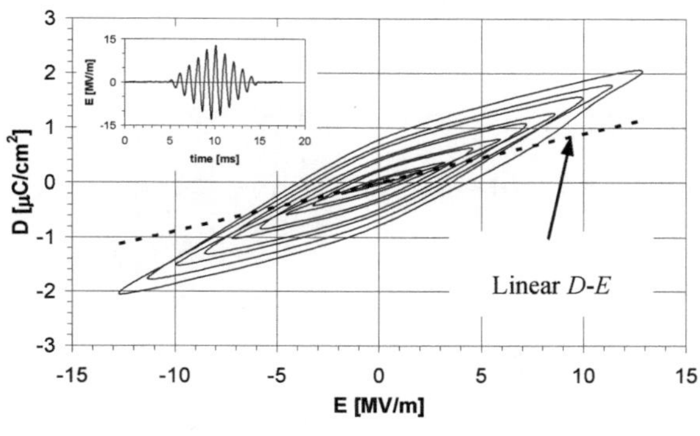

Figure 3. Experimental hysteresis loop. In the top-left inset is shown the excitation electric field.

4. Ink deposition using DOD technology

The drop-on-demand (DOD) technology was then tested to deposit the piezoelectric ink. A commercial piezoelectric microdispensing device (Microfab MJ-AB-80) with a 80-μm diameter orifice was used. This device permits the drop ejection of fluids with viscosities lower than 40 cP. In order to meet this requirement, the piezoelectric ink was dispersed in ethylene glycol diacetate dissolving the polymeric binder and creating a PZT powder suspension. The weight ratio of the ink-ethylene solution was about 1:3. The PZT suspension was then deposited onto copper substrate with the microdispensing device driven with a squared pulse train waveform by means of a programmable function generator and a custom designed power amplifier. The microdispensing device and the deposition substrate are shown in Fig. 4.

Each driving pulse has an amplitude of 60 V, a rise and fall times of 3 μs and lasts for 20 μs. The driving frequency was set to 80 Hz while the number of consecutive pulses in a train was varied in order to vary the number of the ejected droplets.

Figure 4. Microdispensing device over a deposition substrate.

Examples of 25-, and 50-drop depositions are shown in Fig. 5. It can be observed that the deposition shape is well defined with a homogeneous content of PZT powder. Figure 5b shows a single drop deposition, with a magnification of 40x. The diameter of the deposited single drop is about 175 μm.

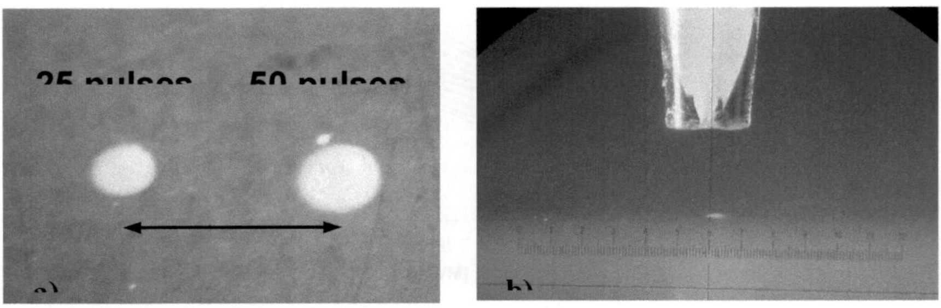

Figure 5. Piezoelectric ink depositions for 25, 50 drops (**a**). Single drop deposition with a magnification of 40× (**b**).

Acknowledgment Italian MIUR for partial funding under the project PRIN 2007 (20078ZCC92).

References

1. S.A. Rocks, Q. Tredez, H.J. Almond, C.P. Shaw, R.A. Dorey, "Bottom up fabrication of a nickel–lead zirconate titanate piezoelectric microcantilevers", Materials Letters, [vol. 63], pp. 88–90, January, 2009.
2. K. Sivanandan, A.T. Achuthan, V. Kumar, I. Kanno, "Fabrication and transverse piezoelectric characteristics of PZT thick-film actuators on alumina substrates", Sensors and Actuators A, [vol. 148], pp. 134–137, November, 2008.

3. V. Ferrari, D. Marioli, A. Taroni, "Theory, modeling and characterization of PZT-on-alumina resonant piezo-layers as acoustic-wave mass sensors", Sensors and Actuators A, [vol. 92], pp. 182–190, August, 2001.
4. M. Renaud, K. Karakaya, T. Sterken, P. Fiorini, C. Van Hoofa, R. Puers, "Fabrication, modelling and characterization of MEMS piezoelectric vibration harvesters", Sensors and Actuators A, [vol. 145–146], pp. 380–386, July–August, 2008.
5. N. Ramakrishnan, P.K. Rajesh, P. Ponnambalamc, K. Prakasan, "Studies on preparation of ceramic inks and simulation of drop formation and spread in direct ceramic inkjet printing", Journal of Materials Processing Technology, [vol. 169], pp. 372–381, December, 2005.

COMPARATIVE BIOAFFINITY STUDIES FOR IN-VITRO CELL ASSAYS ON MEMS-BASED DEVICES

C. RESS[1], L. ODORIZZI[1], C. COLLINI[1], L. LORENZELLI[1], S. FORTI[2], C. PEDERZOLLI[2], L. LUNELLI[2], L. VANZETTI[2], N. COPPEDÈ[3], T. TOCCOLI[3], G. TARABELLA[3] AND S. IANNOTTA[3]

[1]*FBK, Materials and Microsystems Center – BIOMEMS Unit, Via Sommarive 18, 38123 Povo-Tn-Italy*

[2]*FBK, Materials and Microsystems Center – M2B2 Unit, Via Sommarive 18, 38123 Povo -Tn-Italy*

[3]*IFN-CNR, Institute of Photonics and Nanotechnology, Via Alla Cascata 56/C, 38123 Povo-Tn-Italy*

Abstract. Nowadays, bioaffinity studies of materials are playing a key role in the realization of microdevices for a wide number of biological applications in cell-based assays such as drug screening, cell tracking and transfection. The aim of this research is to evaluate the bioaffinity of different substrates often employed in microfabrication techniques, in order to check the optimal materials aimed at enhancing cell growth or improving cell confinement. Bioaffinity has been tested in the presence of nanostructured TiO_2 films produced by Pulsed Microplasma Cluster Source (PMCS). To explore the difference in affinity among these materials, SKOV-3 cells have been cultivated over them by performing studies in terms of morphological appearance and cell density. Both of these aspects have been assessed with fluorescence microscopy and also the surface chemistry has been investigated by means of X-ray Photoelectron Spectroscopy (XPS).

1. Introduction

Bioaffinity represents a fundamental parameter for the appropriate choice of the most suitable materials to develop BioMEMS-based devices aimed at *in-vitro* cell assays. Since there is an increasing attention towards nanoparticle-assembled materials in the fabrication of biocompatible microsystems [1], the presence of a nanostructured TiO_2 film on different substrates has been evaluated throughout a comparative bioaffinity study.

P. Malcovati et al. (eds.), *Sensors and Microsystems: AISEM 2009 Proceedings*, Lecture Notes in Electrical Engineering 54, DOI 10.1007/978-90-481-3606-3_13,

2. Materials and methods

2.1. *Candidate substrates for bioaffinity tests*

Different candidate substrates have been chosen among those most employed for the development of MEMS-based devices (Table 1). As shown in the schematic sketch (Fig. 1), a comparison between these materials (zone 1) and a nanostructured TiO_2 film (zone 2) has been carried out by depositing the last one on half of each substrate.

Table 1. Substrates and relative microfabrication technologies employed for bioaffinity study (Note: a thin layer of chromium has been used as adhesion promoter for Au/Pt layers).

Substrates	Microfabrication technology
Ti (120 nm)	Sputtering
Si_3N_4 (100 nm)	PECVD
SiO_2 (18 nm)	PECVD
Au (150 nm)	Electron beam
Pt (150 nm)	Electron beam

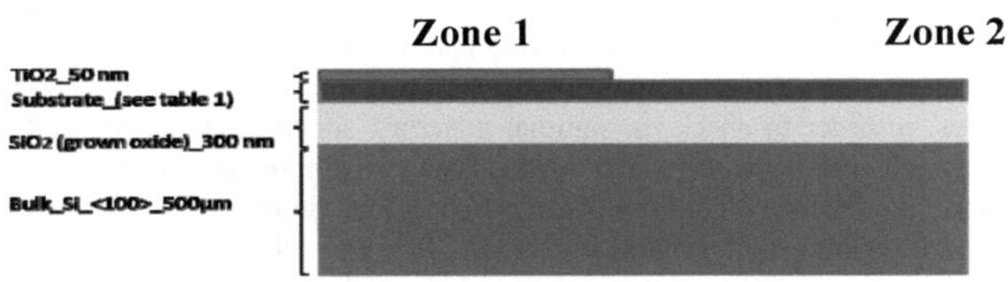

Figure 1. Cross-section of the deposited materials.

The deposition of nanocrystalline TiO_2 clusters has been performed by means of a Pulsed Microplasma Cluster Source (PMCS) [2]. In this technique, clusters are produced by condensation of plasma in a buffer gas after a discharge between two electrodes.

2.2. *Surface chemistry analysis and bioaffinity studies*

The surface chemistry of the materials listed in Table 1 has been analysed by means of XPS (tilt 30°). The relative atomic concentrations have been determined of both substrates (zone 1) and nanostructured TiO_2 film (zone 2). Morphological appearance and cell density have been assessed by cultivating SKOV-3 human ovarian carcinoma cell line on different substrates (15,000 cells/cm^2) and on tissue culture polystyrene (TCPS) as control. After 24 h, cells have been labelled with calcein acetoxymethylester (calcein AM, 0.5 µM, 10 min) and observed by means of fluorescence microscopy analysis (Fig. 2).

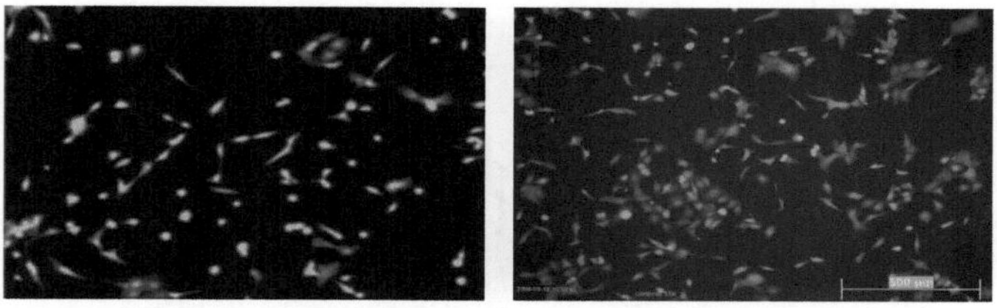

Figure 2. Fluorescence microscopy images of SKOV-3 cells cultured for 24 h on TiO_2 (*left*) and tissue culture polystyrene (TCPS) as control (*right*), after staining with calcein AM. ×10 magnification.

3. Results

3.1. *X-ray photoelectron spectroscopy (XPS) analysis of the substrates*

The atomic concentrations of TiO_2 film and of the selected materials have been determined by XPS analysis. A representative chemical composition of all analyzed nanostructured TiO_2 films deposited on the different substrates is reported in Table 2. Chemical composition of substrates resulted as expected in terms of atomic concentration. However, XPS analysis revealed the presence of carbon contaminations with percentage values ranging from 16 to 50 (Table 3).

Table 2. Atomic concentration of TiO_2 film deposited by PMCS as determined by XPS surface analysis (tilt 30°), expressed in percentage.

Nanostructured film	O (%)	C (%)	Ti (%)
TiO_2	44.4	39.7	15.8

Table 3. Atomic concentration of the analyzed materials as determined by XPS surface analysis (tilt 30°), expressed in percentage.

Substrates	O (%)	C (%)	N (%)	Si (%)	Ti (%)	Au/Pt (%)
Ti (Sputtering)	43.1	41.9	–	–	15.0	–
Si_3N_4 (PECVD)	33.9	16.4	12.3	37.3	–	–
SiO_2 (PECVD)	44.4	23.9	–	31.7	–	–
Au (Electron beam)	19.8	50.7	–	–	–	29.5
Pt (Electron beam)	22.7	42.3	–	–	–	35.1

3.2. *Cell density analysis by means of fluorescence microscopy*

In order to evaluate the effect of the nanostructured TiO_2 film on cell adhesion, both quantitative (which considers the total viable cell density) and morphological analysis (which considers only well adherent spread cells) have been performed (Fig. 3a and b, respectively).

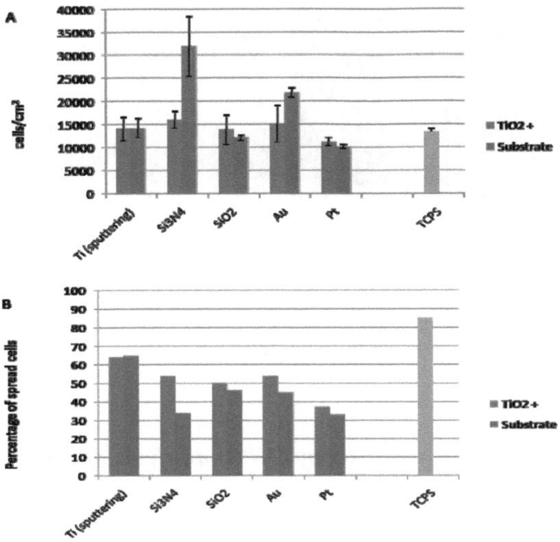

Figure 3. Quantitative (**a**) and morphological (**b**) analysis of SKOV3 cells after 24 h. Cell density (cells/cm^2) and percentage of spread cells are represented.

At 24 h cell density on TiO_2 films and substrates resulted comparable or even higher (i.e. Si_3N_4) to that observed on TCPS. However, if only well adherent spread cells are analyzed (Fig. 3b), it appears that the presence of TiO_2 deposited by PMCS (TiO_2+) increases the percentage of spread cells, improving the biocompatibility of our candidate materials for MEMS-based devices.

4. Conclusions

In this work preliminary results about material surface chemistry and their cell biocompatibility have been presented. Concerning the surface properties, in order to better control surface quality and to assure repeatability of the chemical surface properties the carbon contamination could be removed by optimizing the cleaning protocols. The short time SKOV3 culture analysis (24 h) has demonstrated that the presence of TiO_2 increases the percentage of spread cells suggesting that this nanostructured material could promote cell adhesion. This parameter plays a key role in adherent cell electroporation, as it could improve the overall transfection efficiency. In addition, surface roughness will be analyzed by means of Atomic Force Microscopy (AFM) to evaluate the function of surface morphology on the biological response. Moreover, long term experiments (48–72 h) will be performed to evaluate proliferation and functionalization processes of the TiO_2 surfaces with appropriate organic molecules will be investigated, in order to control the chemical groups interacting with cells and to further enhance the surface biocompatibility.

Acknowledgments This work has been supported by CARITRO Foundation Rovereto, Italy under the project: CELTIC 2006 – "Development of an integrated system based on innovative nano-microfabrication technologies for in vitro- diagnostic assays".

References

1. D. M. Brunette, P. Tengvall, M. Textor, P. Thomsen (editors), "Titanium in medicine: material science, surface science, engineering, biological responses and medical applications". Berlin: Springer, 2001.
2. T. Toccoli, A. Boschetti, L. Guerini, S. Iannotta, S. Capone, P. Siciliano, A. Taurino, "Growth of Titanium Oxide films by Seeded Supersonic Beams for Gas and VOC Sensing Applications", Sensors Journal IEEE [3], 2003.

EFFECT OF THE LAYER GEOMETRY ON INK-JET SENSOR DEVICE PERFOMANCES

A. DE GIROLAMO DEL MAURO[1]*, F. LOFFREDO[1], G. BURRASCA[1], E. MASSERA[1], G. DI FRANCIA[1] AND D. DELLA SALA[2]

[1]*Dipartimento Tecnologie Fisiche e Nuovi Materiali Sezione Nanomateriali e tecnologie per dispositivi (FIM MAT-NANO), ENEA Centro Ricerche Portici, Piazzale E. Fermi, 80055 Portici (NA), Italy*

[2]*Dipartimento Tecnologie Fisiche e Nuovi Materiali Sezione Nanomateriali e tecnologie per dispositivi (FIM MAT-NANO), ENEA Centro Ricerche Casaccia,via Anguillarese, 00123 S. Maria di Galeria (RM) Italy*

Abstract. In this work, we report on ink-jet manufactured chemical sensors based on polymer-carbon black composites. In particular, in order to study the correlation between ink-jet pattern geometry and device response sensors characterized by sensing layers consisting of polystyrene/carbon black composites lines of various shape and surfaces were fabricated. The morphology of films, the electrical characteristics and time stability of sensors were investigated.

1. Introduction

Ink-jet printing is a particularly attractive technique especially for the controlled solution deposition of small quantities of functional materials in conformity with specific patterns [1]. The technique operates ejecting single drops of an ink through very fine nozzles (with a 20–100 μm diameter), at ambient temperature and with no contact or vacuum [2–4]. The only constraint is the requirement to have fluids with suitable viscosity (<20 cP) and surface tension (in the range 28–350 mN m^{-1}) [3].

Recently, different groups have investigated the possibility to use ink-jet technique for chemical sensor applications [5], mainly using conducting polymers such as poly (thiophene) [6] and polyaniline [7] as the sensing film. In a previous work [8], we showed that it is possible to manufacture chemiresistive sensors all made by ink-jet printing. Indeed, sensor devices where both the sensing layer (a conductive surface of polymer/carbon black composite) and transducer (inter-digitated silver contacts) were completely fabricated by ink-jet and characterized versus various VOCs.

Ink-jet is a particularly attractive technology because it opens a way to design a new class taking advantage from the specific features of the manufacturing technique: patterning capability of small quantities of functional materials, efficient

P. Malcovati et al. (eds.), *Sensors and Microsystems: AISEM 2009 Proceedings*, Lecture Notes in Electrical Engineering 54, DOI 10.1007/978-90-481-3606-3_14,

use of materials, possibility to use flexible substrate, low cost of process and industrialization opportunity.

In order to optimize the geometry of sensing layer obtained by ink-jet, in this work we studied the correlation between pattern geometry and device response. Sensors based on polystyrene/carbon black composites lines or surface were fabricated changing ink-jet parameters (drop overlapping, number of overlapped printed layers, the temperature of substrate). In particular, we examined the effect of these ink-jet parameters on the device performances.

2. Experimental section

2.1. *Materials*

The Carbon Black (CB), used in preparing the composites, was Black Pearls 2000, a furnace black material with 1,500 m^2/g specific surface area, with 12 nm average particle size and with 150 g/l density. The 1-methyl-2-pyrrolidinone (NMP), solvent of preparation of composites, and atactic polystyrene (PS) were supplied from Sigma-Aldrich and were used as received.

2.2. *Preparation and characterization of PS/CB ink suspension*

In order to manufacture the ink, the PS (80 mg) was dissolved in hot NMP in a 0.5 wt% concentration. The filler (20 mg) was dispersed in this polymeric solution by ultrasonic bath for 90 min. The final PS/CB suspension was filtered using a 0.2-μm Teflon filter to safely remove agglomerates above the 0.2 μm. This ink, suitably processed to get viscosity (≈2 mPas) and surface tension compatible with ink jet printhead, has been characterized by Dynamic Light Scattering; about 220 nm mean particle dimension (Fig. 1, left) and 70 days time stability (Fig. 1, right) have been found.

2.3. *Sensing devices*

As substrate was used alumina with 5×5 mm^2 dimensions and 0.5 mm thickness. The electrical contacts geometry has been realized by a lift-off photolithographyc process followed by e-beam Au evaporation [9]. Ink-jet single drop on this substrate has about 170 μm diameter. Table 1 lists the ink-jet parameters, the experimental conditions used and the optical and photographic images of the sensing devices prepared. Afterward, each sensor was mounted on a commercial electronic TO8 case by soldering gold wires to the pins and tested in a climatic chamber towards specific concentration of acetone vapors.

3. Results

I–V characterization shows an ohmic behaviour for all devices while the sensor characteristic electrical impedance was stable for a 65 days period (see Fig. 2).

The dynamic responses of sensors based on surface printed at 40°C characterized by different thickness at acetone vapors are reported in Fig. 3.

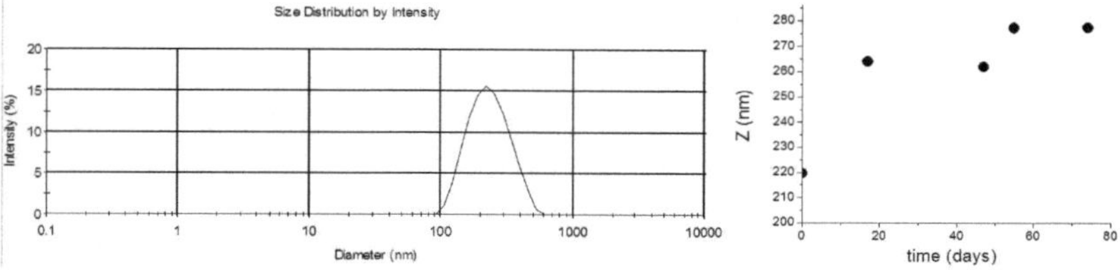

Figure 1. Dynamic Light Scattering (*left*) and time stability (*right*) for polystyrene/carbon black ink suspension in NMP.

It is observed that, decreasing the thickness of sensing material, the sensors show better reversibility and higher responses.

The results show that the dynamic response of sensor based on different geometry is strongly influenced by base resistance.

Table 1. Ink-jet parameters and experimental conditions used to print polystyrene/carbon black in NMP sensing layer.

Geometry	Substrate T(°C)	Layers	Drop overlap	Optical images of films (100×)	Device photo	Series
LINE	50	1	50%			L1_T50
	50	3	50%			L3_T50
SURFACE	40	1	50%		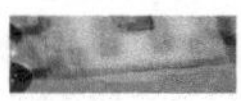	S1_T40
	50	1	50%			S1_T50
	50	1	10%			SD1_T50
	40	3 and 5	50%		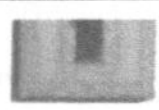	S3_T40 and S5_T40

Critical factors are the number of layers in the line type sensor and the overlapping between the drop in surface type sensor. For the former it is possible to improve the signal/noise ratio increasing the number of printed layer maintaining short response times.

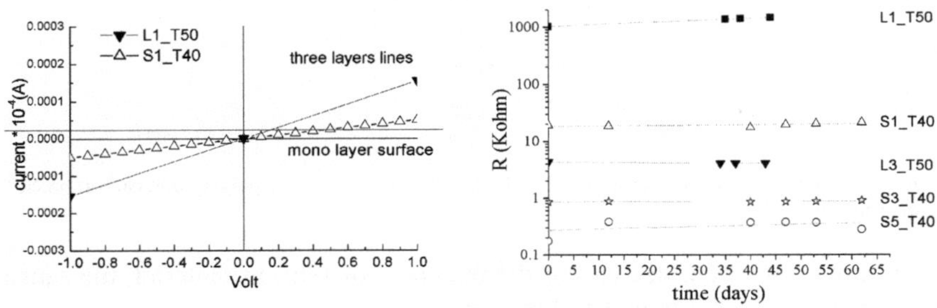

Figure 2. I–V characteristic (*left*) and time stability (*right*) of the devices.

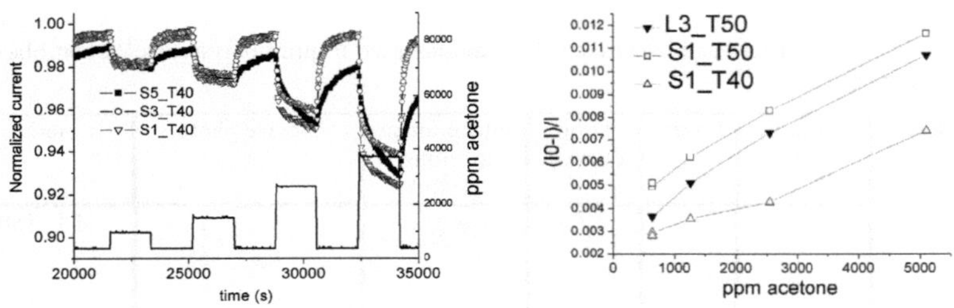

Figure 3. Electrical responses to acetone at 20°C.

4. Discussion and conclusions

We studied and optimized ink-jet parameters (drop overlapping, geometry and number of sensing layers, temperature of substrate) to obtain sensor devices based on PS/CB surfaces and lines.

The comparison between the responses towards acetone vapors shows it is possible to improve sensor performances using a simple linear geometry for the sensing layer. This permits to reduce the time of deposition, sensor device size and its response times. For this type of geometry, the quality of response is clearly affected by ink parameters deposition as the percentage of drop overlapping and number of layers. The work is in progress to design other sensor devices geometries capable to optimize the performances of ink-jet sensor devices.

References

1. R. F. Service, Science, 304, 675 (2004).
2. P. Calvert, Chem. Mater., 13, 3299 (2001).
3. B. J. de Gans, P.C. Duineveld, U. S. Schubert, Adv. Mater., 16, 204, (2004).

4. E. Tekin et al., J. Mater. Chem., 14, (2004).
5. A. Bietsch, J. Zhang, M. Hegner, H. P. Lang, C. Gerber, Nanotechnology, 15, 873 (2004).
6. M. F. Mabrook, C. Pearson, M. C. Petty, IEEE Sens. J., 6, 1435, (2006).
7. K. Crowley, E. O'Malley, A. Morrin, M. R. Smyth, A. J. Killard, Analyst, 133, 391–399, (2008).
8. F. Loffredo, A. De Girolamo Del Mauro, G. Burrasca, V. La Ferrara, L. Quercia, E. Massera, G. Di Francia, D. Della Sala, Sens. Actuat. B. (Submitted)
9. L. Quercia, F. Loffredo, M. Bombace, I. Nasti, G. Di Francia, Sens. Actuat. B, 111–112/166–170, (2005).

DEVICES

OPTICAL FLOWMETER SENSOR FOR BLOOD CIRCULATORS

M. NORGIA[1], A. PESATORI[1] AND L. ROVATI[2]

[1]*Dipartimento di Elettronica e Informazione, Politecnico di Milano, Milano, Italy*
[2]*Dipartimento di Ingegneria dell'Informazione, Università di Modena e Reggio Emilia, Modena, Italy*

Abstract. This work describes the design and realization of a new low-cost instrument for real time measurement of blood flow, especially developed for dialysis machine monitoring. The instrument is realized by means of a self-mixing interferometer and a dedicated data elaboration system.

1. Introduction

The blood flow monitoring is a key measurement for the control of the correct working conditions in hemodialysis and extracorporeal blood circulator systems. Commercial instruments are based on ultrasound sensors [1, 2], requiring high-cost crystal actuators, working in the megahertz frequency range. Other kind of sensors are inductive flowmeters, which require special cuvettes, not used in disposable equipments for their cost [2]. Some optical instruments have been proposed for flow measurements [3], but they are difficult to implement at the very low cost required by the application. In this work we propose an optical sensor based on the self-mixing interferometry [4], realized with low-cost components and managed by a dedicated data processing system.

2. The realized self-mixing interferometer

Any laser source, when subjected to back-injection of a small fraction of the emitted power, exhibits an amplitude modulation [3, 4], which depends on the phase of the back-reflected optical field, and hence on the remote target distance. The amplitude modulation signal can be detected by the monitor photodiode typically placed inside the laser package. This effect has been used for realizing different kind of sensors [4–6].

For the measurement of a blood flow inside a plastic duct, we implemented an optical setup as described in Fig. 1. The setup has been studied and experimentally tested in order to maximize the back-injection signal. The first lens collimates the laser beam, while the second lens creates a focus inside the blood flow.

P. Malcovati et al. (eds.), *Sensors and Microsystems: AISEM 2009 Proceedings*, Lecture Notes in Electrical Engineering 54, DOI 10.1007/978-90-481-3606-3_15,

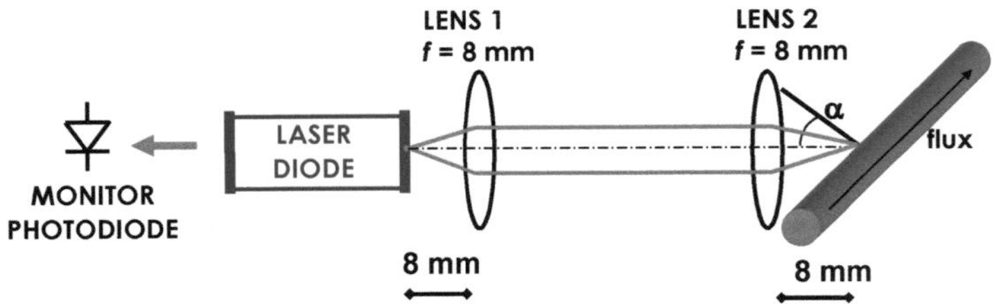

Figure 1. Schematic of the optical sensor.

The signal from the monitor photodiode is amplified by a trans-impedance amplifier with bandwidth 1.25 MHz, and acquired by a digital oscilloscope equipped with FFT module.

The laser beam is directed at an angle α with respect to the normal to the liquid flux. When the liquid moves at a speed v, the back-diffused light exhibits a Doppler shift equal to

$$f_{\text{Doppler}} = \frac{2}{\lambda} \cdot v \cdot \sin\alpha \tag{1}$$

This Doppler shift is read by the self-mixing interferometer and is measured as an amplitude modulation of the laser power, directly by the monitor photodiode.

3. Measurement results and data processing

We conducted several measurements over a polycarbonate tube with diameter 4.3 mm. The scattering liquid employed in the first experiment is a water solutions of Intralipid (Fresenius Kabi SPA, Verona, Italy). This suspension is often used in the biomedical optics community as a scattering medium for phantoms that mimic turbid tissues [7, 8]. The goal of the system is the measurement of the mean velocity of the liquid. The peristaltic pump used in the experiment can reach 400 ml/min, corresponding to about 45 cm/s as average liquid speed. The signal acquired from the monitor photodiode has been elaborated in real time by an FFT transform. Figure 2 shows the spectra for different liquid speeds (from 0 to 45 cm/s), acquired at an angle $\alpha = 25°$.

Taking into account the need for a real-time elaboration, we decided to implement a very simple data processing algorithm. The measured Doppler spectrum has been considered as a Probability Density Function (PDF) of the velocity distribution, for the particles suspended in the fluid. Let's consider the physical phenomenon: each particle back-diffuses a fraction of the impinging optical power, inducing a Doppler shift proportional to its velocity; moreover the contribution of each particle can be considered as uncorrelated from the other ones (it can be summed up in power). Therefore, the mean value of the particles velocity can be calculated as the center of gravity of the probably density function $p(x)$.

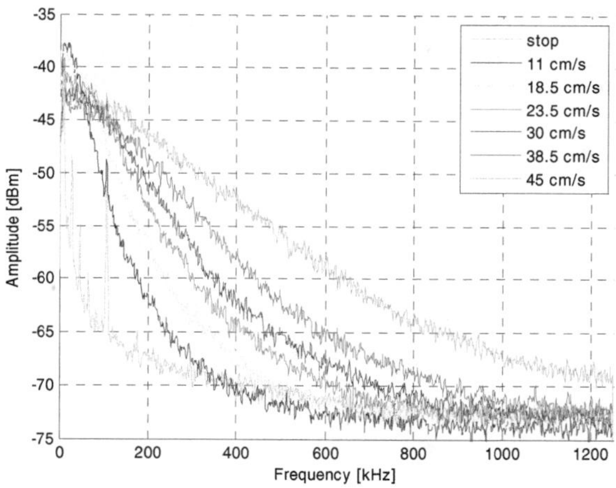

Figure 2. Spectra of the acquired signal at different velocity.

Hence, the spectral center of gravity is obtained as

$$\bar{f} = \frac{\int_0^\infty p(f) \cdot f \cdot df}{\int_0^\infty p(f) \cdot df} \cong \frac{\sum F(f) \cdot f}{\sum F(f)} \tag{2}$$

where *F*(*f*) is the power spectrum vector, calculated as the square module of the FFT signal. This elaboration algorithm is simple to implement, because it requires only two sums on the previously calculated FFT spectrum. The power spectrum, however, does not represent exactly the velocity distribution of the particles, because the contribution of each particle is weighted by the level of its back-injected optical power. We have to take into account the different attenuations undergone from each back-reflection. Moreover, the focus position is really important, because it determines the portion of fluid that gives the major contribution to the back-scattered radiation. The best solution for our setup is the beam focus position at the center of the tube, about 2 mm inside the liquid flux. The average velocity estimated by this algorithm is reported in Fig. 3. The linearity of the first realization of the sensor is about 5%, corresponding also to the value of repeatability: a performance comparable with commercial ultrasound measurement systems.

4. Conclusions

In this work we proposed a new low-cost optical sensor, for real-time blood-flow measurement. For the correct and robust working of the instrument we studied a dedicated data-handling system, which can be implemented in a low-cost embedded electronics. The first realized prototype has an accuracy of about

5%, estimated from repeated measurements. Future works can improve the accuracy of the system, by means of a custom realization of a fixed setup, and a better estimation of the optical attenuation as a function of the signal frequency.

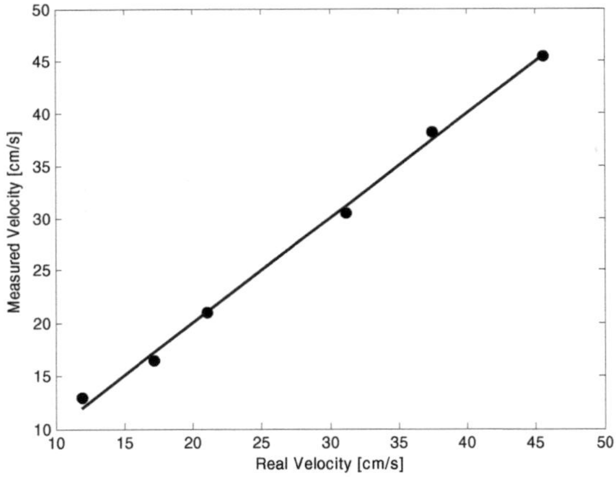

Figure 3. Sensor characterization: measured vs real velocity.

Acknowledgments The authors wish to thank DATAMED S.r.l. for the active collaboration and support to the research project.

References

1. The Flow and Liquid Level Handbook, Vol. 29, 1995. "Ultrasonic Doppler Flow-meters," Omega Engineering, Inc., Stanford, CT.
2. J. Webster, editor-in-chief, "The measurement, instrumentation, and sensors handbook", chapter 76, II series, CRC Press LLC, 2000.
3. S. Donati, "Electro-Optical Instrumentation – Sensing and Measuring with Lasers", Prentice-Hall, USA, 2004.
4. G. Giuliani, M. Norgia, S. Donati, T. Bosch, "Laser diode self-mixing technique for sensing applications" J. Opt. A: Pure Appl. Opt., vol. 4, no. 6, pp. S283–S294, 2002.
5. M. Norgia, S. Donati, D. D'Alessandro, "Interferometric measurements of displacement on diffusing target by a speckle tracking technique", IEEE J. Quan. Electron., vol. 37, no. 6, pp. 800–806, giugno 2001.
6. M. Norgia, G. Giuliani, S. Donati, "Absolute distance measurement with increased accuracy using laser diode self–mixing interferometry in a closed loop", in stampa su IEEE Trans. Instrument. Measure., dicembre 2007.
7. S.T. Flock, S.L. Jacques, B.C. Wilson, W.M Star, M.J.C. van Gemert, "Optical Properties of Intralipid: A phantom medium for light propagation studies," Laser. Surg. Med., vol. 12, pp. 510–519, 1992.
8. H.G. van Staveren, C.J.M. Moes, J. van Marle, S.A. Prahl, M.J.C. van Gemert, "Light scattering in Intralipid-10% in the wavelength range of 400–1100 nanometers," Appl. Opt., vol. 30, pp. 4507–4514, 1991.

UV LASER BEAM PROFILERS BASED ON CVD DIAMOND

M. GIROLAMI, P. ALLEGRINI, G. CONTE AND S. SALVATORI
Electronic Engineering Department, University Roma Tre Via della Vasca Navale 84 – 00146 Rome, Italy

Abstract. 1D and 2D detector arrays have been realized on 270 μm thick 10 × 10 mm^2 CVD-diamond samples. The relatively high resistivity of diamond specimens in the dark (>3 × 10^{12} Ωcm) allowed the fabrication of photoconductive "sandwich" pixel-detectors. For both the structures, a semitransparent light-receiving back-side contact was used for detector biasing. Cross-talk between pixels was limited by using intermediate guard contacts connected at the same ground potential of the pixels. Each pixel photocurrent was conditioned by a read-out electronics composed by a high sensitive integrator and a Σ-Δ ADC converter. The overall 500 μs conversion time allowed a data acquisition rate up to 2kSPS. The measured fast photoresponse of the diamond samples in the nanosecond time regime suggests to use the proposed devices for fine tuning feedback of high-power pulsed-laser cavities, whereas the solar-blindness guarantees high performance in UV beam diagnostics also under high intensity background illumination.

1. Introduction

Diamond is well known for its wide range of extreme properties: ultra-low electrical conductivity, very high thermal conductivity, chemical inertness, and transparency in the visible and infrared spectral regions. Due to such physical and chemical properties, diamond is one of the most suitable candidate for the detection of high power UV laser emission sources [1]. In this field, the technology of laser beam-profiling is evolving with the advent of more applications for high power lasers. The development of UV beam-profilers, used to improve the emission performances of excimer lasers by fine tuning of the laser cavity, appears mandatory to increase the applications in VLSI and MEMS technologies [2]. In order to make CVD-diamond an adequate tool for laser beam analysis, 1D and 2D detector arrays have to be fabricated. Main aim of this paper is to report on the realization and testing of a polycrystalline diamond detector for UV beam monitoring. Multistrip and pixel-array have been used for 1D and 2D photodetectors, respectively. A dedicated conditioning electronics has been used to sample, independently, the signal generated by each pixel. Despite of a multiplexer-based acquisition system [3], this method allows to reconstruct the beam profile in a real-time fashion, eventually being capable of providing data faster than the repetition rate of the laser pulses.

P. Malcovati et al. (eds.), *Sensors and Microsystems: AISEM 2009 Proceedings*,
Lecture Notes in Electrical Engineering 54, DOI 10.1007/978-90-481-3606-3_16,

2. Experimental

10×10 mm^2 MWCVD diamond specimens, $200 \div 300$ μm thick, were selected for the present study, cleaned by dipping in hot sulphochromic solution and rinsed in *aqua regia*. Sandwich contacts were realized by evaporating silver on the top and bottom faces of the diamond films by thermal evaporation. The thickness of the bottom electrode, used for device biasing, was 100 nm, whereas the top contact was only 50 nm thick, in order to be semi-transparent to the impinging UV light. Standard photolithography was used to define the array structures shown in Fig. 1a, b. The 1D detector consisted of a multistrip structure with 32 fingers, 5 mm in length, 80 μm wide; the 2D detector was a 64-pixel array (500×500 μm^2 each).

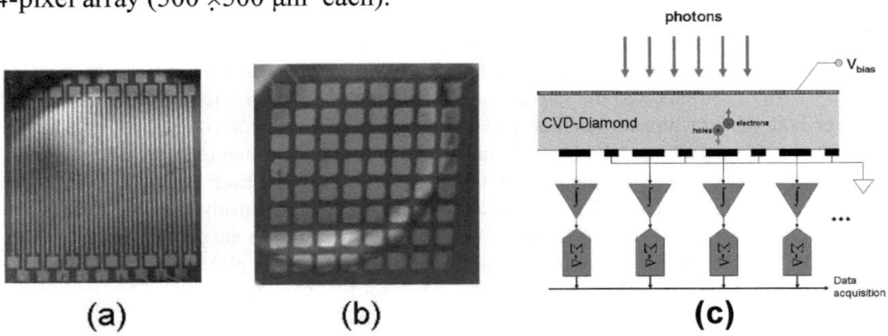

Figure 1. Micrograph of the 1D strip-array detector (**a**), of the 2D pixel-array detector (**b**) and operating principle of both sensors (**c**).

For UV-beam monitoring measurements, sensor pads were wire-bonded to a dedicated "daughter"-board where the front-end electronic circuits, based on the DDC11X current input analog-to-digital converters [4], were also mounted. In particular, each pixel of the sensor was conditioned by a dedicated electronics in which a high-precision switched capacitor integrator was followed by a high-resolution Σ-Δ analog-to-digital converter, as shown in Fig. 1c. A computer-interface main board was designed to include the circuitry used to control the front-end daughter-board and the interface to a personal computer (PC) for data transfer. The interface-board had the capacity to collect the samples converted from each channel to be downloaded to the PC. Moreover, depending on user's preferences, the digital-control circuitry established the integration time of analog integrators in the $50 \div 1{,}000$ μs range for the on-board clock frequency f_{CK} of 10 MHz.

The characterization process was performed in two different ways: in the former, a low-intensity continuous-wave UV spot light was used for the 1D and 2D beam-profile measurements; in the latter, the 2D array-sensor was used for pulsed excimer laser monitoring.

3. Results and discussion

To evaluate the sensor capability to give information on the profile of a UV spot, a monochromatic UV-light source was selected (5 ×5 mm^2 spot size). The impinging photon intensity, around 1 μW at λ = 235 nm, allowed an easy discrimination of the photocurrent signal in comparison to the "background" value induced by the dark conductivity and the offset errors of the input channels. A 500 μm pinhole was used for a preliminary "calibration" procedure of the input channels. Such a sliced spot light allowed to irradiate a small portion of the detector area. The sensor was then mounted on a micrometric *X-Y* moving stage having 10 μm of resolution. The calibration procedure started from the alignment of the centroid spot on a first strip line.

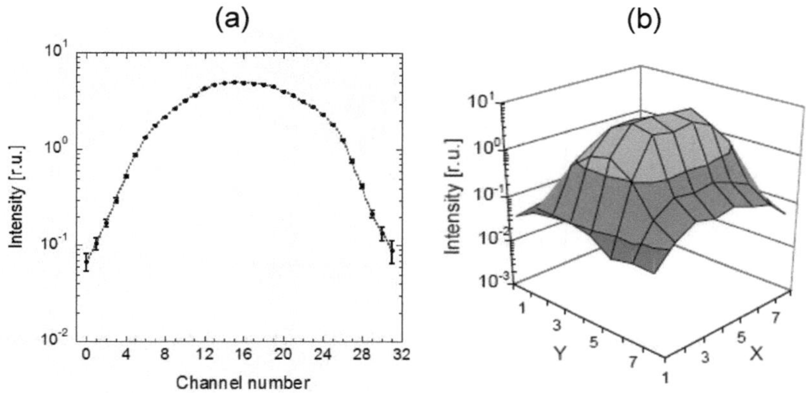

Figure 2. Beam profile acquired in the vertical *Y* axis for the 1D strip-array detector (**a**) and beam profile of the *X-Y* cross-section of the UV spot acquired by the 2D pixel-array detector (**b**).

Successive translations of the array along the *Y* direction, by steps equal to the distance between strips, allowed to evaluate the normalization factors to take into account during the ordinary "parallel" acquisition of sensor signals. As an example for the strip-line 1D sensor, in Fig. 2a we report the beam profile along the vertical direction *Y* as the result of the mean value of 100 samples (50 ms of overall acquisition time) for each pixel-signal. It is worth noting the relative low standard deviation values (error bars) of the ADC output codes we gained (also assured by copper-clad shielding of the front-end electronics board). For the 2D 8 ×8-pixels sensor we adopted a calibration procedure similar to that previously described under the same 500 μm ∅ spot light. Obviously, the micrometric-stage was now used to move the detector along both the two *X* and *Y* directions. Finally, the overall 5 ×5 mm^2 UV-light was used to measure the beam profile along both the two horizontal and vertical axis of the beam section. An example of the typical sensor photoresponse is reported in Fig. 2b. The beam profile obtained along the *Y* direction, by integrating that obtained at *X* channels, fairly reproduces that shown in Fig. 2a. The 2D sensor array, bonded to the front-end electronics, was then used to evaluate the time evolution of an excimer laser beam intensity (λ = 193 nm) along the *X-Y* axis. In Fig. 3, the signal acquired for the four central sensor pixel is shown: due to the non-uniform spatial distribution of the impinging light, the four collected signals are slightly different. The intensity change the laser shows after its first turn-on is well evidenced, and it increases with time before reaching a steady-state condition. The noisy behaviour of the response after turning on is mainly due to two factors: the laser intensity is different at each pulse, especially at higher repetition rate, when the laser efficiency strongly decreases; moreover, a noise contribution due to electro-magnetic interference introduced by the thyratron high frequency generator cannot be excluded.

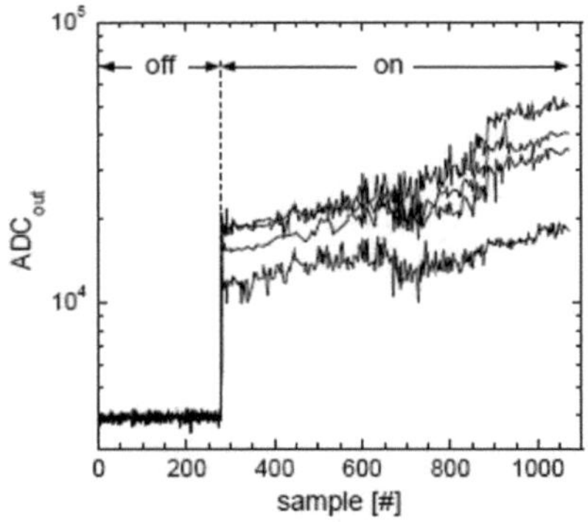

Figure 3. Signals acquired by four central pixels of the 2D pixel-array detector when the sensor is irradiated by the excimer laser used for tests.

4. Conclusions

Polycrystalline diamond turned out to be an extremely appealing material for the realization of high performance UV beam-profile meters. The outstanding properties of CVD-diamond in terms of photoresponse speed and high acquisition rate (2 kSPS) of the read-out electronics allowed to collect and process single excimer laser pulses, so to enable a real-time diagnostics of UV sources.

References

1. G. Mazzeo, S. Salvatori, M. C. Rossi, G. Conte, M. C. Castex, "Deep UV pulsed laser monitoring by CVD diamond sensors", Sensors and Actuators A, [113], pp. 277–281, 2004.
2. M. C. Gower, "Industrial applications of laser micromachining", Optics Express, [vol. 7, no. 2], pp. 56–67, 2000.
3. G. Mazzeo, S. Salvatori, G. Conte, V. Ralchenko, V. Konov, "Electronic performance of 2D-UV detectors", Diamond and Related Materials, [16], pp. 1053–1057, 2007.
4. DDC11X series datasheets, Texas Instruments Inc, www.ti.com.

PHOTOCONDUCTIVE POSITION SENSITIVE CVD DIAMOND DETECTORS

M. GIROLAMI, P. ALLEGRINI, G. CONTE AND S. SALVATORI

University Roma Tre, Department of Electronic Engineering, Via della Vasca Navale 84 – 00146 Rome (Italy)

Abstract. Different contact structures were studied to realize CVD-diamond sensors able to detect light beam displacements. Photons irradiate the sample back-side where a mesh contact is used for device biasing. Photogenerated carriers are collected at the top-side where metal strip lines are realized. Depending on the layout of such contacts, beam displacement is measured by means of the elaboration of acquired strip signals. A very compact front-end electronics was used for signal conditioning. As for an image sensor array, the signal of each strip is integrated in the same time-period. Analog to digital conversion is performed in a second time period by means of high-resolution Σ-Δ converters and digital filtering. During such a second time-period, strip line photocurrents are integrated by a second set of ideal integrators allowing signal acquisition in a continuous-mode fashion. Limited by the 500 μs conversion time of ADCs, the detection system allows up to 2 kSPS data output acquisition rate.

1. Introduction

In the past few years, the demand for high resolution position sensors for ultraviolet high intensity light sources has significantly grown up, due to the developments in microfabrication technologies, such as VLSI and MEMS: in such a context, UV sensors play an important role for in situ laser beam diagnostics. Other application fields of UV light monitoring are: bacteriological water treatment, molecular biology, astronomy and material technology. Conventional semiconductors (e.g. silicon and III-V compounds) are still largely used in industry for the realization of UV detectors, mainly for the low cost of the materials and the proven and reliable technologies, that make up for the low energy bandgap. CVD diamond overcomes these intrinsic limitations, due to its wide bandgap (5.5 eV) and thereby its high UV versus VIS selectivity; moreover, insensitivity to radiation damage due to its high cohesive energy makes diamond sensors suitable for operation in harsh environments and high intensity UV irradiation. Even if damaging has been observed in polycrystalline diamond at lower levels than those expected from the graphitization threshold [1], CVD diamond has been proposed as the elective material for the development of sensors in the deep UV spectral range [2]. In such a context, we present two simple but performing prototypes for the detection of UV beam displacements,

P. Malcovati et al. (eds.), *Sensors and Microsystems: AISEM 2009 Proceedings*,
Lecture Notes in Electrical Engineering 54, DOI 10.1007/978-90-481-3606-3_17,

achieving a spatial resolution down to tens of μm, about one order of magnitude better than other recently proposed CVD diamond UV sensors [3].

2. Experimental

High quality MWCVD diamond 10 × 10 mm^2 samples were selected to realize 1D and 2D position sensitive detectors. The 1D sensor (Fig. 1a) has a multistrip structure (8 Ag strips, 5 mm × 72 μm, 500 μm pitch) as top contact, while the 2D sensor (Fig. 1b) has a leaf-shaped top contact pattern (32 Ag strips, 96 μm wide with variable length, 600 μm pitch): the leaf design allows to simplify the front-end electronics and to collect the adequate amount of information about the beam positon, so avoiding the troubles related to the use of a common 2D pixel detector (e.g. complex read-out circuitry and signal redundancy). Both sensors have a semitransparent 50 nm-thick Ag rectangular pad as back contact: when UV photons impinge on the back contact, electron-hole pairs are created inside the bulk material ("sandwich" configuration), and collected by the front electrodes by means of a uniform electric field, obtained applying a bias voltage. Each strip is conditioned by a dedicated electronics (a high precision switched-capacitor integrator followed by a high resolution Σ-Δ ADC) and isolated from nearest by means of a guard connected to ground. The overall 500 μs conversion time afforded a data acquisition rate up to 2 kSPS. To test the devices under UV irradiation, a monochromatic light source has been used ($P_{INC} \approx 1$ μW @ λ = 235 nm); the beam selected spot-size was 5 × 5 mm^2. Each sensor has been mounted on a 10 μm-resolution micrometric X-Y moving stage in order to perform UV beam displacement measurements.

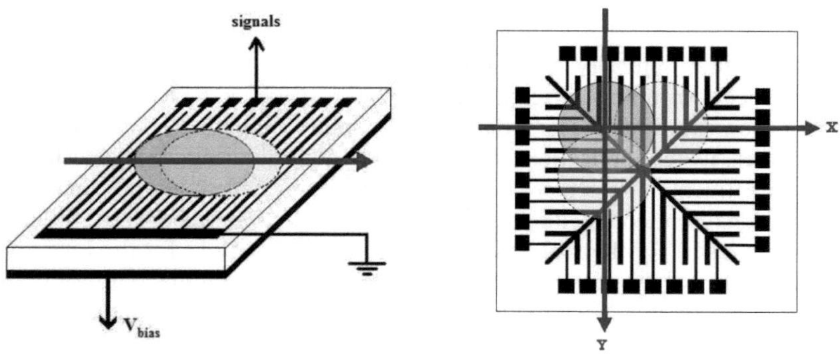

(a) (b)

Figure 1. 1D strip array sensor (**a**) and top view of the 2D leaf-shaped sensor (**b**).

3. Results and discussion

Spatial resolution of the 1D strip detector is in the worst case (spot focused on the pitch centre) equal to half the pitch (250 μm). In order to deepen this point,

UV spot has been gradually shifted (50 μm steps) between two adjacent strips. Processing at each step the 8-channels signal, eight different beam profiles have been obtained (Fig. 2a): the beam centroid displacement at each step is plain, so ensuring at least a 50 μm position resolution. The spot centroid is calculated as the weighted mean of the collected strip signals N_i:

$$Centroid(\#\text{Ch}) = \frac{\sum_i N_i \cdot i}{\sum_i i} \qquad (1)$$

where weights i are the strip positions (channels). In Fig. 2b spot centroid position versus beam displacement is reported: noticeable the linear relationship, with a 5 μm maximum error (max. distance of the experimental values from the best fit curve). Such a small error suggests a finer resolution than 50 μm: to confirm this point, it will be necessary an evaluation of the minimum beam shift that yields a 8-channels signal variation above the noise level.

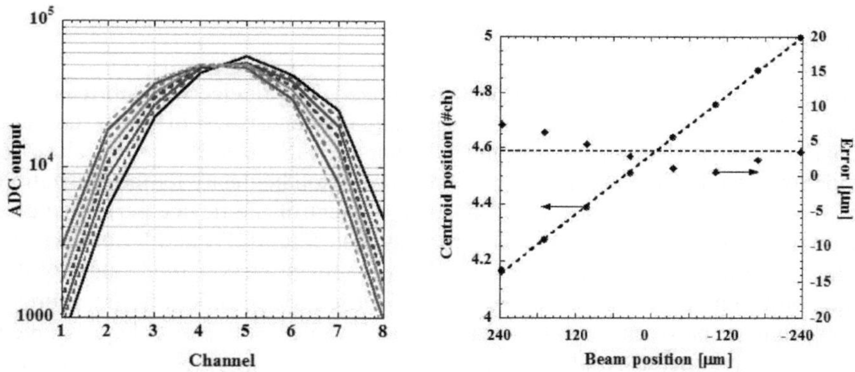

(a) (b)

Figure 2. 1D PSD: Eight different UV beam profiles (**a**) and spot centroid position versus beam displacement (and related error) (**b**).

As regards 2D PSD measurements, UV spot has been gradually shifted along X axis, collecting and processing at the same time both the X- and Y-strip (channel) signals. Noticeable in Fig. 3a the spot centroid versus beam displacement along X axis linear relationship, with a 8 μm maximum error (the larger error compared to 1D PSD is probably due to a wider pitch). As expected, spot centroid position along Y axis is rather constant with the beam displacement (Fig. 3b). Similar results have been obtained shifting UV beam along Y axis.

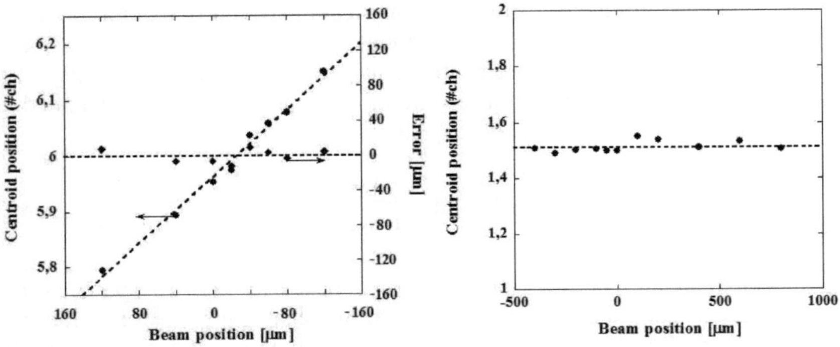

Figure 3. 2D PSD, beam moving along X axis: spot centroid position versus beam displacement along X axis (and related error) (**a**) and along Y axis (**b**).

4. Conclusions

A PSD with high lateral resolution was realized. The device was fabricated on high quality CVD diamond films, showing characteristics in terms of dark conductivity, activation energy and UV versus VIS selectivity very close to those of natural diamond. The use of a fine line photolithography for the realization of the contact structure has led to an improvement of the device performances, allowing to obtain a measurement error less than 5 µm (1D) and 8 µm (2D). In these conditions the resolution is related to both the polycrystalline nature of the diamond film and the distance between strips used for detection. In the future, a compromise between high sensitivity (achievable with thick large-grained samples) and spatial resolution (achievable by means of fine-grained samples and reducing the pitch) will need to be found.

References

1. M.D. Whitfield, S.P. Lansley, O. Gaudin, R.D. McKeag, N. Rizvi and R.B. Jackman, “Diamond photodetectors for next generation 157-nm deep-UV photolithography tools”, Diam. Relat. Mater., [10], pp. 693–697, 2001.
2. G. Mazzeo, S. Salvatori, V. Ralchenko and G. Conte, “Design and test of deep-UV position sensitive detectors”, Sensors 2004 – Proc. IEEE, [3], pp. 1554–1557, Oct. 2004.
3. K. Tsuji, K. Hayashi, J.H. Kaneko, F. Fujita, A. Homma, Y. Oshiki, T. Sawamura and M. Furusaka, “Development of high resolution position sensitive UV detector based on highly oriented polycrystalline diamond”, Diam. Relat. Mater., [14], pp. 2035–2038, 2005.

OPAQUE-GATE PHOTOTRANSISTORS ON H-TERMINATED DIAMOND

P. CALVANI, M. C. ROSSI AND G. CONTE

University Roma Tre, Department of Electronic Engineering Via della Vasca Navale, 84 – 00146 Rome, Italy

Abstract. Opaque-gate ultraviolet sensitive transistors were fabricated on H-terminated polycrystalline diamond. Butterfly shaped structures with different geometric ratios were realized. Observed trends with the gate unbiased demonstrated behavior as p-channel normally-off transistors, switched-on by the impinging UV light. Linearity with the UV beam power was also observed with over-gap radiation. Under steady state illumination, a linear increase of the photocurrent was found when the gate is biased at different voltages in the saturation regime. The operative generation-charge transport mechanism of fabricated devices is discussed.

1. Introduction

Because of its wide bandgap (5.47 eV) diamond is expected to be an ideal candidate for ultraviolet applications. Moreover, due to the high thermal conductivity and availability of large area polycrystalline diamond substrates, with reduced surface roughness, the fabrication of beam shapers to be exploited for UV excimer laser monitoring has been proposed [1]. Selectivity, sensitivity and high speed operation are stringent requirements for such application, as well as optical gain. These requirements can be met by semitransparent Schottky-gate field effect transistors (FET). The first optically activated FET (OPFET) was demonstrated by Lansley et al. [2]. More details on such devices were presented successively [3]. The device operated as a photodiode in terms of turn-on characteristics and as a FET in terms of signal amplification. Optical gain equal to 4 at 200 nm was measured. In view to develop UV laser beam shapers, we report the fabrication of OPFETs working in the short channel regime and as p-channel normally-off transistors. Device performance with an UV-opaque gate structure is presented and results observed with the gate biased or unbiased are discussed.

2. Experimental details

Thermal grade polycrystalline diamond specimens, by Element Six, 1 cm^2 wide and 0.25 mm thick were used for this work. Before hydrogenation at 800°C, substrates were cleaned by a strong acid oxidation with boiling (100–250°C)

P. Malcovati et al. (eds.), *Sensors and Microsystems: AISEM 2009 Proceedings*, Lecture Notes in Electrical Engineering 54, DOI 10.1007/978-90-481-3606-3_18,

mixtures of perchloric, nitric and sulphuric acid to obtain O-terminated surfaces. Samples were then rinsed in water, treated in ultrasound, dried and then transferred to the Astex S1500 vacuum chamber for surface hydrogenation. The fabrication of butterfly shaped devices started with the realization of the source and drain contacts by evaporation of 150 nm of silver directly on the H-terminated diamond surface through windows prepared by UV photolithography. The gate electrode was realized by flash-evaporation of 150 nm of Aluminum for an accumulation type channel. The gate ratio, width/length, was 12.5 μm/μm. The electric insulation among devices was performed by Oxygen/Argon plasma ion etching thus replacing the conductive H-terminated diamond surface with O-terminated zones. To shine light on the devices active area, estimated to be $2 \times 200\ \mu m^2$, a Xe lamp was coupled to a Jobin-Yvon H10 monochromator to obtain UV radiation at 225 nm which was focused on the device active region. The UV beam intensity was estimated by means of a calibrated pyroelectric radiometer Molectron PR200. Devices response in the dark and under illumination was analyzed by evaluating the drain current, I_{DS}, as a function of gate-source voltage (V_{GS}) and drain-source voltage (V_{DS}) by using a HP4140B picoammeter. Measurements were conducted at room temperature.

3. Results and discussion

I_{DS}–V_{DS} characteristics of a UV-FET measured in the dark for different gate bias is reported in Fig. 1a, whereas the same device under UV illumination is shown in Fig. 1b. The UV beam power was estimated to be 55 $\mu W/cm^2$. As it is apparent, the device behaves like a p-channel normally off transistor, showing short channel effects as demonstrated by the linear increase of the current in the saturation regime both, in the dark and under illumination.

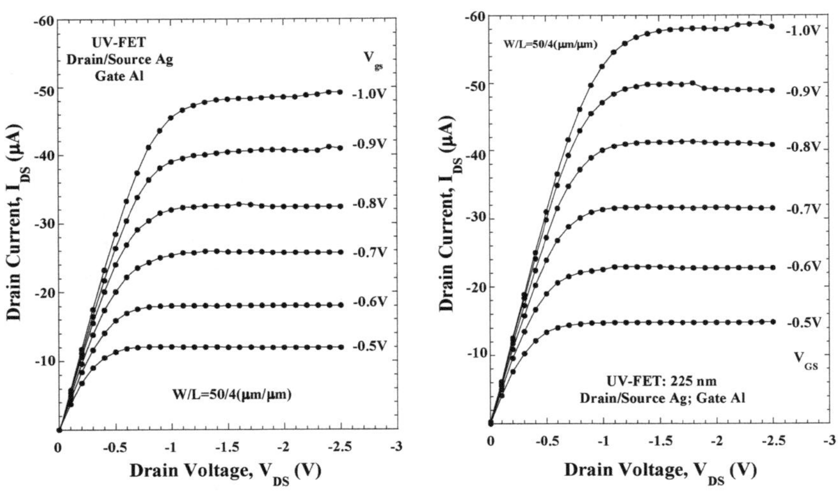

Figure 1. MESFET characteristics (**a**) in the dark; (**b**) 55 $\mu W/cm^2$, UV illumination.

The channel conductance (g_D) increases rapidly with V_{GS} showing enhancement mode behavior. Leakage current at zero V_{GS} is few pA. The UV light clearly enhances the channel conductance. g_D changes between 40 μS and 70 μS and approaches the pinch-off for values around $V_{DS} = -1.5$ V at $V_{GS} = -1$ V. As it is apparent in Fig. 2, the photocurrent increases linearly with V_{GS} leading to a constant transconductance (g_m), with improvement from 18 mS/mm to 22 mS/mm under illumination, whereas $V_{th} = -0.34 \pm 0.01$ V remains constant.

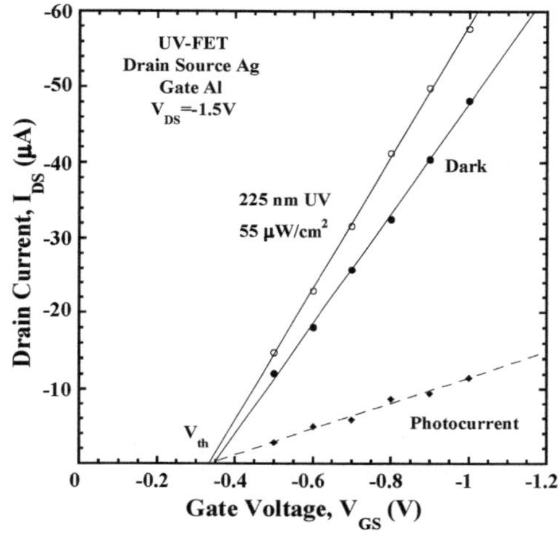

Figure 2. I_{DS}–V_{GS} characteristic @ $V_{DS} = -1.5$ V, in the dark and under UV illumination. The photocurrent is also reported. Lines are the best fit to data.

The response to different UV beam powers with the gate floating is reported in Fig. 3. Due to the device geometry and Aluminum gate thickness, charge carriers generation is possible only in between the H-terminated gate-drain region, estimated to be 200 μm^2. The device response shows initially a constant channel conductance $g_d = 10$ μS up to $V_{DS} = -0.5$ V, thus initiating to decrease reaching the channel pinch-off. Channel length modulation appears at higher powers. The charge carriers inside the channel increases linearly with the UV beam power as shows Fig. 3b. This trend addresses the device to act as a simple photoconductor in the saturation regime.

As charges in the channel below the gate are produced mainly by diffusion of those generated by absorption in the volume source-gate and gate-drain (i.e. about 400 μm^2), and within few microns in depth from the diamond surface at the wavelength used, it is expected that only a minor number of photo-generated holes will results accumulated below the gate in such a way inducing a band bending that increases with the impinging beam power. In this optics fabricated devices behaves like an opaque-gate MESFET, as proposed for GaAs/AlGaAs power MESFET [4]. In our case, UV light stimulates a change in the surface holes-density between source and drain, through carriers diffusions below the gate similar to the electrical bias application, enabling a channel current to flow.

As already found in reported papers, the device act as a photodiode in terms of turn-on behavior and as a FET in terms of signal amplification.

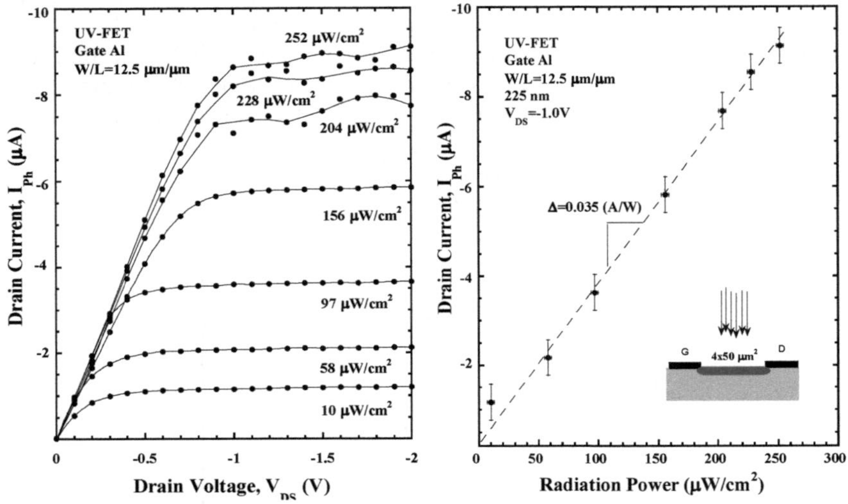

Figure 3. (**a**) Photocurrent versus the drain-source voltage measured with the floating gate; (**b**) photocurrent versus the beam power at $V_{DS} = -1$ V. The dashed curve is the linear fit to data.

The best fit to I_{DS}-Radiation Power data leads to a slope equal to 35 mA/W. Taking into account the total illuminated active area of 400 μm^2 for each device, we obtain an evaluation for the responsivity at 225 nm, equal to 8.75 mA/W.

4. Conclusions

Optically activated opaque-gate FET were fabricated on H-terminated commercial polycrystalline diamond [5]. Floating gate operation under UV illumination appears to be similar to electrical stimulated gate behaviors. A linear response to 225 nm UV light was observed with high responsivity, in such a way addressing the device behaving like a photoconductor in terms of turn-on operation. Observed trends suggested such device to be suitable for operation as light activated power switch.

References

1. S. Salvatori, M. Girolami, P. Allegrini, G. Conte, S. Salvatori, Proc. 7th IEEE Sens. Conf., pp. 270–273 (2008).
2. H. J. Looi, J. S. Foord, R. B. Jackman, Appl. Phys. Lett., [vol. 72], pp. 353–355, 1998.
3. S. P. Lansley, H. J. Looi, M. D. Whitfield, R. B. Jackman, Diam. Rel. Mater., [vol. 8], pp. 946–951, 1999.
4. S. B. B. Pal, R. U. Khan, IEEE Trans. Elec. Dev., 45 (1998) 78.
5. Thermal grade substrate by Element Six, http://www.e6sales.com/

FABRICATION AND CHARACTERIZATION OF A SILICON PHOTODETECTOR AT 1.55 MICRON

M. CASALINO[1], L. SIRLETO[1], M. GIOFFRE'[1], G. COPPOLA[1], M. IODICE[1], I. RENDINA[1] AND L. MORETTI[2]

[1]*Consiglio Nazionale delle Ricerche – IMM sez. Napoli, Via P. Castellino n. 111, Italy*
[2]*Seconda Università degli Studi di Napoli, Via Vivaldi n. 43, 81100 Caserta, Italy*

Abstract. In this paper the realization and the characterization of a new kind of resonant cavity enhanced photodetector (RCE), fully compatible with silicon microelectronic technologies and working at 1.55 µm, are reported.

1. Introduction

In the range of the fiber optic communications O band (1,260–1,360 nm) and/or C band (1,530–1,565 nm) where the silicon is transparent due to its cut-off wavelength of about 1,100 nm, internal photoemission effect (IPE) allows to realize a Silicon based photodetector (PD) [1]. Silicon infrared photodiodes based on the IPE are already used in the infrared (3–5 µm wavelength) imaging systems [2], unfortunately, due to the metal losses, their quantum efficiency is very low. In this paper, with the aim to enhance the quantum efficiency of photodetector based on the IPE, a new kind of RCE photodetector is realized and characterized. We note that the resonant cavity structure is not a new concept in photodetector design, while it is quite new to incorporate into RCE structure a photodetector based on the IPE. A theoretical and numerical analysis of this structure was provided by the same authors in [3], in this paper the experimental validation of previous theoretical and numerical analysis is demonstrated.

2. Device fabrication

The sketch of the device is shown in Fig. 1. The samples were fabricated starting from a slightly doped (10^{14} cm^{-3}) p-type bi-polished 100-µm-thick silicon wafer, acting as $\lambda/2$ resonant layer. On the back, an a-Si:H/Si_3N_4 Bragg mirror is deposited by a Plasma Enhanced Chemical Vapor Deposition (PECVD) system. A measured reflectivity of about 0.99 in a 200-nm-wide range spanning from 1,400 to 1,600 nm is obtained. The collecting ohmic contact and the Schottky contact were both realized on the top of the sample. Moving our attention to the Schottky contact we have compared the performances of two devices having

P. Malcovati et al. (eds.), *Sensors and Microsystems: AISEM 2009 Proceedings*, Lecture Notes in Electrical Engineering 54, DOI 10.1007/978-90-481-3606-3_19,

different Schottky metals: copper and aluminum. These metals were thermally evaporated (200 nm) and patterned by lift-off.

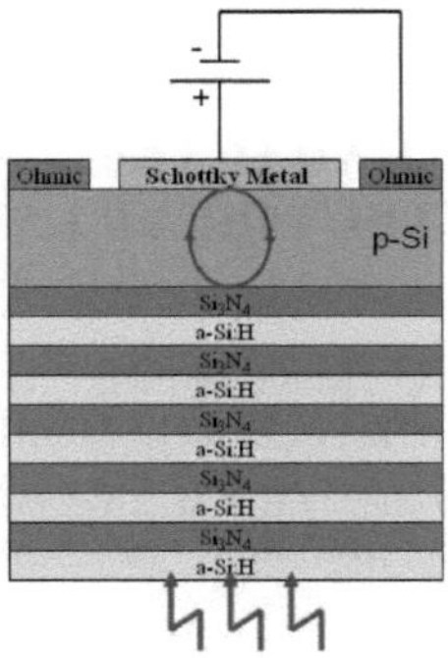

Figure 1. Schematic cross section of the proposed RCE Schottky photodetector.

3. Measurements

3.1. *Electrical measurements*

Density current-voltage (J-V) measurements of Al/p-Si and Cu/p-Si diodes, obtained by a parameter analyzer, were carried out. The Schottky barrier height (SBH) of the devices having aluminum and copper as Schottky metal, deduced fitting the experimental J-V characteristic by canonical equation of the Schottky diode, are 0.81 ± 0.01 eV and 0.72 ± 0.01 eV, respectively.

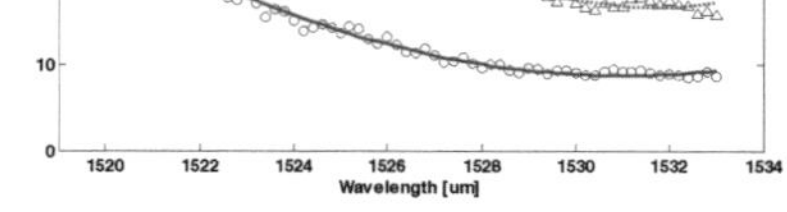

Figure 2. Measured responsivity versus wavelength, and their curve fitting, for the fabricated Al/p-Si device without Bragg reflector.

3.2. *Optical measurements*

In order to experimentally demonstrate the responsivity enhancement, a comparison between two typologies of Schottky diodes, with and without Bragg reflector, was made. Measurements on the device having aluminum as Schottky metal were carried out in the range of 1,520 and 1,533 nm (step of 0.2 nm), with no bias applied and with a reverse voltage of 10 mV. For this device, even if a responsivity modulation was expected, no resonance peaks were obtained (Fig. 2). In our opinion, because the Al/p-Si interface has a potential barrier very close to photon energies, in the considered wavelength range, the responsivity contrast is low and therefore it is covered by noise. Responsivity at 1,520 nm is about 25 nA/W without applied bias, and 60 nA/W for 10 mV of reverse voltage applied. By performing the same measurements onto the device with the Bragg reflector, no enhancement was observed. This is probably due to the high potential barrier which destroy any improvement due to the cavity. Finally, measurements on the device having copper as Schottky metal were carried out into the range of 1,545–1,558 nm (step of 0.05 nm). Figure 3 shows the room temperature responsivity versus the wavelength in two conditions: no applied voltage (solid) and low reverse bias of 10 mV (dashed). If no bias is applied, a peak responsivity of about 0.8 μA/W was measured. We observe three distinct peaks. The measured free spectral range of 3.3 nm agrees the theoretical. Moreover, the measured cavity finesse F is 2.9. Applying 10 mV of reverse bias, while the free spectral range and finesse are unchanged, an enhancement of peak responsivity up to 1.6 μA/W, due to an improvement of the photocarriers collection, is achieved. Figure 4 shows the room temperature responsivity measurements for the Cu/p-Si device fed on Bragg reflector side. For no reverse voltage applied a peak responsivity and finesse of about 2.3 μA/W and 4.7, are obtained respectively, while the peak responsivity increases up to 4.3 μA/W at 10 mV due to an improvement of the photocarriers collection. As expected the Bragg presence does not modify the free spectral range, which is measured to be about 3.3 nm. Moreover, in Fig. 4 theoretical results compared to the experimental, are shown. The SBH calculated fitting the optical responsivity curve is 0.71 ± 0.02 eV, in strong agreement with the results returned by the electrical characterization. We note that an enhancement of the peak responsivity, due to the presence of Bragg reflector, of about three times is obtained.

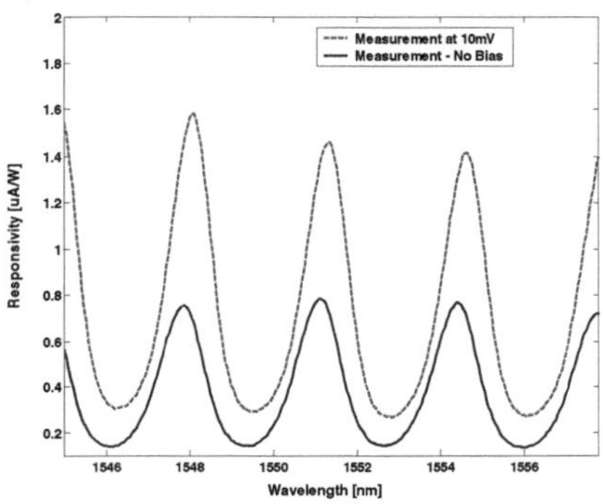

Figure 3. Measured responsivity versus wavelength for the fabricated Cu/p-Si device without Bragg reflector.

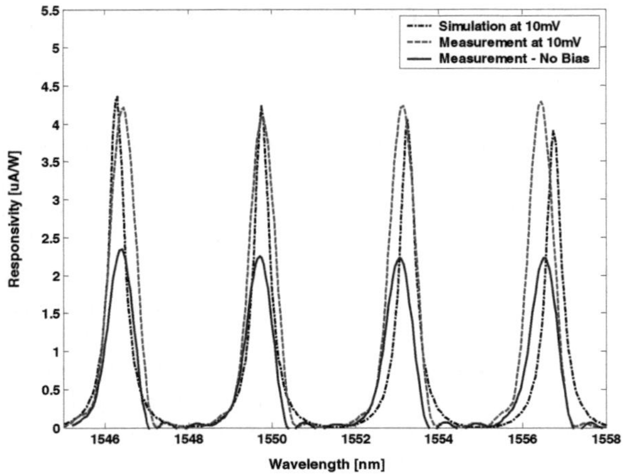

Figure 4. Measured responsivity versus wavelength for the fabricated Cu/p-Si device fed on a-Si:H/Si_3N_4 Bragg reflector.

4. Conclusions

In conclusion the aim of this paper was not to propose an optimized device, but to provide the proof of concept of the responsivity enhancement obtained by a new kind of RCE structure based on IPE and an appropriate choice of the Schottky metal. Our paper has the merit to experimentally prove the possibility to enhance the IPE by using RCE structure incorporating Schottky copper device. Even if the measured device responsivity is not attractive for application, we believe that our results are encouraging and open the way to the investigation of more complex microcavity having higher finesse (for example ring resonator), which could be very promising in order to get a significant improvement in responsivity. The main advantages of the proposed photodetector are that it should be intrinsically fast and its fabrication process is simple and completely compatible with standard VLSI technologies.

References

1. M. S. Unlu and S. Strite, Appl. Phys. Rev., 78 (1995), 607.
2. W. F. Kosonocky, F. V. Shallcross, T. S. Villani, IEEE Trans. Electron Dev., ED-32 (1985), 1564.
3. M. Casalino, L. Sirleto, L. Moretti, I. Rendina, Semicond. Sci. Technol., 23 (2008), 075001.

ACTIVE AREA DENSITY OPTIMIZATION TECHNIQUE FOR HARVESTER PHOTODIODES EFFICIENCY MAXIMIZATION

E. DALLAGO, M. FERRI, P. MALCOVATI AND D. PINNA
Department of Electrical Engineering, University of Pavia, Via Ferrata 1, 27100, Pavia, Italy

Abstract. In this paper we present a test chip in standard CMOS 0.35 um with different photovoltaic structures. In particular we realized several integrated micro solar cell to characterize an harvesting supply system thus to implement a power efficiency optimization technique based on the active area geometry. The photodiodes can be used as stand alone supply solar cells as well as an integrated harvester for micro-systems realized on the same silicon.

1. Introduction

Modern and future ultra-deep-submicron technologies [1] allow very low power consumption Microsystems to be realized, with increasing analog functions and computational capabilities. Despite the technology scaling, the slow growth of the battery energy density [2] limits the lifetime and the weight reduction of portable and nomadic devices as well as of sensor network nodes. Therefore, harvesting systems are becoming the new challenge in both research and comercial communities. Macro solar harvesting systems represent the most developed commercial applications. Usually, photodiodes and phototransistor [3] are used as low cost photodetectors, but, if properly operated, they can be used both as a stand alone, high efficiency micro solar cell and as solar energy source for an integrated microsystem diffused on the same silicon die. In this paper we present several micro solar cell structures [4], implemented on the same test chip, and the design approach to maximize the efficiency of such harvesters. The test chip is realized with a standard 0.35 um CMOS technology. In particular, we realized several photodiodes, with different dimensions of the n-well and of the p-diffusion geometry, featuring different density of active area.

1.1. *Integrated micro solar cells*

Figure 1 shows the different p-diffusion geometries realized to maximize the active area density, and the relative dimensions on-chip. Table 1 summarizes the dimensions and the equivalent active area for each structure. The presented work is focused on the characterization of the structures in terms of geometry dependent efficiency and the relative performances improvement.

P. Malcovati et al. (eds.), *Sensors and Microsystems: AISEM 2009 Proceedings*, Lecture Notes in Electrical Engineering 54, DOI 10.1007/978-90-481-3606-3_20,

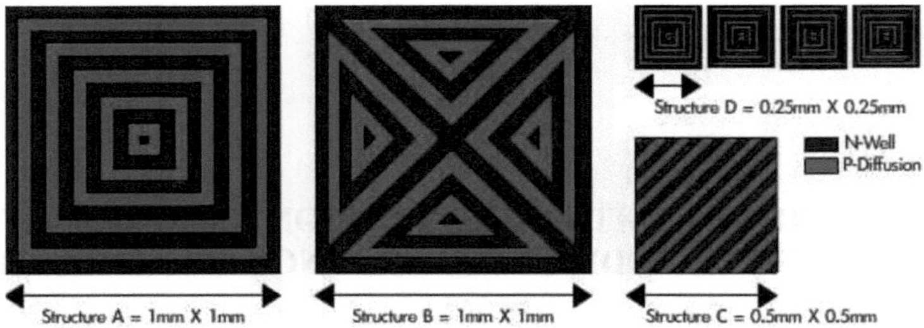

Figure 1. Geometries of the realized micro solar cells.

The equivalent circuit model for each solar structure consists of a couple of pn junction, as shown in Fig. 2. The upper one (between p-diffusion and n-well) is the desired harvester for on-chip micro-systems, while the deeper one is a parasitic diode, whose junction is composed by the n-well and the low doped p-substrate. These diodes can be used for several purpose:

- As high efficiency micro solar cell
- As harvester for integrated microsystem on the same chip
- As photodiode or phototransistor based light sensor

The work presented in this paper is focused on the first two applications, and, in particular on the characterization of the different structures on the basis of geometry dependent efficiency.

When the micro solar cell is used as stand-alone device, it is possible to exploit the photo-generated current of both junctions, considering them as independent photodiodes connected in parallel (when the substrate is short-circuited to the p-diffusion).

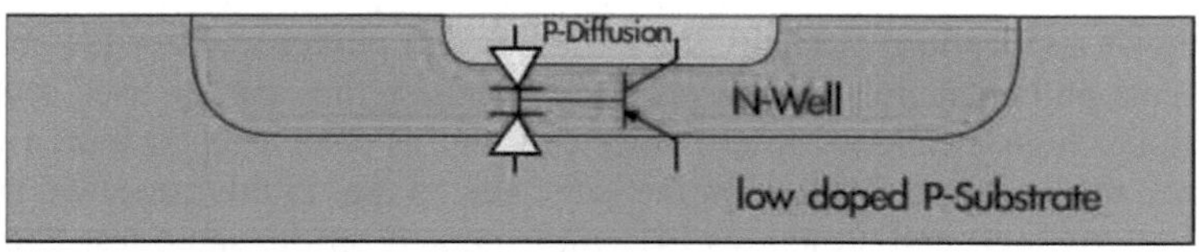

Figure 2. Circuit model of realized solar structures.

Table 1. Realized integrated solar cell parameters.

Type	Harvester active area (mm²)	Harvester active perimeter (μm)	Parasitic active area (μm²)	Parasitic active perimeter (μm)	Number of structures on silicon
A	2.43929e-5	642,546	1e-6	4,000	1
B	2.89457e-7	827,845	1.9826e-6	5,632.2	1
C	7.24565e-8	207,431	5.11719e-7	2,861.65	1
D	1.78246e-7	47,073	6.23e-8	998.4	4

As the standard CMOS technology is based on the reverse biasing isolation, in Harvester application the p-substrate and the n-well must be short circuited, thus to ensure to bias the substrate to the lower voltage. As there was not any chance to change the process parameter, the presented structures could be used as light sensors just in a restricted area of application (optical switches for example).

2. Results

In order to characterize the micro solar cell as a stand alone deice, it is useful to connect the substrate to the – diffusion, thus connecting the two diodes in parallel. The current contributions of both diodes are, therefore, added. The output power of the structure C, obtained with an illumination of 100 W/m², is shown in Fig. 3. The conversion efficiency of the integrated micro solar cells, depending on the p-diffusion geometry, is reported in Table 2. Device C features an efficiency as large as 17%. In order to avoid contributions fro other devices on the chip, both terminal of all the diodes not being tested are short-circuited to the substrate.

The largest photo-generated current contribution is given by the parasitic diode, since its junction is deeper and the substrate is less doped than the p-diffusion, thus leading to a higher efficiency than in the corresponding harvester. As expected, geometry C, featuring the largest relatives active area density, is the most efficient diode structure.

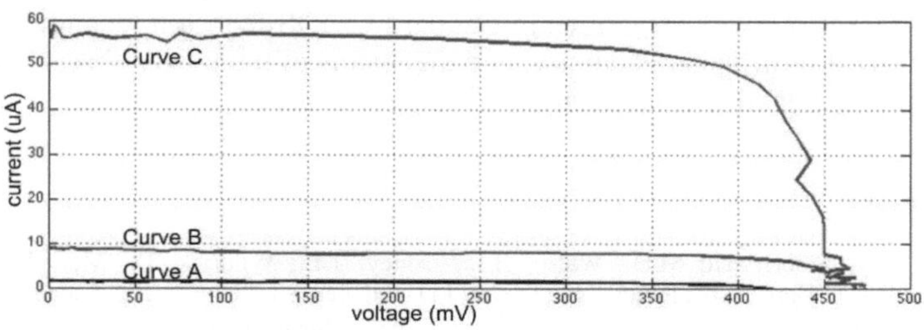

Figure 3. Power characterization of structure C: Curve A is relative to the Harvester configuration, Curve B represents the power characterization of the upper diodes with floating parasitic diodes, Curve C is relative to stand alone solar cell.

Table 2. Efficiencies Table.

Type	Harvester with floating substrate diode	Harvester with short-circuited substrate diode	Parasitic with floating harvester diode	Parasitic with short-circuited harvester diode	Harvester and substrate diodes in parallel
A	2.91	0.44	9.45	8.69	9.45
B	3.02	0.44	15.12	15.12	15.12
C	2.91	1.04	17.64	17.64	17.01
D	2.26	0.56	11.9	11.97	12.6

However, in order t use the diode structures as harvesters to supply an integrated micro-system, realized on the same silicon, it is necessary to hold the p-substrate to the lowest voltage value. Therefore, the parasitic diode must be short-circuited (i.e. the substrate must be connected to the n-well and not to the p-diffusion). The resulting structure is not just equivalent to a couple of diodes, but it has to be modeled with a PNP transistor. This is due to the interaction of the currents of the two junction (transistor effect). In particular, when the junction between p-substrate and n-well is short-circuited, it contributes negatively to the current of the harvester diode. This causes a loss of efficiency, reducing the photogenerated current. For example, with device C, for 100 W/m² the current is reduced from 10 μA, obtained when the substrate is floating, to 2.5 μA. To validate the PNP transistor model, we extracted the Ebers-Moll parameter in the absence of light. When the substrate diode is short-circuited, it is necessary to consider the transistor effects also in light irradiation conduction. In view of the obtained results, with the proposed micro solar cells, it is possible to actually design an integrated microsystem with a significant power budget, considering that the incident power in an usual spring day is, at least, 600 W/m².

3. Conclusions

In this paper we present an active area density optimization technique for harvester photodiodes power efficiency maximization. In particular we realized several micro solar cell structures in standard CMOS 0.35 μm technology, with three different p-diffusion geometry, and the results of the power characterization in the main applicative configuration.

References

1. Ho, H.L.; Steigerwalt, M.D.; Walsh, B.L.; Doney, T.L.; Wildrick, D.; McFarland, P.A.; Benedict, J.; Bard, K.A.; Pendleton, D.; Lee, J.D.; Maurer, S.L.; Corrow, B.; Sadana, D.K., "A 0.13 μm high-performance SOI logic technology with embedded DRAM for system-on-a-chip applicationl", Electron Devices Meeting, 2001. IEDM Technical Digest. International, 2–5 Dec. 2001 Page(s):22.3.1.
2. Gao, J.; Yu, L., "Use of battery ohmic testing to improve network reliability and decrease battery maintenance cost", Telecommunications Energy Conference, 2007. INTELEC 2007. 29th International, Sept. 30 2007–Oct. 4 2007 Page(s):194 – 202
3. Moriizumi, T., Takahashi, K., "Theoretical analysis of heterojunction phototransistors", *Electron Devices, IEEE Transactions on Volume 19, Issue 2, Feb 1972 Page(s):152–159*
4. Neamen, D.A., "Semiconductor Physics and Devices", 1992

ALL-FIBER HYBRID FIBER BRAGG GRATINGS CAVITY FOR SENSING APPLICATIONS

D. PALADINO[1], G. QUERO[1], A. CUTOLO[1], A. CUSANO[1],C. CAUCHETEUR[2] AND P. MEGRET[2]

[1]*Optoelectronic Division-Engineering Department, University of Sannio, Benevento, Italy*

[2]*Electromagnetism & Telecommunication Unit, Faculté Polytechnique de Mons, Belgium*

Abstract. By now, fiber Bragg gratings (FBGs) represent a well assessed technology in sensing field. Unfortunately, to make FBGs sensitive to surrounding refractive index, hosting fiber structuring is needed. In last years, also tilted FBGs (TFBGs) have been proposed as promising technological platform for sensing applications. However, complex spectral features limit the practical exploitation of this technology. It would be extremely useful to merge the peculiar spectral characteristics of both grating types. To address this issue, here, we propose a hybrid cavity involving two unbalanced uniform FBGs written at both sides of a TFBG. The proposed configuration provides a wavelength gated reflection signal. Such a structure preserves uniform FBG advantages in terms of interrogation methods and allows the possibility to contemporarily measure multiple parameters.

1. Introduction

Over the last decade, fiber Bragg gratings (FBGs) have been utilized as optical sensors to measure a wide range of physical parameters including temperature and strain [1]. Their wavelength encoding, flexibility of design, and relative low cost make FBGs ideal devices for a multitude of different sensing applications. However, to make them sensitive to the surrounding refractive index (SRI), hosting fiber structuring is needed with consequent weakening of the grating structure [2]. In contrast, tilted FBGs (TFBGs) – refractive index modulation blazed with respect to the fiber axis – are intrinsically SRI sensitive [3]. In TFBGs, a core and several cladding mode resonances appear simultaneously in their transmission spectrum [4]. This has several advantages: the cladding mode resonances are sensitive to external perturbations (SRI, bending, etc.) while the core mode resonance is only sensitive to temperature and mechanical strain. Up to now, two main types of techniques have been proposed to demodulate the transmitted spectrum of a TFBG subject to an external perturbation: global monitoring of the area delimited by the cladding modes [3] and local monitoring of selected cladding mode resonance shifts [5]. To correctly operate, spectral measurements on a few tens of nanometers are required, severely complicating

P. Malcovati et al. (eds.), *Sensors and Microsystems: AISEM 2009 Proceedings*, Lecture Notes in Electrical Engineering 54, DOI 10.1007/978-90-481-3606-3_21,

the demodulation technique of such sensing elements. It would be extremely useful to opportunely combine uniform FBGs and TFBGs in a single in-fiber structure to obtain strain, temperature, bending, and SRI sensitivity for multi-parametric measurements maintaining the advantages of FBG interrogation. To this aim, here we demonstrate that a hybrid cavity – involving two unbalanced FBGs and an interposed TFBG – can be used for sensing purposes, offering simple multi-parametric measurements working in reflection mode.

2. Device description

The proposed hybrid cavity is made by an unbalanced interferometer composed of two uniform FBGs in the middle of which a weakly TFBG is inscribed. The lateral FBGs are 0.5 mm long, but they have a different maximum reflectivity: about 30% and 100%, respectively. Experimental results concerning two different cavities will be reported: one involves a 4 mm long 4° TFBG, while the other one a 3 mm long 2° TFBG. The uniform FBGs are approximately centred at 1,577.7 and 1,574.7 nm, respectively. A schematic diagram of the investigated structure is reported in Fig. 1a. Such an interferometric cavity can be interrogated in reflection from the non-strong FBG side. The experimental spectral response of the cavity involving the 2° TFBG is shown in Fig. 1b. The reflected spectrum is given by interference fringes due to the Fabry-Pérot cavity formed by the two uniform FBGs with superimposed the attenuation bands – only those overlapping with the Bragg mirrors spectra – due to the cladding modes resonances of the TFBG within the cavity. This way, the hybrid interferometric cavity preserves the versatility of uniform FBGs with respect to measurements in reflection for strain and temperature sensing applications In addition, compared to a single TFBG, the structure preserves the spectral SRI and bending sensitivities, but strongly reduces the useful signal bandwidth.

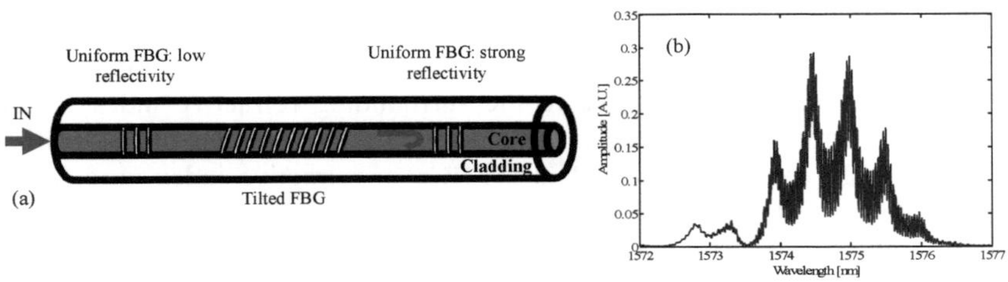

Figure 1. (**a**) Scheme of the all-fiber hybrid cavity; (**b**) reflected spectrum by the 2° TFBG cavity.

3. Experimental results

To demonstrate the device multi-parametric sensitivity and discrimination, SRI, bending, temperature, and strain characterizations have been carried out.

3.1. *SRI and bending characterization*

Figure 2a shows the spectrum of the 4° TFBG hybrid cavity for different SRIs. As well known [3], as the SRI increases the cladding mode resonances red shift due to the increase in the mode effective indices. In addition, approaching their cut-off condition, the attenuation bands progressively disappear starting from the ones at lower wavelengths (higher order cladding modes) and are gradually replaced by a broadband low level attenuation due to the coupling with a continuum of radiation modes. The same spectral effects can be observed for the cladding mode resonances involved in the cavity spectra reported in Fig. 2a. As direct consequence, the total reflected power progressively decreases. This suggests a simple demodulation technique based on reflected power monitoring. Figure 2b shows the reflectance changes – normalized to the reflectance in air ($\Delta R/R_{air}$). Note that the cut-off condition depends on the modes order, and thus it is possible to design the cavity in order to work within the desired SRI range. Similarly, reflected power decrease has been also pointed out for increasing bending along the cavity. In particular, Fig. 2c shows the spectrum of the 2° TFBG hybrid cavity subject to different curvature radii. Figure 2d, instead, shows the correspondent normalized (to the free bending state) reflectance changes. As observable from the reported spectra, no experimental wavelength shift has been observed during SRI and bending characterizations.

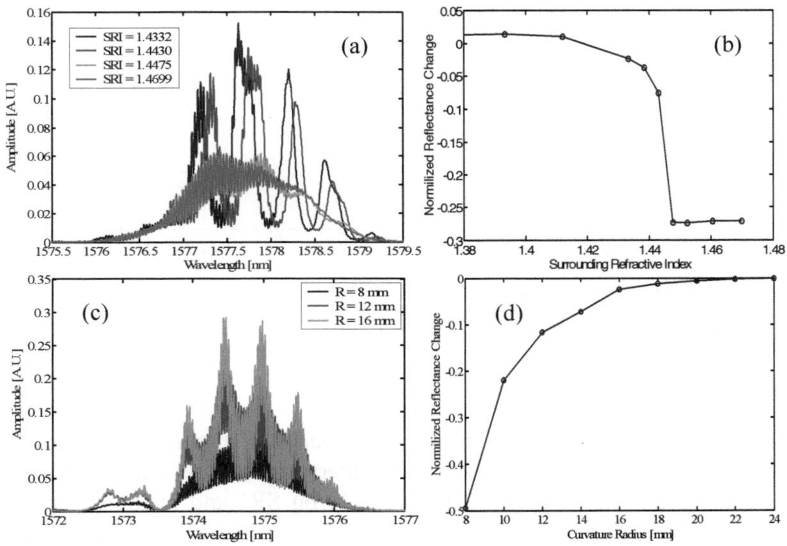

Figure 2. Cavity spectra (**a**) and normalized reflectance change (**b**) for different SRIs; Cavity spectra (**c**) and normalized reflectance change (**d**) for different bending states.

3.2. *Temperature and strain characterization*

Figure 3a shows the hybrid cavity spectrum (4° TFBG) as the temperature changes. As observable, temperature increases only red shift the reflected spectrum. As consequence, uniform FBGs demodulation technique based on wavelength shift monitoring could be adopted. In particular, Fig. 3b shows the cavity central wavelength versus temperature, revealing sensitivity of 9.5 pm/C. Very similar spectral behavior has been registered versus strain (applied mass to tension the fiber structure). Figure 3c shows the spectra of the same hybrid cavity for different applied masses, whereas Fig. 3d shows the correspondent cavity central wavelength – sensitivity of 8.5 pm/g. Note that no reflectance changes have been registered versus temperature and strain.

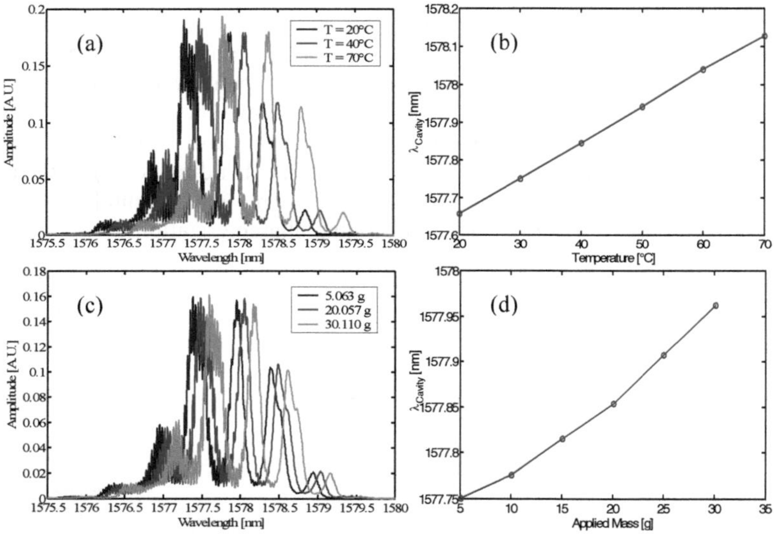

Figure 3. Cavity spectra (**a**) and cavity central wavelength (**b**) for different temperatures; Cavity spectra (**c**) and cavity central wavelength (**d**) for different strain states.

4. Discussion and Conclusions

The results demonstrate simultaneous SRI, bending, temperature, and strain sensitivity. In particular, standard FBG interrogation methods could detect temperature and strain changes independently on SRI and bending variations, while intensity based measurements could detect external SRI and bending changes independently on temperature and strain variations. In addition, the narrow reflected signal could easily allow the multiplexing of several of these cavity structures by wavelength division multiplexing.

References

1. A. Mendez, "Fiber bragg grating sensors: A market overview", Third European Workshop on Optical Fibre Sensors – SPIE 6619, 2007.
2. A. Cusano, A. Cutolo, M. Giordano, "Fiber bragg gratings evanescent wave sensors: A view back and recent advancements", Sensors 21, 113–152, 2008.
3. G. Laffont, P. Ferdinand, "Tilted short-period fibre-Bragg-grating-induced coupling to cladding modes for accurate refractometry", Meas. Sci. Technol. 12, 765–770, 2001.
4. T. Erdogan, J. E. Sipe, "Tilted fiber phase gratings", J. Opt. Soc. Am. A 13, 296–313, 1996.
5. C.-F. Chan, C. Chen, A. Jafari, A. Laronche, D. J. Thomson, J. Albert, "Optical fiber refractometer using narrowband cladding-mode resonance shifts", Appl. Opt. 46, 1142–1149, 2007.

AN OPTICAL PLATFORM BASED ON FLUORESCENCE ANISOTROPY FOR C REACTIVE PROTEIN AND PROCALCITONINE ASSAY

F. BALDINI, A. GIANNETTI, F. SENESI, C. TRONO, L. BOLZONI AND G. PORRO

[1]*Istituto di Fisica applicata "Nello Carrara" IFAC-CNR, via Madonna del Piano 10, 50019 Sesto Fiorentino (FI), Italy*

[2]*Datamed s.r.l., Via Papa Giovanni XXIII 45, 20090 Rodano (MI), Italy*

Abstract. A plastic biochip was developed for the detection of C reactive protein (CRP) and procalcitonin (PCT). It consists of a polymethylmethacrylate (PMMA) chip shaped in order to achieve several flow microchannels. Sandwich assays using the appropriate monoclonal antibodies were implemented with the capture antibody immobilized on the PMMA surface and the target antibody labelled with a fluorophore. A laser diode excites the fluorescent sensing layer. Thanks to the anisotropy of the fluorescence the emitted light travels along the thickness of the plastic material. The fluorescence coming out from the chip is collected by 1 mm plastic optical fibre and detected with a spectrum analyser.

1. Introduction

Optical biosensors find wide applications in the development of point of care testing (POCT) instrumentation. The basis of POCT devices is the need of physicians to have a fast and reliable response to formulate the right diagnosis or to decide the correct therapy, avoiding to deliver the samples to the central laboratories and to wait for a period of time, generally several hours long, to achieve the results of the analysis. In the case of inflammation, the rapid discrimination of viral and bacterial sepsis in intensive care patients, or the fast identification of the origin of infections are mandatory. In this case, the simultaneous measurements of analytes such as C-reactive protein (CRP), procalcitonin (PCT), tumour necrosis factor α, myeloperoxidase, interleukine-6, interleukine-8, interleukine-10 and neopterin can be extremely important for physicians. Among the above mentioned biomarkers CRP and PCT are the parameters analysed in this paper.

Following an acute-phase stimulus, the circulating concentration of human CRP rises from normal levels of about 1–10 mg L^{-1} up to 20-40 mg L^{-1} for a viral infection or even up to about 500 mg L^{-1} in case of a bacterial infection, also for periods 24–48 h long [1]. As for PCT, in microbial infections and in various forms of severe systemic inflammation, its circulating levels increase several fold, and this increase often correlates with the severity of the condition and with mortality [2]. During severe infections (bacterial, parasitic and fungal)

P. Malcovati et al. (eds.), *Sensors and Microsystems: AISEM 2009 Proceedings*,
Lecture Notes in Electrical Engineering 54, DOI 10.1007/978-90-481-3606-3_22,

with systemic manifestations, PCT levels may rise considerably. The circulating concentration of human PCT in healthy patients is less than 0.1 μg L^{-1}, but in the presence of infection, this values can rise up to 100 μg L^{-1}.

2. Methodology

2.1. *The optical device*

The heart of the device is a polymethylmetacrylate (PMMA) biochip, containing microchannels through which the analysed sample flows. Figure 1 show the photo of the PMMA-based chip, produced by injection moulding, with 13 different channels. The dimension of a single microchannel is (100 μm high, 600 μm width, 10 mm long). The optical setup for the interrogation of a single microchannel is described in Fig. 2 [3–5]. The excitation source is a laser diode emitting at 635 nm. A self made plastic cylindrical piano-convex lens located on the cover of the chip leads to an improvement of the excitation efficiency. Thanks to the fluorescence anisotropy [6], a large portion of the fluorescence emitted by the sensing layer immobilised on the bottom of the PMMA cover travels along the thickness of the PMMA up to its end-face where it is collected by means of a plastic optical fibre (core diameter 1 mm) connected to a Hamamatsu optical spectrum analyzer. The end-face of the cover is polished at 60° to facilitate the alignment with the collecting fibre. The particular profile of the cover with the presence of air gaps, as it can be seen from Fig 1, allows a physical separation of the fluorescent signal coming from the different channels, so that each channel is optically separated from the others.

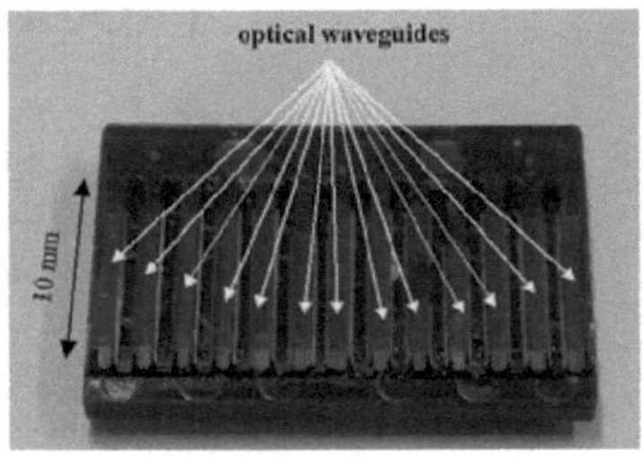

Figure 1. The optical biochip.

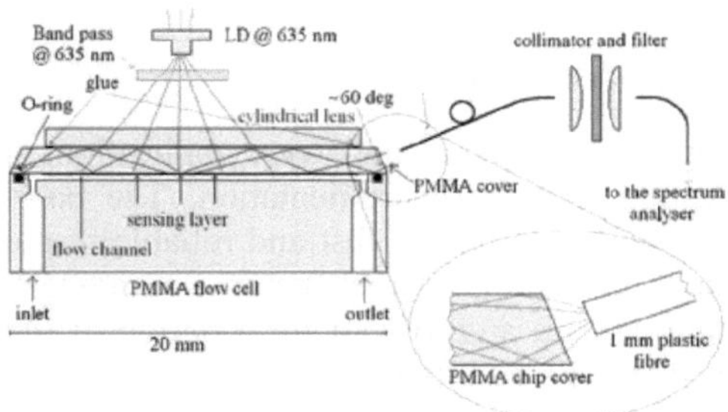

Figure 2. Sketch of the optical system for the chip interrogation.

2.2. *Chemical protocols*

Sandwich assays for CRP and PCT spiked in serum were performed. The PMMA cover of the chip was functionalised by depositing 0.01 mM Eudragit LI00 on the microchannel. The carboxylic groups, now available on the chip surface, were used to covalently immobilise the capture antibody. In the CRP assay, C5-clone and DY647-labeled C7-clone were used as capture antibody, immobilised on the PMMA surface, and target antibody, respectively. PCT assay was carried out using the antibody PROC1 3G3 clone and the labeled DY647-PROC4 6C6 clone as the capture and target antibody, respectively.

The sandwich assay was performed flowing the antigen in buffer at different concentrations for 10 min, washing with buffer for 3 min and flowing, again for 10 min, the target antibody. A final washing step was performed for 3 min with the buffer. All these steps were performed at a flow rate of 10 μL min^{-1}. The fluorescent spectra were then recorded stopping the flow.

3. Results and discussions

Figure 3 shows the fluorescence emission spectra collected with the optical spectrum analyser for different concentrations of CRP ranging from 1 μg L^{-1} up to 500 mg L^{-1} and the related calibration curve. The spectra of the emitted fluorescence are given by the optical intensity for $\lambda > 650$ nm. Each spectrum is the average of ten spectra acquired with an integration time of 10 s. The fluorescence signal detected with the serum solution, which is the zero level of the assays, is due to the autofluorescence of the PMMA biochip.

Each point of the calibration curves is the emitted fluorescence evaluated as the sum of the optical intensity detected by the spectrum analyser between 650 and 720 nm. The standard deviation is evaluated on the ten recorded different spectra. The fitting with the logistic function $y = F_{max} + \dfrac{F_{min} - F_{max}}{1+(x/x_0)^p}$ is also shown where F_{Max} and F_{Min} are the asymptotes of the sigmoidal curve, x_0 is the value of the concentration for which the fluorescence signal is equal to the 50% of the dynamic range (generally denoted with IC_{50}) and p is a coefficient which is related to the slope of the sigmoidal curve

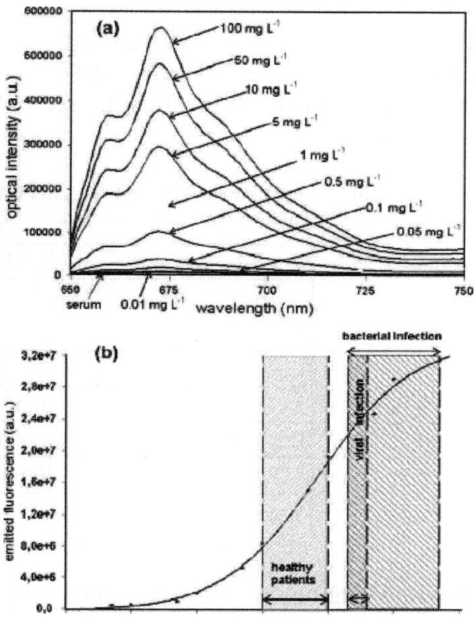

Figure 3. Spectra **(a)** and calibration curve **(b)** for the CRP spiked in serum at different concentrations. The ranges of clinical interests are shown.

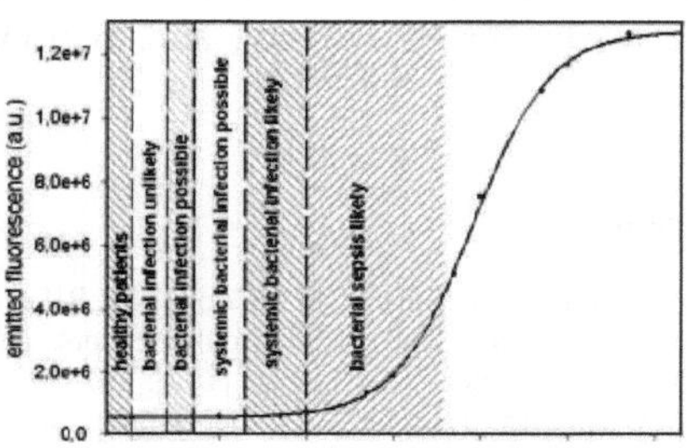

Figure 4. Calibration curve for the PCT spiked in serum at Mifferent concentrations. The ranges of clinical interests fire shown.

for x = IC_{50}. Figure 4 shows the calibration curve achieved for different concentrations of PCT ranging from 1 μg L^{-1} up to 50 mg L^{-1}. Table 1 shows the working ranges and limit of detections for both CRP and PCT. The limits of detection (LOD) and of quantification (LOQ) were evaluated following the IUPAC rules, as three and ten times of the standard deviation of the blank measurement, i.e. the measurement achieved with serum.

4. Conclusions

From these results, it is apparent that the CRP assay satisfy quite well the clinical requirements, whereas the PCT assay requires an (improvement, at the moment in progress, in order to achieve the limit of detection required by clinicians.

Table 1. Achieved performances of the CRP and PCT assays. Parameters of the logistic functions as well as LOD, LOQ and working range are given.

	CRPO	PCTO
IC_{50} (mg L^{-1})	6.67	0.79
F_{max}	33,607,000	12,736,000
F_{min}O	–63,000	540,000
p	0.62	0.97
LOD (mg L^{-1})	0.0008	0.002
LOQ (mg L^{-1})	0.0045	0.006
Working range (mg L^{-1})	0.2–220	0.8–7.6

Acknowledgments This research study was supported by the (European Community within the framework of the five-year EU funded project under IST priority, CARE-MAN (HealthCARE by Biosensor Measurements and Networking).

The authors wish to thank EXBIO Praha, Czech Republic, for labelling the detection antibodies, and the Helmholtz Zentrum München for the provision of the mAbs for PCT assay, and, particularly, Petra Kramer for her fruitful advices.

References

1. P.M.Ridker, J.E.Buring, N.R.Cook, N.R.Rifai, "C-reactive protein, the metabolic syndrome, and risk of incident cardiovascular events: an 8-year follow-up of 14,719 initially healthy American women," Circulation [107], 391–397, January (2003).
2. W.Weglohner, J.Struck, C.Fischer-Schulz, N.G.Morgenthaler, A.Otto, C.Bohuon, A.Bergmann, "Isolation and characterization of serum procalcitonin from patients with sepsis", Peptides, [22], 2099–2103, December, 2001.
3. F.Baldini, A. Carloni, A.Giannetti, G.Porro, C.Trono, "A new optical platform for biosensing based on fluorescence anisotropy", Anal. Bioanal. Chem., [391], 1837–1844, July, 2008.

4. F.Baldini, A.Carloni, A.Giannetti, G.Porro, C.Trono, "An optical PMMA biochip based on fluorescence anisotropy: Application to C-reactive protein assay", Sensors Actuators B, [139], 64–68, May, 2009.
5. F.Baldini, L.Bolzoni, A.Giannetti, M.Kess, P.M.Krämer, E.Kremmer, G.Porro, F.Senesi, C.Trono, "A new procalcitonin optical immunosensor for POCT applications", Anal. Bioanal. Chem., [393], 1183–1190, February, 2009.
6. L.Polerecky, J.Hamrle, B.D.MacCraith, "Theory of the radiation of dipoles placed within a multilayer system", Appl. Opt., [39], 3968–3977, August, 2000.

GOLD COATED LONG PERIOD GRATINGS IN SINGLE AND MULTI LAYER CONFIGURATION FOR SENSING APPLICATIONS

A. IADICICCO[1], S. CAMPOPIANO[1], D. PALADINO[2], A. CUTOLO[2], A. CUSANO[2] AND W. BOCK[3]

[1]*Department of Technology, University of Naples "Parthenope", Centro Direzionale Napoli, Isola C4, 80143, Napoli, Italy*

[2]*Optoelectronic Division, Engineering Department, University of Sannio, Corso Garibaldi 107, 82100, Benevento, Italy*

[3]*Centre de recherche en photonique, Université du Québec en Outaouais, 101 rue Saint-Jean-Bosco, Gatineau, QC J8X 3X7, Canada*

Abstract. Long Period fiber Gratings (LPGs) fabricated by electrical discharge have found large interest as emerging devices in the field of optical communications and optical fiber sensors. Besides, the integration of UV LPGs with noble metal nano-layers, such as Au or Ag, seem to act as a promising sensing platform partly due to their potential applications for chemical and biological sensing. This work, for the first time to our knowledge, reports on the spectral characteristics of arc-induced LPGs coated with nano-scale gold coating in single and multi layer configurations. In particular, gold layers at nano-scale have been deposited on LPGs by adopting evaporation technique in a vacuum chamber with good uniformity, optically smooth interfaces and different thicknesses. Also, multi-layer configurations involving pre and post deposition of polymeric thin film have been investigated.

1. Introduction

In recent years, considerable attention has been focused on in-fiber Long Period Gratings (LPGs) for a wide range of applications in optical communications and sensing systems, such as in-fiber band rejection filters and various kinds of sensors for temperature, strain and Refractive Index (RI) measurements [1]. At the same time, the sensing characteristics of LPGs also encourage the development of new methods for LPG fabrication: CO_2 laser pulses or electrical arc discharge have been proposed as a simpler and lower-cost approach if compared with UV-induced LPGs [2, 3]. Besides, the sensitivity of LPGs to environmental parameters and the easy tailoring of the sensing performances of this class of devices motivate the ever increasing research activities aimed to transform this technology in a valid platform for chemical and biological sensing applications. In particular, polymeric coatings have been proposed for specific chemical sensing and as simple tool capable to tailor the LPG sensitivity characteristics [4–7]. Moreover, recently,

P. Malcovati et al. (eds.), *Sensors and Microsystems: AISEM 2009 Proceedings*, Lecture Notes in Electrical Engineering 54, DOI 10.1007/978-90-481-3606-3_23,

great attention has been focused on the integration of LPGs with noble metal nano-layers, such as Au or Ag, partly due to their potential applications for chemical and biological sensing. Some biosensors and chemosensors schemes have been realized based on Surface Plasmon Resonance (SPR) phenomena [8–11]. This work, for the first time to our knowledge, reports on the spectral characteristics of Arc-Induced LPGs coated with nano-scale gold coating in single and multi layer configurations.

2. Experiment

The LPGs for our experiments were manufactured from single-mode Corning SMF-28 fibers. The arc setup is based on periodic melting of the fiber, while a pulling weight stretches it, thus determining a periodically tapered fiber [12]. The grating period was mainly determined by the moving step of the translation stage that was controlled by a computer and by some other factors such as arc volume, arc intensity, arc duration time (τ), and pulling weight. From optical microscope analysis it was possible to estimate a fiber cross section reduction in the taper waist of about 5%, in the investigated gratings; moreover, it is reasonable to assume that the total fiber diameter and the core diameter stay in the same ratio along the taper [10]. Gold layers at nano-scale have been deposited on LPGs by adopting evaporation technique in a vacuum chamber. To coat the fiber as uniformly as possible, two coating runs were made with the fiber holder rotated by 180° between the coatings. We believe that under these conditions, the film uniformity around the fiber circumference is acceptable. The gold layer thicknesses were estimated by measuring layer deposited on glass flat sample treated simultaneously to the optical fiber. Layer thicknesses of 18 nm and 45 nm have been obtained. Moreover, multi-layer configurations have been obtained by pre-deposition and post-deposition of polymeric thin film of Syndiotactic PolyStyrene (sPS) deposited by dip-coating method in 8% solution by weight of sPS (in chloroform).

3. Results

The presence of nano-scale gold layers modifies the cladding modes distribution and thus their interaction with external medium. While Surrounding RI (SRI) increases the attenuation bands of the gold coated LPG shift towards lower wavelengths whereas an anomalous behavior can be observed for specific SRI values. Figure 1a reports the spectra of a 45 nm gold coated arc-induced LPG with pitch of 400 μm for different SRI values. As the SRI moves from air to 1.4423 the band at 1,270 nm exhibits a phase-shift like effect whereas the band at 1,470 nm registers a blue shift of 36 nm compared to the air case. For further increase of the SRI (from 1.4423 to 1.4488) the phase-shift effect can be observed only for the band at 1,470 nm. For comparison with the uncoated grating, Fig. 1b plots the resonance wavelength of the band around 1,450 nm of the 45 nm gold coated LPG (square marks) with similar uncoated grating (circle marks) whereas the dark triangular marks report the phase-shift wavelength. As evident the uncoated grating exhibits leaky mode regime for SRI higher than 1.458 whereas the 45 nm gold layer seems to shift the leaky mode regime to a slightly lower SRI value (around 1.452). However the 45 nm gold coated grating exhibit phase-shift like spectrum as the SRI moves between 1.448−1.451. This anomalous behavior is currently under investigation to better identify the coupling regime. It could be compared with SPR observations recently reported by Albert et al. [11]. Additionally, in guided regime (SRI lower than 1.442) the metallic film exhibit a shielding effect for SRIs lower than 1.415 whereas a slight SRI sensitivity enhancement is observed as the SRI moves between 1.415 and 1.442. Figure 1b also reports the resonance wavelength of the same grating in multi-layer configuration involving 45 nm gold layer and polymeric sPS layer (star marks). It is evident that the multi-layer configuration exhibits a strong shielding effect. In particular negligible resonance shift and band transmittance changes are measured in the entire investigate SRI range (1−1.469).

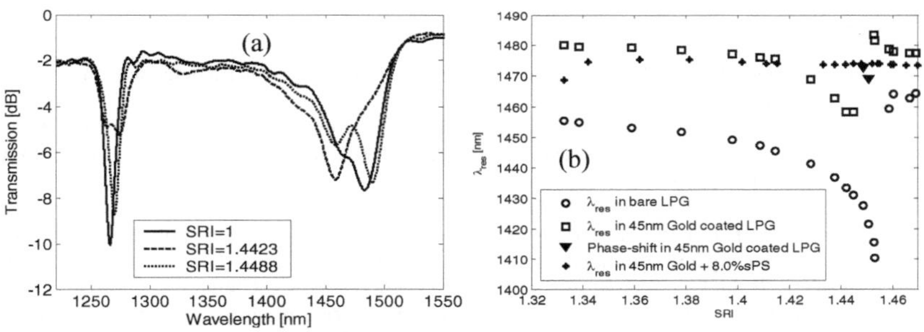

Figure 1. (**a**) Spectra of 45 nm gold coated LPG for several SRIs; and (**b**) resonance wavelength of the bare, 45 nm gold coated and multi-layer configuration (45 nm gold + sPS layer).

Figure 2a instead plots the band resonance wavelength versus the SRI in a similar grating (arc-induced with 400 μm period) when coated with 18 nm gold film. Here, the metallic film strongly affects the cladding mode behavior. In fact, differently from the bare case (circle marks), the metallic coated grating exhibits a red shift (square marks) as the SRI increases from the water value. On the other side a strong increase in the absolute SRI sensitivity is measured for SRI around the 1.34 (close to the water refractive index). Besides, also multi-layer configuration involving pre-deposition of sPS layer and 18 nm gold film have been investigated along a 400 μm LPG. Figure 2b compares the resonance wavelength versus the SRI for single layer deposition. As observable the sPS layer was designed to achieve a modal transition centered at about 1.38 [5] whereas the metallic film shields the transition phenomena. The multi-layer configuration exhibits wavelength shift similar to bare grating whereas the leaky mode regime seem to be achieved for lower SRIs (around 1.442).

In conclusion, gold films along arc-induced LPGs strongly influence the guiding features of the guided modes and thus affects the behavior of the grating versus the SRI. However, deep analyses are requested and currently in progress to better understand the novel guiding regime in order to identify eventually novel sensing performances.

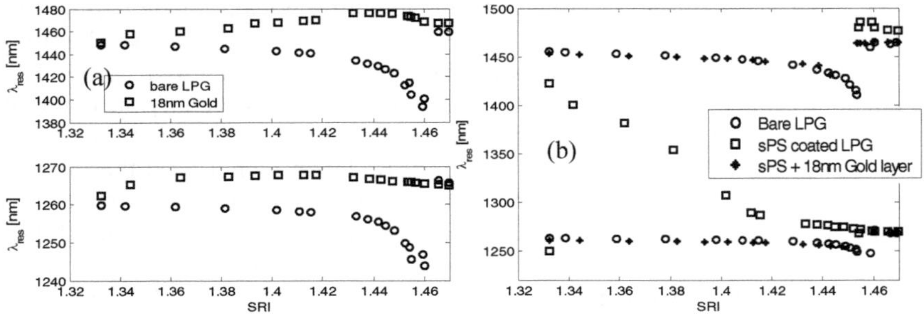

Figure 2. Resonance wavelength of the bare, sPS coated and multi-layer configuration (sPS layer + 18 nm gold film).

References

1. V. Bhatia, "Applications of long-period gratings to single and multiparameter sensing," Optics Express Vol. 4, pp. 457–466, 1999.
2. Y. J. Rao, Y. P. Wang, Z. L. Ran, T. Zhu, "Novel fiber-optic sensors based on long-period fiber gratings written by high-frequency CO2 laser pulses", Journal Of Lightwave Technology, Vol. 21, No. 5, pp. 1320–1327, 2003.
3. S. Nam, C. Zhan, J. Lee, C. Hahn, K. Reichard, P. Ruffin, K. Li Deng, S. Yin, "Bend-insensitive ultra short long-period gratings by the electric arc method and their applications to harsh environment sensing and communication", Optics Express, Vol. 13, No. 3, pp. 731–737, 2005.
4. A. Cusano, P. Pilla, L. Contessa, A. Iadicicco, S. Campopiano, A. Cutolo, M. Giordano,"High sensitivity optical chemo-sensor based on coated long period gratings for sub ppm chemical detection in water", Applied Physic Letters, Vol. 87, 234105, December 2005.
5. A. Cusano, A. Iadicicco, P. Pilla, L. Contessa, S. Campopiano, A. Cutolo, M. Giordano, "Mode transition in high refractive index coated long period gratings", Optics Express, Vol. 14, No. 1, pp. 19–34, January 2006.
6. J. M. Corres, I. del Villar, I. R. Matias, F. J. Arregui, "Two-layer nanocoatings in long-period fiber gratings for improved sensitivity of humidity sensors", IEEE Transactions On Nanotechnology, Vol. 7, No. 4, July 2008.
7. J. Barnes, M. Dreher, K. Plett, R. S. Brown, C. M. Crudden, H. P. Loock, "Chemical sensor based on a long-period fibre grating modified by a functionalized polydimethylsiloxane coating", Analyst, Vol. 133, pp. 1541–1549, 2008.
8. T. Okamoto, I. Yamaguchi, T. Kobayashi, "Local plasmon sensor with gold colloid monolayers deposited upon glass substrates", Optics Letters, Vol. 25, pp. 372–374, 2000
9. S. -F. Cheng, L. –K. Chau, "Colloidal gold-modified optical fiber for chemical and biochemical", Sensing. Anal. Chem., Vol. 75, pp. 16–21, 2003.
10. J. L. Tang, S. Chengb, W. T. Hsub, T. Y.Chiang, L. K. Chaub, "Fiber-optic biochemical sensing with a colloidal gold-modified long period fiber grating", Sensors and Actuators B, Vol. 119, pp. 105–109, 2006.
11. Y. Y. Shevchenko, J. Albert, "Plasmon resonances in gold-coated tilted fiber Bragg gratings", Optics Letters, Vol. 32, No. 3, 2007.
12. W. J. Bock, J. Chen, P. Mikulic, T. Eftimov, "A novel fiber-optic tapered long-period bragg grating sensor for pressure monitoring", IEEE Instrumentation and Measurement Technology Conference, Sorrento, Italy, 24–26 April 2006.

UV SCHOTTKY SENSORS BASED ON WIDE BANDGAP SEMICONDUCTORS

P. ALLEGRINI, P. CALVANI, M. GIROLAMI, G. CONTE AND M.C. ROSSI

University of Roma Tre, Department of Electronic Engineering, Via della Vasca Navale 84 - 00146 Rome, Italy

Abstract. Planar Al/GaN/Ni Schottky diodes were realized on GaN films deposited on sapphire substrates and characterized in the dark and under illumination. The optoelectronic characteristics of GaN photodetectors appear largely influenced by structural defects and impurities, which are clearly detected in photocurrent yield measurements. In particular, an exponential increase of the photocurrent is observed and explained in terms of a barrier lowering photoeffect, hence a light induced shrinking of the space charge region, related to carrier trapping at defects and impurities. Trapping events are also responsible for a dispersive behavior of the AC responsivity with the light chopping frequency. Such effects point out the importance of a proper selection of bias voltage and working frequency for GaN photodetector operations.

1. Introduction

In recent years, the increasing interest in the ultraviolet imaging for scientific instrument and astronomy applications has triggered a research effort to develop radiation detectors based on wide band gap materials, showing solar blind spectral response. Among suitable devices, photodetectors based on AlGaN alloys, show high sensitivity, large UV/visible contrast, extremely low dark current and negligible damage caused by operation in harsh environments [1–3]. The photoresponse of these devices can be tuned over the whole UV spectral range from 200 to 360 nm by changing the Al content in a Al/GaN/Ni planar Schottky diodes. However, GaN optoelectronic applications requires a deeper comprehension of fundamental questions about defect related detrimental effects. The purpose of this article is to concentrate on this aspect affecting the optoelectronic performance of GaN photodetectors.

2. Experimental

GaN samples were grown on sapphire substrate by Molecular Beam Epitaxy (MBE) technique according to a procedure reported elsewhere [4]. For device fabrication samples were cleaned in acid before contacts realization. Planar arrays of ohmic Al contacts and Schottky Ni contacts were thermally evaporated

P. Malcovati et al. (eds.), *Sensors and Microsystems: AISEM 2009 Proceedings*, Lecture Notes in Electrical Engineering 54, DOI 10.1007/978-90-481-3606-3_24,

on the sample slab in order to test photoconductor, Schottky photodiode and MSM Schottky detector structures. In each case about $1mm^2$ free surface was left for device illumination. Electrical characteristics were measured in the dark by using an HP4140B picoammeter. Current-voltage measurements were performed in a helium cryostat in the temperature range 150–300 K. Spectrally resolved photocurrent measurement was performed by standard lock-in technique using a 300 W xenon arc lamp passing through a monochromator scanning from long to short wavelength, unless otherwise specified, in order to minimize PPC effects. Detector responsivity values was also measured with the same set up at different chopping frequency.

3. Results and discussion

Al/GaN/Ni planar Schottky diodes with ideality factor n = 1.4 and more than four order of magnitude of rectification have been realized on sapphire substrates. As reported in Fig. 1, the I–V characteristic have been measured at different temperature (a reverse dark current decrease of about one order with temperature is detected in the range 150–300 K), allowing the evaluation of the Schottky barrier height through the equation holding for a reversely biased Schottky diode:

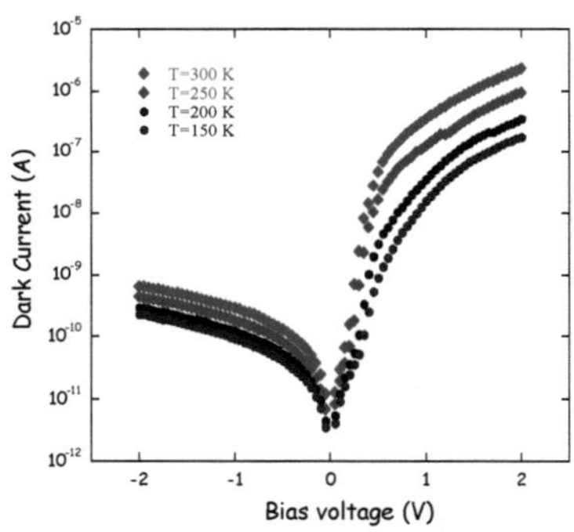

Figure 1. I–V characteristics of GaN Schottky diode at different temperatures in the range 150–300 K.

$$J_R \sim A^*T^2 \exp(-q\Phi_{B0}/kT)\exp[q(qE/4\pi\varepsilon_s)^{0.5}/kT] \quad (1)$$

Here J_R is the reverse current density, A^* is the Richardson constant, Φ_{B0} is the zero field barrier height, ε_s is the dielectric constant and E is the applied electric field given by:

$$E=[2qN(V+V_{bi}-kT/q)] \quad (2)$$

Here V_{bi} is the Schottky junction built in voltage and V the applied bias. Similarly to previous results [5], the barrier height of the Ni/GaN contact was evaluated from the I–V characteristic measurements at different temperatures, which turns out to be 1.2 ± 0.1 eV. It is noteworthy that such a value can be affected by the upward surface band bending caused by the presence of charged surface states [6]. Ni/GaN/Ni back to back Schottky structures have been also realized and tested both in the dark and under illumination, showing the expected relation $I = I_0 \tanh (V/2V_T)$ (not shown). Spectrally resolved photocurrent spectra of GaN detectors fully cover the UV range, as shown in Fig. 2, exhibiting a good UV/visible contrast, i. e. the ratio between the peak responsivity and that below the bandgap at about 3 eV. At this energy the detector discrimination is about three order of magnitude. A rather sharp transition is observed at the optical band edge around 3.3 eV, where also an exciton peak is present. Such a band gap value suggests a main zinc-blende film structure (β phase with $E_g(\beta) = 3.2$ eV). A contribution from electronic transition between levels induced by nitrogen vacancies could also play a role in this spectral range. At higher energy, a continuously increasing photoresponse is found due to the rise in the density of states of the valence and conduction band, hence in the number of the band to band transitions. A small but well structured sub band gap photosignal is also detected. In particular, the contribution of defect and/or impurity related absorption bands can be clearly distinguished in Fig. 2 at about 2.2, 2.6 and 2.8 eV and usually called yellow (Y), green (G) and blue (B) luminescence bands, respectively. Such contributions were even observed GaN nanowiskers [7]. A much discussed defect has been the one giving rise to the so-called yellow emission band in luminescence spectra of GaN. The presence of this emission has often been correlated with low quality of the material. The identity of the emission is not known yet, but it's clear that the emission has slow decay time (0.1–1 μs), many orders of magnitude slower than the one observed for the band-gap excitons (of the order 100 ps at low temperatures). Under illumination, a small, but non-negligible signal is also detected at V = 0 followed by a non linear voltage dependence of the photocurrent, as shown in Fig. 3. In particular, an exponential increase of the photocurrent is observed, which can be explained in terms of a barrier lowering photoeffect, hence a light induced shrinking of the space charge region, induced by carrier trapping. Modulation of the space charge regions around defects induced by illuminations have been previously reported in AlGaN photoconductive detector [8]. Qualitative band diagram of the MSM detector under applied bias is reported in Fig. 3,

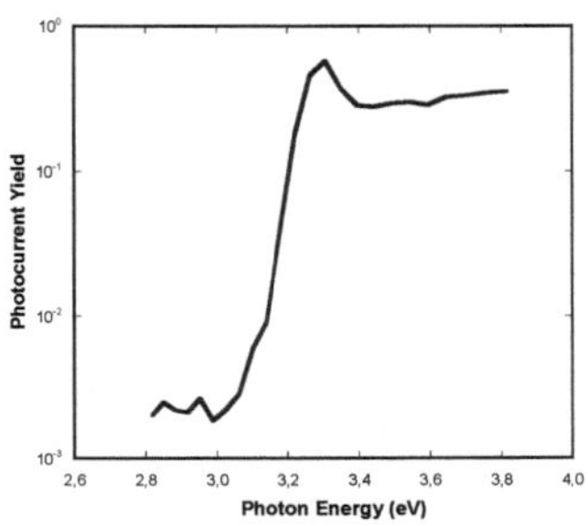

Figure 2. Spectrally resolved photocurrent yield of typical GaN photodiode in the 2.8–4 eV photon energy range.

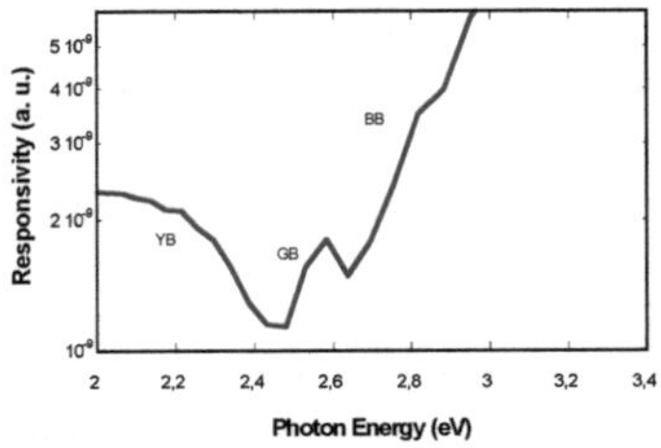

Figure 3. Sub band gap details of spectrally resolved photocurrent yield of typical GaN photodetector.

evidencing the light induced barrier lowering and space charge modulation. According to this simple model, photogenerated holes can be easily collected at the cathode, whereas electrons can be trapped by surface defects close to the anode, where they experience an energy barrier, which is lowered by electrons accumulation. The detected photocurrent is dominated by this effect which completely hide the primary photocurrent related to the space charge layer of the reversely biased diode. According to this simple model the photocurrent can be expressed as:

$$I_{ph} = I_{00} \exp [(\beta\ \mathrm{sqrt}(V) + \Delta\Phi_L)/nV_T] \tag{3}$$

where I_{00} is a pre-exponential factor, β is the Schottky barrier lowering coefficient, V_T is the room temperature thermal energy, n is the diode ideality factor and $\Delta\Phi_L$ is the light-induced barrier lowering due to the accumulation/trapping of photogenerated electrons at the rectifying interface. Best fitting of experimental data gives the Schottky barrier lowering coefficient $\beta = 2.54\ V^{-1/2}$ and $I_{00} \exp (\Delta\Phi_L/nV_T\ ) = 9.4\ 10^{-6}$ A. This suggests that GaN based devices a significant role is played by surface trapping states modeled as charged states able to reduce the surface band bending under illumination. The presence of trapping states and their effect on the photogenerated carrier kinetics plays also a role in determining a strongly sublinear responsivity dependence on the illumination intensity, in agreement with previous observations [8, 9]. A further evidence of the presence of defects is given by the AC responsivity dependence on the chopping frequency. A sort of "gain (responsivity) – bandwidth product" law is observed, with an almost constant responsivity at low frequencies, followed by a faster decay eventually approaching an inverse frequency dependence law (1/f). A corner frequency separating the two regions can be found. Such a effect points out the importance of a proper selection of the working frequency, a lower frequency resulting into a larger detector selectivity. This dispersive behavior reflects the energy distribution of defects with different response times which is also responsible for extremely slow decay and persistence of photocurrent observed when the illumination is switched off. Similar effects of trapping-related persistent photocurrent and long response time were previously observed and tentatively related to sample preparation conditions in continuous film [10] and to size effects in GaN nanowiskers [11] (Figs. 4 and 5).

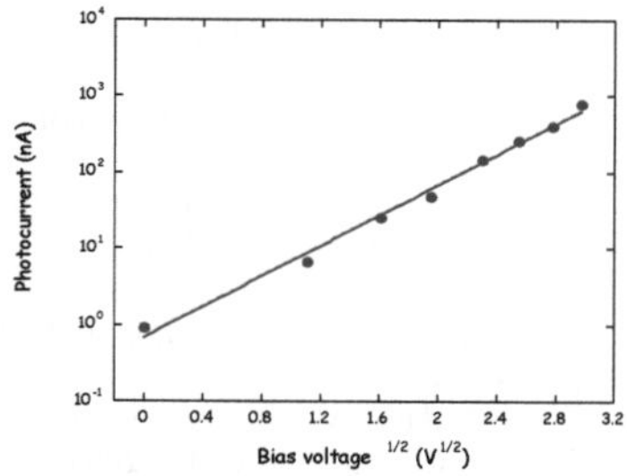

Figure 4. Voltage dependence of the detector photocurrent.

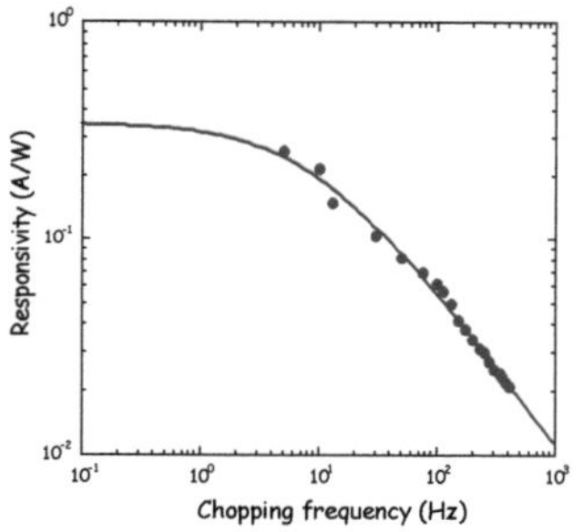

Figure 5. Responsivity dependence on chopping frequency, showing a typical low pass behaviour.

4. Conclusion

We have reported on the realization and characterization of GaN ultraviolet photodetectors with Al/GaN/Ni planar Schottky diode structure. Diodes have shown more than four order of magnitude of rectification with ideality factor n = 1.4. The spectral response, either close to and within the gap regions have clearly shown the signature of charged structural defects and impurities, possibly located close to the surface. In particular, the observed exponential increase of the photocurrent have been explained in terms of a barrier lowering photoeffect, and consequently a light induced shrinking of the space charge region, related to carrier trapping. Trapping processes were also responsible for a dispersive behavior of the AC responsivity with the light chopping frequency. Such effects point out the importance of a proper selection of bias voltage and working frequency for GaN photodetector operations.

References

1. E. Monroy, F. Omnès, F. Calle, Semicond. Sci. Technol 18 (2001) R33.
2. E. Monroy, F. Calle, J. L. Pau, F. J. Sanchez, F. Omnes, B. Beaumont, P. Gibart, E. Munoz, J. Appl. Phys. 88 (2000) 2081.
3. Z. C. Huang, J. C. Chen, D. Wichenden, J. Cryst. Growth 170 (1997) 362.

4. M. E. Lin, Z, Ma, F. Y. Huang, Z. F. Fan, L. H. Allen, H. Morkoc, Appl. Phys. Lett. 69 (1996) 1556.
5. T. Sawada, Y. Izumi, N. Kimura, K. Suzuki, K. Imai, S.-W.Kim, T. Suzuli, Appl. Surface Sci. 216 (2003)
6. S. Chevtchenko, X. ni, Q. Fan, A. A. Baski, H. Morkoc, Appl. Phys. Lett. 88 (2006) 122104.
7. A. Cavallini, L. Polenta, M. Rossi, T. Richter. M. Marso, R. Meijers, R. Calarco, H. Lüth, Nano Lett. 6 (2006) 1548.
8. E.Monroy, F. Calle, J. A. Garrido, P. Youinou, E. Munoz, F. Omnès, B. Beaumont, P. Gibart, Semicond. Sci. Technol. 14 (1999) 685.
9. M. De Vittorio, B. Poti, M. T. Todaro, M. C. Frassanito, A. Pomarico, A. Passaseo, M. Lomascolo, R. Cingolani, Sensors and Actuators A 113 (2004) 329.
10. M. Salis, A. Anedda, F. Quarati, A. J. Blue, W. Cunningham, J. Appl. Phys. 97 (2005) 33709.
11. R. Calarco, M. Marso, T. Richter, A. A. Aykanat, R. Meijer, A. v. d. Hart, T. Stoica, H. Lüth, Nano Lett. 5 (2005) 981.

DESIGN AND REALIZATION OF A NOVEL PIXEL SENSOR FOR COLOR IMAGING APPLICATIONS IN CMOS 90 NM TECHNOLOGY

G. LANGFELDER, A. LONGONI AND F. ZARAGA

Electronics and Information Department, Politecnico di Milano, via Ponzio 34/5 20133, Milano, Italy

Abstract. Toward the complete integration of color sensors in CMOS technologies, a novel color sensitive device, the Transverse Field Detector, is proposed. The TFD color detection principle is based on the creation of a transverse, V-shaped electric field configuration in a Silicon active layer. The electric field is generated only by means of surface biasing/ collecting electrodes. Taking advantage on the dependence of the Silicon absorption length with respect to the incoming wavelength, each of the surface contacts collect photo-carriers down to a different depth. In this way three spectral functions are obtained at the three electrodes, without the use of any color filter. Newly developed pixel structures and a preliminary Active Pixel readout circuitry design are presented.

1. Introduction

Digital imaging in CMOS technology is typically performed by means of a spatial sampling device, formed through a matrix of basic picture elements (pixels). Silicon quantum sensors have been used so far as intrinsically "gray-scale" sensors, thus to reconstruct the color of the incoming radiation a color filter array (CFA) is superimposed on top of the sensor. Each color element is formed by four pixels, typically with a Bayer RGGB CFA, and the different transmittances of the filters constituting the CFA allow obtaining three different spectral functions (namely Red, Green and Blue) of the incoming radiation, which allows color reconstruction [1, 2]. This commonly used solution has some drawbacks: the CFA absorption reduces the overall quantum efficiency (QE); errors and artifacts may arise as interpolation is required to reconstruct color information at the single pixel position; moreover, the deposition of CFA is not part of a standard CMOS technology process flow. Solutions that are not based on the deposition of CFA have been proposed in the literature [1, 3, 4] but either they can solve only some of the listed drawbacks or they aren't fully compatible with CMOS standard processes.

Toward the complete integration of color sensors in CMOS technology, a novel color sensitive CMOS device, the Transverse Field Detector (TFD) has been recently proposed [5]. The TFD color detection principle is based on the

P. Malcovati et al. (eds.), *Sensors and Microsystems: AISEM 2009 Proceedings*, Lecture Notes in Electrical Engineering 54, DOI 10.1007/978-90-481-3606-3_25,

creation of a transverse, V-shaped, electric field configuration in the Silicon active layer. This electric field is generated only by means of surface biasing/ collecting electrodes. For instance an array of suitably reverse biased N+ source/ drain implants on a low doped P type substrate or epitaxial layer accomplishes this task.

Taking advantage on the strong dependence of the Silicon absorption length with respect to the incoming wavelength in the visible range, each of the surface contacts collect photo-carriers down to a different depth. Carriers' collection may be inferred from the electric field streamlines arising in the depleted region, as schematically depicted in Fig. 1. This is the way the three different spectral functions are obtained at the three electrodes, without the use of any color filter. No reduction in QE is caused using this solution. No interpolation is required to reconstruct the color at each position, as each pixel is color sensitive itself. Moreover, no further step with respect to a CMOS process flow is added.

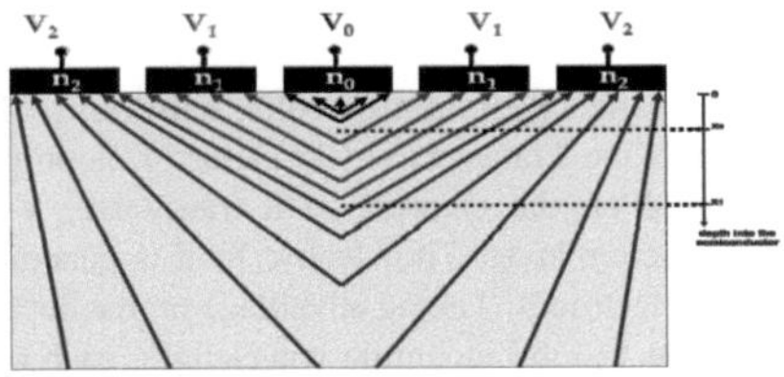

Figure 1. TFD basic working principle: an array of electrodes is built on top of a Silicon active layer. Through a suitable biasing, a transverse electric field configuration, here represented by the streamlines arrows, is obtained. Different contact pairs collect carriers generated up to different depths, thus each contacts pair has a different spectral response.

The paper is organized as follows: in section 2 a description of TFD prototypes realized in a CMOS standard 90-nm process are described, and an improved device, with smaller dimensions is presented. Section 3 reports the layout design of a mini-matrix of Active Pixel (AP) TFD, where the AP is suitably designed to cope with the TFD requirements.

2. 90-nm CMOS standard TFD prototypes

First prototypes of the TFD, realized in a CMOS 90-nm standard technology, demonstrate the basic working principle and the color detection capabilities of this sensor. The structure is schematically depicted in Fig. 2a, which reports a three-colors pixel with a width of 5.2 μm. The biasing/collecting electrodes are designed through reversely biased n-type regions, built directly on a lowly doped (~10 Ωcm) epitaxial layer. The outer N-type region is an N-well rather than an N+ implant, to obtain a better inter-pixel isolation. Each n-type electrode is isolated from punch through by means of grounded P+ implants. Experimental results obtained with this sensor were presented in [5, 6].

To further reduce the pixel dimension a novel test structure has been designed, which schematic cross section is reported in Fig. 2b. With respect to the previous structure, the transverse electric field is here obtained only through three single contacts rather than using contacts pairs. The electric field configuration symmetry is interrupted using a P-well aside the TFD active area. This P-well is need in the AP electronics design to separate the lateral N-well of the TFD from the N-well hosting the PMOS, thus it is not an added structure. The overall active area of the pixel is 3.5 μm, with a foreseen Active Pixel fill factor around the 35%, including the electronics and the contact areas. The electronics of the AP will be described in section 3.

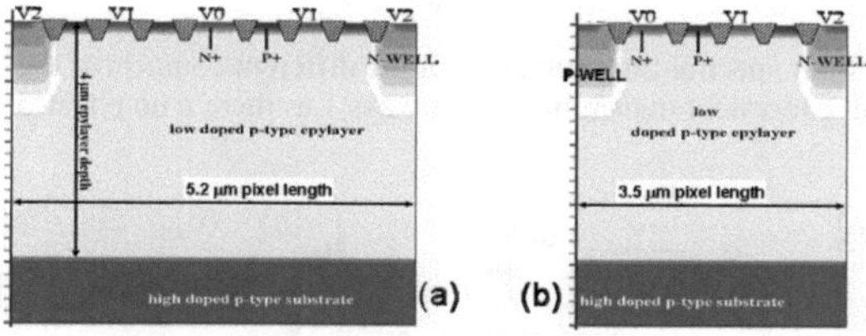

Figure 2. Schematic cross section of two version of the active area of a TFD three colors pixel: (**a**) fully symmetric structure which exactly matches the working principle expressed in Fig. 1; (**b**) asymmetric structure in which the transverse electric field is laterally blocked by a P-well.

Figure 3 reports the *equivalent filter spectral transmittances* ([6]) obtained at the three surface contacts in two different biasing conditions. From the solid lines it can be seen that the innermost electrode has the highest equivalent transmittance in the blue region, the intermediate electrode has an equivalent transmittance peak around 500 nm in the green part of the visible spectrum, and the outermost electrode has an equivalent transmittance which is maximum in the red region. From the dashed lines it can be inferred how a change in the biasing values allows tuning the equivalent filter transmittances. It should be remarked that no technology rule of the used 90 nm process is violated during the sensor design.

3. Design of an active pixel for the TFD

Typical Active Pixel Sensors are based on the three-transistor (3T) readout circuit or on derived topologies [7]. In these AP configuration, the anode voltage decreases during the integration time, as the photo-charge directly integrates on the reversely biased diode capacitance. The TFD basic working principle on the contrary requires a constant voltage at each anode to hold on the electric field configuration during the readout phase.

To comply with this requirement a specific AP electronics has been developed, which is based on a single transistor charge amplifier. The AP is formed by the sensitive area, four transistors (4T) and a feedback capacitance per color channel. The AP circuit is shown in Fig. 4a. Using a process with several metal layers, the capacitances can be built on top of the transistors area.

A first mini matrix (20 pixels) of APS TFD has been designed in the same 90 nm technology: overall three color pixel dimensions are 15 × 15 um (TFD active area is here 7 × 8 μm). The three capacitances are ~4 fF each. The fill factor (ratio between the sensitive area and the whole active pixel area) is ~50%. Part of the mini matrix layout is shown in Fig. 4b.

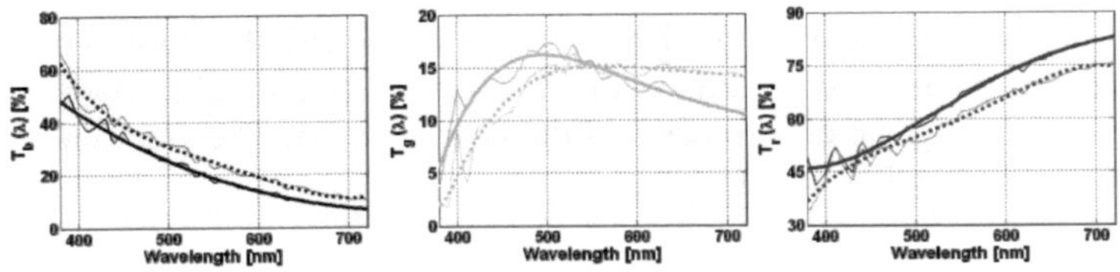

Figure 3. Equivalent filter spectral transmittances of the different contacts of the TFD in two biasing conditions: the overall equivalent transmittance is always 1 as there's no physical filter in the device.

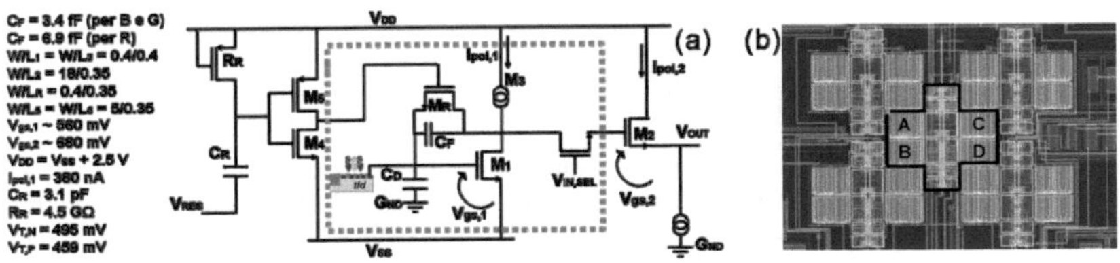

Figure 4. Schematic (**a**) and layout (**b**) of the AP circuit for the TFD: when the reset MOS M_R is opened, photo-charges integrate on the feedback capacitance C_F through the feedback mechanism on the transistor M_1. When M_R is closed the circuit is reset. Only devices enclosed in the dashed line are part of the AP: the other circuitry (a charge pump and the output bus) are part of the whole matrix.

4. Conclusions

The Transverse Field Detector is a recently proposed pixel device capable to perform colour imaging without color filters. In this paper, improvements on the pixel structure which allow reducing the pixel dimensions down to 3.5 μm are presented. A dedicated Active Pixel circuitry has been designed which relies on the use of a single transistor charge amplifier per pixel. The designed mini-matrix has a pixel fill factor around 50%.

References

1. B. E. Bayer, "Color Imaging Array", US Patent n. 3 971 065, July 1976.
2. A. El Gamal, H. Eltoukhy "CMOS Image Sensors", IEEE Circuits and Devices Magazine, pp 6–20, May–June 2005.
3. Foveon Inc., "Color separation in an active pixel cell imaging array using a triple-well structure", US Patent n. 5 965 875, October 1999.
4. P. B. Catrysse, B. A. Wandell, A. El Gamal, "An Integrated Color Pixel in 0.18μm CMOS technology," 2001 International Electron Devices Meeting – Technical Digest, 559, 2001.
5. A. Longoni, F. Zaraga, G. Langfelder, L. Bombelli, "The Transverse Field Detector (TFD): a novel color sensitive CMOS device", Electron Device Letters, [29] n. 12, November 2008.
6. G. Langfelder, F. Zaraga, A. Longoni, "Tunable spectral responses in a color sensitive CMOS pixel for imaging applications", submitted to IEEE Trans. on Electron Devices, January 2009.
7. J. Janesick, F. Gunawan, T. Dosluoglu, J. Tower, N. McCaffrey, "Scientific CMOS Pixels", Experimental Astronomy vol. 14, 33–43, 2002.

TECHNOLOGY AND *I–V* CHARACTERISTICS OF FULLY POROUS *PN* JUNCTIONS

N. BACCI, G. BARILLARO AND A. DILIGENTI

Dipartimento di Ingegneria dell'Informazione: Elettronica, Informatica, Telecomunicazioni, Università degli Studi di Pisa, via G. Caruso 16, 56126, Pisa, Italy

Abstract. Fully-porous *pn* junctions were fabricated, by anodization of *p/n* substrates, and electrically tested. Process parameters, current density and illumination, were chosen to take into account the different morphological and mechanical characteristics of porous silicon layers resulting from anodization of the *p*- and *n*-type silicon. *I–V* curves were measured as a function of the aging time. Interestingly, it was found that fully-porous junctions still have a rectifying behavior, which is better evident as the porous silicon oxidizes with aging. The validity of this hypothesis was confirmed by measuring the *I–V* characteristic of a sample subjected to controlled aging, the latter performed by rapid thermal oxidation on freshly grown structures.

1. Introduction

In preceding papers [1, 2] *pn* junctions surrounded by a porous silicon (PS) layer have been proposed as gas sensors (Porous Silicon Surrounded Diode – PSsD). In PSsDs, gas adsorption by the peripheral porous layer has been shown to have noticeable effects on the electrical behavior of the *pn* crystalline junction. It has been supposed that gas-induced changes in PS electrical properties could be accountable for the modification of the electrical behavior (i.e. *I–V* curve) of the crystalline junction in contact with the PS-itself. However, since the PS layer was formed by anodizing both *p*- and *n*-type silicon, the sensing characteristics could be also partially related to the fully-porous *pn* junction surrounding the crystalline junction.

In order to investigate the behavior of a fully-porous junction, anodization of *p/n* wafers was carried out, and *I–V* curves of the resulting porous structure were measured as a function of the aging time. Electrical measurements show that fully-porous junctions still behave as a diode, with a reverse current fairly lower than the forward current. The rectifying behavior of the diode slowly improves with the aging, over a temporal period of several days, most likely due to oxidation of the PS surface. A controlled aging, performed by thermal oxidation soon after the PS formation, seems to validate such a hypothesis.

P. Malcovati et al. (eds.), *Sensors and Microsystems: AISEM 2009 Proceedings*, Lecture Notes in Electrical Engineering 54, DOI 10.1007/978-90-481-3606-3_26,

2. Experimental

The starting material was an *n*-type substrate ($N_D \sim 10^{15}$ cm^{-3}), (100) oriented, with a *p*-type epitaxial layer ($N_A \sim 10^{15}$ cm^{-3}), 4 µm thick, on top. Conventional implantation and diffusion steps were employed to get a shallow, high-doped, p^+ layer at the wafer surface. In this way a *p/n* wafer with higher surface-doping p^+ and junction-depth of 2.4 µm was obtained. Two *p/n* wafers with different surface-doping were considered for fully-porous junction fabrication, the latter performed by anodization in a HF aqueous solution (11% wt.) with addition of an apolar surfactant (NCW-1001):

1. Surface-doping concentration of $2 \cdot 10^{19}$ cm^{-3}. In this case, the presence of the p^+ layer over the *p* region leads to two porous layers with significantly dissimilar morphology. The mechanical stress at the interface between these two layers results in a partial detachment of the porous p^+ layer. To avoid such a problem, before PS formation, the wafer surface was electropolished for removing the p^+ layer, by using a current density of 80 mA/cm^2 for 8 s, under front-side illumination (250 W). After the electropolishing step, the remaining *pn* junction was etched with a constant current density, 1 mA/cm^2 for 1,800 s, using an illumination power of 50 W. The process outcome is reported in Fig. 1a, which shows a fully-porous *pn* junction consisting of a microporous PS layer of about 0.25 µm (the residual *p*-type region), and a macroporous layer of about 1.8 µm, resulting from the *n*-type material.
2. Surface-doping concentration of 10^{17} cm^{-3}. In order to avoid the electropolishing step, a substrate with lower surface doping was employed, thus reducing PS morphology differences between high and low doped *p* regionsa. Fully-porous junctions were formed by a single electrochemical etching step, using a current density of 1 mA/cm^2 for 3,600 s, and an illumination power of 125 W. The process parameters (higher lamp power and longer etching time)

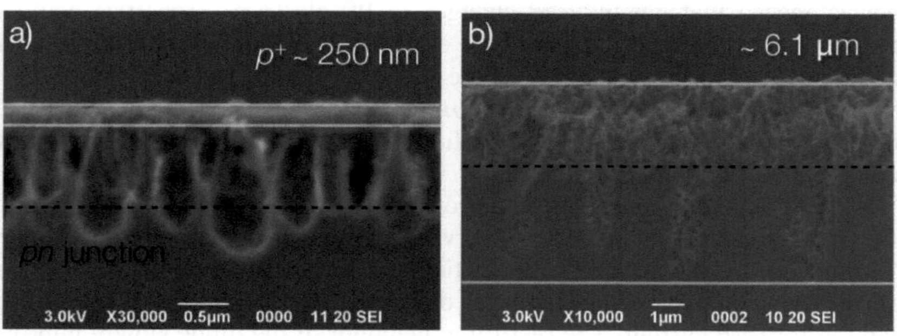

Figure 1. SEM cross-section of fully porous *pn* junctions obtained from: (**a**) type-(i) substrates (highly doped surface) partially etched prior to PS formation; (**b**) direct anodization of type-(ii) substrates (lower doped surface).

were chosen to get a thicker PS layer, with smaller macropore dimension and increased porosity, in the *n*-type material. The result of the etching is shown in Fig. 1b.

In both cases, (i) and (ii), in order to perform electrical tests, an aluminum (Al) film was thermally evaporated on top of the porous *p* layer through a metal mask, which allowed to obtain different metal dots with an area of about 7 mm^2. After a further evaporation of an aluminum layer on the back-side, the samples were mounted on a PCB board by using a silver paste. Thin copper wires, fixed with silver paste, were used for electrical connections with Al dots.

3. Results

I–V measurements, performed at atmospheric pressure and room temperature, showed a rectifying behavior for both kinds of samples, (i) and (ii). Figure 2a shows the *I–V* curves of all on-chip fully-porous diodes, measured at two different aging times, for a type-(i) sample. Each device has a rectifying behavior, even if a spreading in the *I–V* curve of the different devices is clearly visible. Moreover, the *I–V* curves, and, in turn, the diode parameters (e.g. reverse saturation current I_0, exponential factor *n*, parallel and series resistance), were dependent on the sample aging. In particular, a significant reduction of I_0 and an increase of the series resistance were observed (Fig. 2a). In Fig. 2b the behavior of I_0 versus the aging time is reported: the saturation current diminishes with the aging time and seems to stabilize after about 2 weeks. A similar time-dependent behavior is also evident for type-(ii) diodes (Fig. 3a). In this case, a reduction of the reverse saturation current I_0 of one order of magnitude, together with a slight increase of the series resistance, occurs in less than 2 weeks.

A crystalline *pn* junction with a series and a shunt resistor, the latter consisting of conduction-paths on the PS inner surface, can be used to model such a time-dependent behavior. Room temperature oxidation gradually removes superficial shunt-paths, revealing the rectifying behavior.

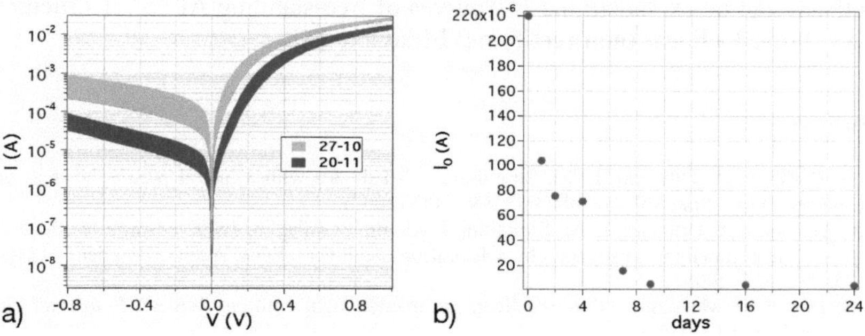

Figure 2. (**a**) *I–V* curves and (**b**) saturation current vs aging time, for type-(i) diodes, measured at different aging times.

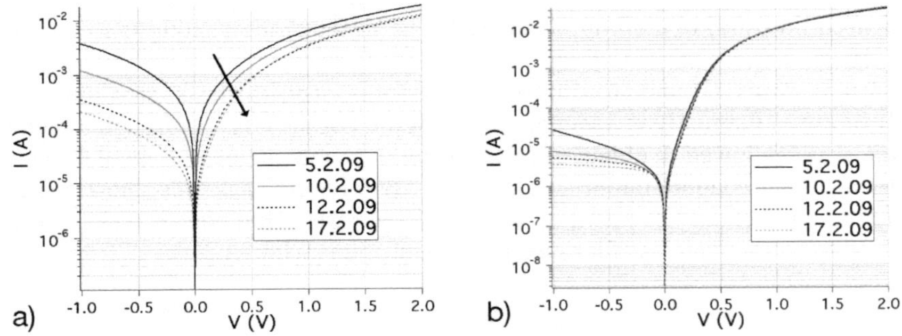

Figure 3. *I–V* curves of a type-(ii) diode for a period of 2 weeks: (**a**) a non-oxidized sample; (**b**) sample oxidized using an RTO process.

To check this hypothesis a controlled thermal oxidation (RTO) was performed (200°C for 1 h) on a type-(ii) sample. Figure 3b shows *I–V* curves of such a diode for a 2-weeks period, where a residual aging effect is still visible on the I_0 current.

4. Conclusions

RTO experiments confirm that the aging-effect of fully-porous junctions can be ascribed to a modification of the conducting properties of the PS inner surface. Such junctions can be viewed as formed by many *pn* crystalline paths, connected in parallel, whose surface behaves as a shunt resistor. As the nature of the surface changes (growth of native silicon dioxide), the *surface* current decreases and the rectifying behaviour prevails. There is in literature only one example [3] of PS diodes and, according to some authors [4], the rectifying properties must be attributed to the Al/PS interface. However, on the basis of many *I–V* measurements performed on PS free-standing membranes, obtained from *p* silicon, this hypothesis can be excluded: the *I–V* curves of free-standing Al/PS/Al structures always showed a linear (non-rectifying) behaviour [5].

References

1. G. Barillaro, A. Diligenti, L. M. Strambini, "p$^+$-n diodes with a lateral porous layer as gas sensors", Phys. Stat. Sol. (a), 204(5), 1399–1403 (2007).
2. G. Barillaro, A. Diligenti, L. M. Strambini, E. Comini, G. Faglia, "NO2 adsorption effects on p+-n silicon junctions surrounded by a lateral porous layer", Sens. Actuat. B: Chem., 134(2), 25, 922–927 (2008).
3. B. Das, S. P. McGinnis, "Porous silicon pn junction light emitting diodes", Semicond. Sci. Technol., 14 988 (1999).
4. M. Theodoropolou, P. K. Karahaliou, C. A. Krontiras, S. N. Georga, N. Xanthopoulos, M. N. Pisanias, "Transient and ac electrical transport under forward and reverse bias conditions in aluminum/porous silicon/*p*-cSi structures", J. Appl. Phys. 96, 12 (2004).
5. N. Bacci, G. Barillaro, A. Diligenti, "Conduction measurements on free-standing PS membrane", unpublished results.

FAST GATING OF SINGLE-PHOTON AVALANCHE DIODES FOR PHOTON MIGRATION MEASUREMENTS

A. DALLAMORA[1], A. TOSI[1], F. ZAPPA[1], S. COVA[1], A. PIFFERI[2], A. TORRICELLI[2], L. SPINELLI[2], D. CONTINI[2] AND R. CUBEDDU[2]

[1]*Dipartimento di Elettronica e Informazione, Politecnico di Milano, Piazza Leonardo da Vinci 32, I-20133 Milano, Italy*

[2]*IIT, ULTRAS-INFM-CNR and IFN-CNR, Dipartimento di Fisica, Politecnico di Milano, Piazza Leonardo da Vinci 32, I-20133 Milano, Italy*

Abstract. Near-Infrared (NIR) picosecond pulsed light shined in biological tissues offers the opportunity for non-invasive imaging. We aimed at developing a winning photo-detector-electronics pairing for a broad field of multiple-wavelengths faint-signal optical investigations, like functional brain imaging. We present an electronic instrumentation based on silicon Single-Photon Avalanche Diode (SPAD) and fast-gating front-end electronics, in a time-correlated single-photon counting set-up. The high detection efficiency allows the acquisition of very faint optical signals on a wide spectral range. Furthermore, the fast-gating circuitry enables the detector very quickly (400 ps), thus allowing the rejection of very intense light scattered from more superficial layers of the head, preceding useful faint signal scattered from the brain. We attain photon-counting dynamic ranges up to 10^7 with photon-timing resolutions of 95 ps, thus allowing the detection of photons delayed up to 6 ns from the laser stimulus.

1. Introduction

The key point in some important biomedical optical measurements, which make use of single-photon detectors, is the possibility of quickly switch-on and off the detectors in order to avoid photon collection out of a precise time gate.

Using time-correlated single-photon counting (TCSPC) techniques [1], the maximum count rate can be limited both by the saturation threshold of the electronics and by the "pile-up" distortion effect. The latter limit is essentially due to the probability that more than one photon can arrive on the detector within the same measurement period, since only the first one can be detected because of the dead time of the photodetector. In applications such as functional near-infrared spectroscopy (fNIRS) [2], optical mammography and molecular imaging, there is the need to discriminate between early arriving and late arriving photons to increase the contrast of scattering and absorbing perturbation embedded in the sample under investigation. For example, in fNIRS the novel approach of small source-detector separation increases the collection of photons, the contrast and the spatial resolution [3], but the strong presence of the "early photons" (i.e. photons reflected from skin, skull and meninges) heavily limits

P. Malcovati et al. (eds.), *Sensors and Microsystems: AISEM 2009 Proceedings*, Lecture Notes in Electrical Engineering 54, DOI 10.1007/978-90-481-3606-3_27,

the performances, since their number overcomes of few orders of magnitude that of the "late photons" (i.e. photons back diffused from brain). Early photons force one to reduce the optical power injected to avoid the count rate saturation, but to attain good performances at small source-detector separation the optical power injected must be high (below the safety limits) to enhance the number of late-photons which carry the information.

Our solution is the use of a time-gated silicon Single-Photon Avalanche Diode (SPAD) [4], enabled at a well defined delay with respect to the arrival time of the early-photons guaranteeing their rejection (see Fig. 1). SPADs have high quantum efficiency on a wide spectral range (about 50% at 550 nm, 15% at 800 nm). Essentially, they are p–n junctions biased above the breakdown voltage V_{BD}. At this bias a single carrier in the depletion layer can trigger a self-sustaining avalanche by impact ionization process. Therefore, if the primary carrier is photogenerated, the leading edge of the avalanche pulse marks with picosecond resolution the arrival time of the photon. After that, the avalanche must be quenched by lowering the reverse voltage near to V_{BD} or below. Finally, the bias voltage has to be restored to detect another photon.

Quenching circuit can generally be either passive or active [4]. In the passive quenching circuit a resistor, in series with the SPAD, develops a voltage drop when crossed by the avalanche current, thus reducing the bias voltage across the detector. The resistor must have a high value (hundreds kΩ), thus strongly limiting the count rate because of the time constant of the restoring transition. Otherwise, commercially available active quenching circuits (AQC) detect the avalanche and subsequently force the quenching and recovering transitions, allowing high count rates but limiting the transition speed to few nanoseconds.

2. Measurement setup and results

We implemented the fast gating setup of Fig. 2 and we developed our own fast pulse generator to gate on and off the detector in about 400 ps for a time duration (from 200 ps to 510 ns) and a repetition rate (from 40 Hz to 133 MHz) that can be selected by the user. The "dummy" network enables to detect

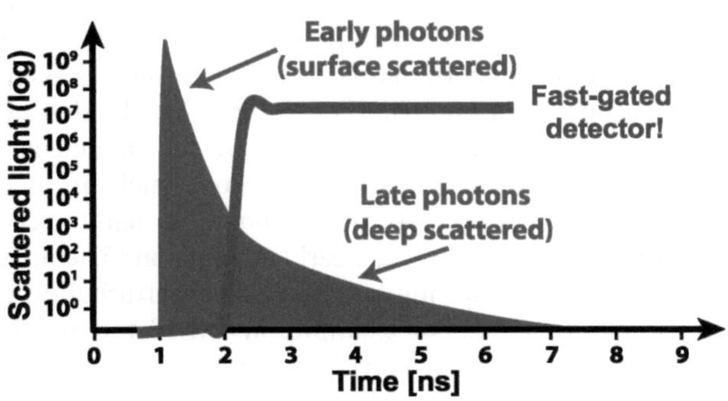

Figure 1. Rejection of early photons by delaying the detector enabling.

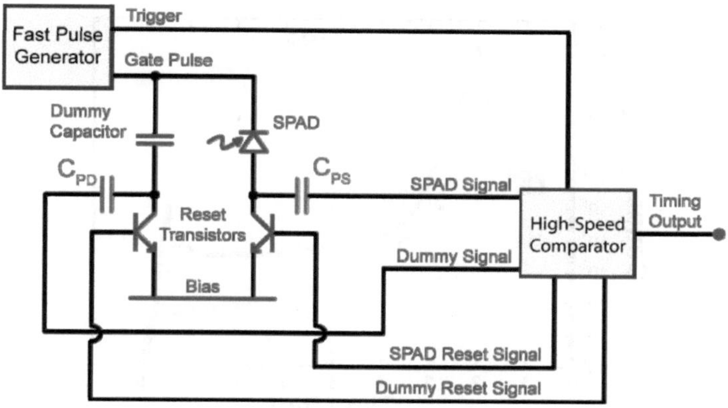

Figure 2. Fast-gating setup with capacitive compensation and active reset transitions.

photons with low time jitter (100 ps) thanks to the differential pick-up circuitry, which avoids the large spurious feed-through spikes on the avalanche signal due to rising- and falling-edge of the gate signal (see Fig. 3) that prevent avalanche detection with standard pick-up networks 0. Moreover, the presence of the reset transistors guarantees fast active reset transitions in order to allow high count rates.

We measured the system instrument response function (IRF) by means of a 800-nm pulsed laser with a repetition rate of 20 MHz, with pulse duration of only few picoseconds. The SPAD was biased 0.5 V under V_{BD} and gated with voltage pulses of 5.5 V (i.e. 5-V excess bias) with 5-ns duration. Starting with the laser pulse in the middle of the ignition interval, the gate window was shifted at steps of 25 ps with respect to the optical pulse. For each delay, we acquired photons for 10 s. When the gate window is sufficiently delayed, the laser peak is out of the window and the pulse tail is measured. The longer is the gate delay, the fainter is the laser tail and therefore the more powerful could be the laser pulse in order to investigate the last part of the laser tail. Figure 4 shows the IRF assembled by combining slices of the gated acquisitions coming

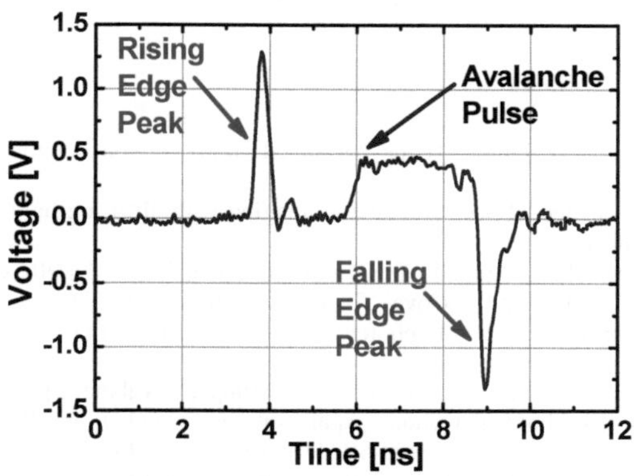

Figure 3. SPAD signal at the comparator input with 200-ps rising/fall transition durations.

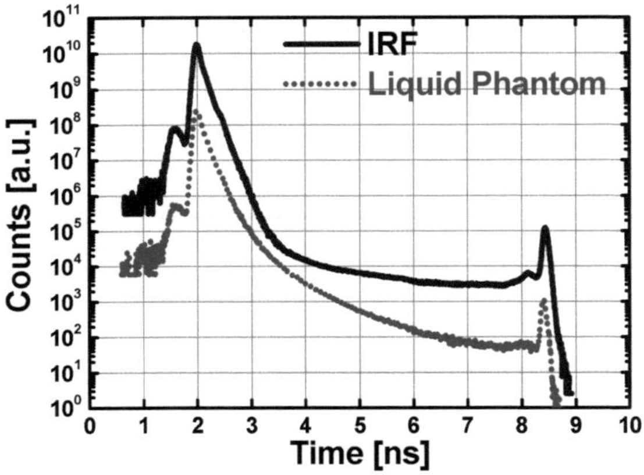

Figure 4. Instrument response function to laser pulse (solid line), phantom reflectance curve (dots).

from different delays. The dynamic range is about 10^7, thus allowing a detection of photons even at 6 ns from the main peak, with time resolution of 100 ps (FWHM). Figure 4 shows also the reflectance curve obtained by launching the laser pulse into a liquid phantom with an absorption coefficient $\mu_a = 0.05\ cm^{-1}$ and a scattering coefficient $\mu_{s'} = 5\ cm^{-1}$ at 2 mm source-detector separation.

3. Conclusions

We developed a photon migration measurement set-up by using a fast-gated SPAD obtaining high time resolution (100 ps) and large photon-counting dynamic range (10^7) in few minutes measurement time. It can be employed for many biomedical applications such as in-vivo spectroscopy.

Acknowledgments The research leading to these results has received funding from the European Community's Seventh Framework Programme (FP7 2007–2013) under grant agreement no. HEALTH-F5-2008-201076.

References

1. D.V. O'Connor and D. Philip, "Time correlated single photon counting," Academic Press, London, 1984.
2. A. Yodh and B. Chance, "Spectroscopy and imaging with diffusing light," Phys. Today, [vol. 48 (3)], pp. 34–40, 1995.
3. A. Pifferi, A. Torricelli, L. Spinelli, D. Contini, R. Cubeddu, F. Martelli, G. Zaccanti, A. Tosi, A. Dalla Mora, F. Zappa and S. Cova. "Time-resolved diffuse reflectance using small source-detector separation and fast single-photon gating," Phys. Rev. Lett., [vol. 100 (13)], art. no. 138101, 2008.
4. S. Cova, M. Ghioni, A. Lacaita, C. Samori and F. Zappa, "Avalanche photodiodes and quenching circuits for single-photon detection," Appl. Opt., [vol. 35 (12)], pp. 1956–1976, 1996.
5. F. Zappa, A. Tosi and S. Cova, "InGaAs SPAD and electronics for low time jitter and low noise," Proc. SPIE Photon Counting Applications, Quantum Optics, and Quantum Cryptography, 65830E, 2007.

PERFORMANCE OF COMMERCIALLY AVAILABLE InGaAs/InP SPAD WITH CUSTOM ELECTRONICS

A. TOSI, A. DALLA MORA, F. ZAPPA AND S. COVA

Politecnico di Milano, Dipartimento di Elettronica e Informazione, P.za Leonardo da Vinci 32, 20133 Milano, Italy

Abstract. InGaAs/InP devices suitable as Single-Photon Avalanche Diodes (SPADs) for photon counting and photon timing applications in the near-infrared provide good detection efficiency and low time jitter, together with fairly low dark-count rate at moderately low temperatures. However, their performance is still severely limited by the afterpulsing effect, caused by carriers trapped into deep levels during the avalanche current flow and later released. We present preliminary experimental characterization of recently-developed InGaAs/InP detectors that can promisingly be operated slightly cooled. We investigate the primary dark-count rate, taking into account both thermal generation in the InGaAs absorption layer and trap-assisted tunnelling in the InP multiplication layer. The fundamental role played by the front-end circuits in minimizing the effects of afterpulsing is assessed and demonstrated.

1. Introduction

During the last years, new InGaAs/InP Single-Photon Avalanche Diodes (SPADs) [1] have become commercially available. These detectors are specifically designed for near-infrared (NIR) single-photon counting applications, such as quantum key distribution (QKD), eye-safe laser ranging (LIDAR), VLSI circuit testing by means of non-invasive optical investigations, singlet oxygen detection for photodynamic therapy (PDT) dosimetry, etc. Other NIR detectors are available (InGaAs photomultiplier tubes, Superconducting Single Photon Detectors), but InGaAs/InP SPADs are a good choice in terms of signal-to-noise ratio and usability in practical applications.

The SPAD is a p–n junction, reverse-biased above the breakdown voltage, in order to exploit the fast and intense avalanche build-up triggered by the absorption of a single photon [2]. The sub-nanosecond rise of the avalanche current marks with high precision the photon arrival-time. A suitable circuit senses the avalanche, quenches the detector by lowering the voltage below the breakdown level, and finally resets the SPAD. InGaAs/InP SPADs exploit the separate absorption and graded multiplication structure (SAGM) [3]. The photons with the wavelength of interest (e.g., 1,550 nm) are absorbed in the narrow-bandgap $In_{0.53}Ga_{0.47}As$ layer (e.g. ~0.75 eV at 295 K), lattice-matched to InP. The photo-generated hole

P. Malcovati et al. (eds.), *Sensors and Microsystems: AISEM 2009 Proceedings*, Lecture Notes in Electrical Engineering 54, DOI 10.1007/978-90-481-3606-3_28,

is then drifted toward a wider bandgap InP region (e.g. ~1.35 eV) in which avalanche multiplication occurs.

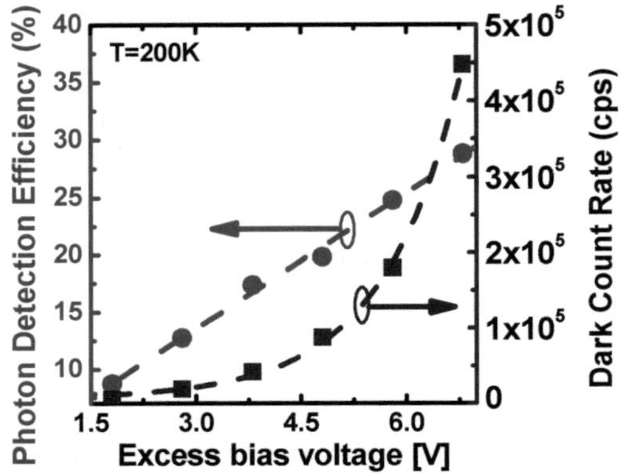

Figure 1. Photon detection efficiency and primary dark count rate of an InGaAs/InP SPAD (old Princeton Lightwave device [4] with 40-µm diameter) dependence on overvoltage (i.e. excess bias).

2. Experimental characterization

The photon detection efficiency depends on both the probability of absorbing a photon and on the probability that the photo-generated electron-hole pair succeeds in triggering the self-sustaining avalanche process. The former depends mainly on the thickness of the absorption InGaAs layer, while the latter rises first linearly with the overvoltage (i.e. the difference between the applied reversed voltage and the breakdown voltage) and tends to saturate to unity at very high overvoltage. High overvoltage is desirable also for reducing the photon timing jitter.

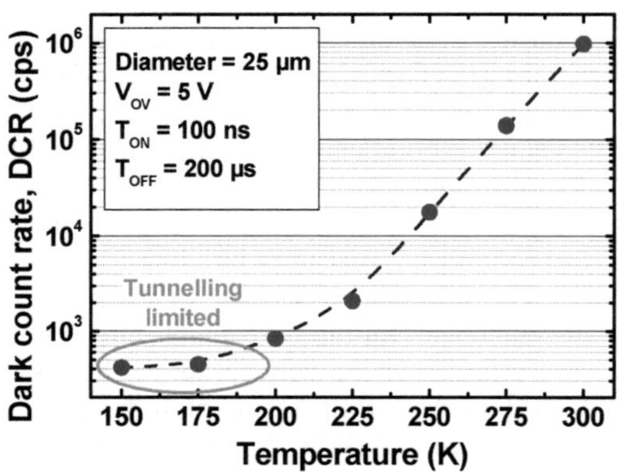

Figure 2. Dependence of the primary dark count rate of a new generation of Princeton Lightwave (PLI) $In_{0.53}Ga_{0.47}As$/InP SPADs on temperature.

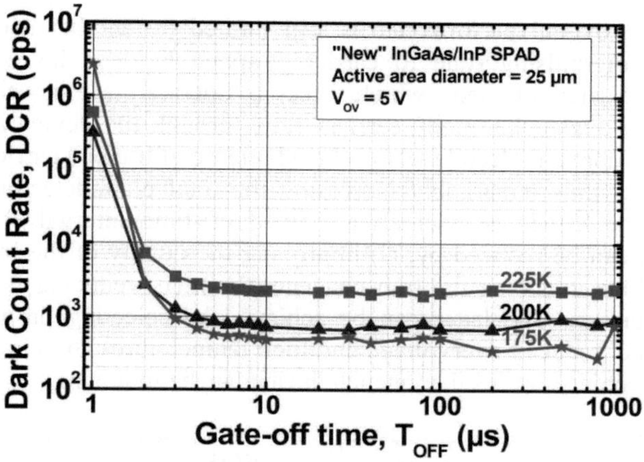

Figure 3. Dark count rate of a new generation of PLI InGaAs/InP SPAD at various temperatures in gated operation (T_{ON} = 20 ns) as a function of the idle interval T_{OFF}.

On the other hand, the primary dark count rate (DCR) rises faster with the overvoltage (almost exponentially) due to the progressive enhancement of tunnel-assisted generation of carriers, as illustrated in Fig. 1. Typically, the active area diameter of InGaAs/InP SPADs ranges from 25 to 40 µm [4], the photon detection efficiency exceeds 25% in the 0.8–1.7 µm wavelength range and the dark count rate is a few kcps (counts per second) when the detector is cooled at 225 K (a temperature easily achievable with a thermo-electric cooler mounted inside the TO-can package of the detector) (see Fig. 2).

The primary dark-counts have two different causes: temperature-assisted processes and field-assisted processes. The temperature-assisted phenomenon is the spontaneous thermal generation of free carriers, which is dominated by the Shockley-Read-Hall (SRH) processes in the depleted regions, mainly due to local defects. SRH generation is intense in the $In_{0.53}Ga_{0.47}As$ absorber because of the small band-gap and can be reduced with high quality lattice. Field-assisted mechanisms include direct band-to-band tunneling and trap-assisted tunneling. Both these generation processes become important when the electric field is high enough (i.e. in the InP multiplier) to give a finite probability of quantum tunneling and depend exponentially on the electric field. The result is that increasing the overvoltage, the DCR increases exponentially (Fig. 1).

In order to fairly compare the old Princeton Lightwave device (Fig. 1) with the new one, we computed the DCR per unit active area (cps/µm^2) at 200 K:

Old generation device:	$DCR_{unit\ area,old} = 120$ cps/µm^2
New generation device:	$DCR_{unit\ area,new} = 1.6$ cps/µm^2

By computing the ratio, we obtain an improvement of a factor of 120/1.6 = 75 from the previous to the last PLI InGaAs/InP SPAD generations.

Presently, the overall performance is still limited by another source of dark counts, the afterpulsing (due to carriers trapped in deep levels during the avalanche and later released), which forces to operate these detectors in gated mode. During avalanche, many carriers flow through the device and some of them can be trapped in deep levels. Delayed release of trapped carriers re-trigger the avalanche (thus remarkably increasing the total photodetector dark count rate) when the SPAD is reset after a fixed off-time interval T_{OFF} (Fig. 3). Afterpulsing can be reduced by: (i) improving the quality of the InP multiplication layer (thus reducing the trap concentration); (ii) reducing the amount of avalanche carriers. The latter can be achieved by selecting proper operating conditions (i.e. reduced overvoltage and high temperature) and circuit solutions (i.e. fast quenching circuits that quench the avalanche in few ns).

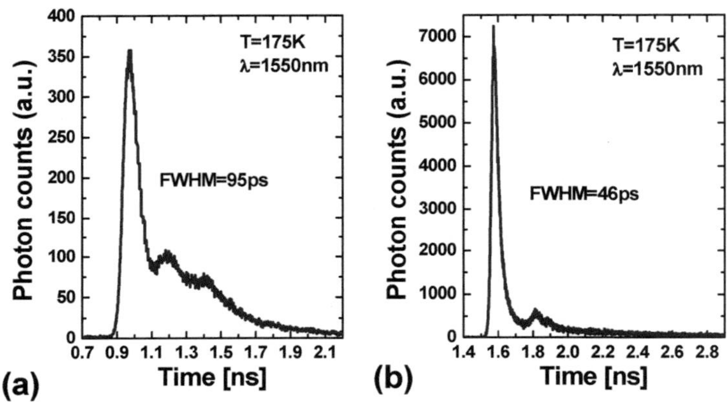

Figure 4. Photon timing jitter of an old PLI InGaAs/InP SPAD as measured with a standard front-end circuit (**a**) and with the differential sensing (**b**).

Finally, we demonstrated the fundamental role played by the front-end circuits, in particular as concerns picosecond photon timing for a SPAD operating in gated-mode with ultrafast gate-on and gate-off transitions. Suitable signal pick-up circuits can be employed for extracting at best the photon-timing information. With a compensated differential read-out circuit, we obtained a time resolution of about 46 ps (Full Width at Half Maximum, FWHM) with an old 25-μm InGaAs/InP SPAD (Fig. 4b), a remarkable improvement in respect of the 95 ps obtained by means of standard pick-up circuits (Fig. 4a). New generation devices show time resolution of about 100 ps (FWHM).

3. Conclusions

We reported the performance of a new generation 25 μm-diameter InGaAs/InP SPAD at 1,550 nm wavelength, with dark count rate of 400 cps at 175 K and just 2,000 cps at 225 K, with afterpulsing showing off only below T_{OFF} = 10 μs.

The important role played by the front-end circuits in minimizing the afterpulsing and maximizing the time resolution effect has been demonstrated.

References

1. M. A. Itzler, R. Ben-Michael, C. F. Hsu, K. Slomkowski, A. Tosi, S. Cova, F. Zappa and R. Ispasoiu, "Single photon avalanche diodes (SPADs) for 1.5 μm photon counting applications," J. Modern Opt., [vol. 54 (2)], pp. 283–304, 2007.
2. S. Cova, M. Ghioni, A. Lacaita, C. Samori and F. Zappa, "Avalanche photodiodes and quenching circuits for single-photon detection," Appl. Opt., [vol. 35 (12)], pp. 1956–1976, 1996.
3. K. Nishida, K. Taguchi and Y. Matsumoto, "InGaAsP heterostructure avalanche photodiodes with high avalanche gain," Appl. Phys. Lett., [vol. 35], pp. 251–252, 1979.
4. Princeton Lightwave Inc., www.princetonlightwave.com

NOVEL VACUUM EVAPORATED CAVITAND SENSORS FOR DETECTING VERY LOW ALCOHOL CONCENTRATIONS

M. TONEZZER[1], G. MAGGIONI[1], M. MELEGARI2 AND E. DALCANALE[2]
[1]*D.I.M.T.I. – University of Trento, via Mesiano 77, 38050 Povo (TN), Italy*
I.N.F.N. – L.N.L., Viale dell'Università 2, 35020 Legnaro, Italy
[2]*D.C.O.I. – University of Parma and ISTM, UdR Parma, Viale Usberti 17/A, 43100 Parma, Italy*

Abstract. High vacuum evaporation (VE) is used for the first time to grow thin films of novel tetraphosphonate, **Tiiii[H, CH3, Ph]**, and tetrathiophosphonate, **TSiiii[H, CH3, Ph]**, cavitands for gas sensing applications. The sensing capabilities of the samples were investigated by exposing **Tiiii**- and **TSiiii**-coated QCMs to very low concentrations of ethyl alcohol. The sensitivity, the speed (t_{50} = 5 s for both the samples) and the detection limit (0.4 ppm for **Tiiii** and 2.5 ppm for **TSiiii**) of the samples were determined, indicating highly competitive sensing capabilities.

1. Introduction

Chemical sensing requires an integrated approach, where both the molecular and the materials properties of the sensing layer must be finely tuned to achieve the desired properties. In the specific case of supramolecular sensing, the material side has been largely neglected, despite its great influence on the ultimate performances of the sensors. In fact, because analyte recognition is mediated by the layer properties of the coated receptors, precise control and accurate characterization of these properties are critical for developing advanced sensing materials. Cavitands are one of the most studied receptors [1–4] for gas/vapor sensing because of their outstanding host-guest properties, which are tunable for the recognizing of different classes of analytes. Standard production methods for obtaining thin solid films from cavitand molecule utilize wet deposition; in particular, spray- and spin-coating methods are commonly employed. Nevertheless, despite promising results for cavitands as chemical receptors, these deposition methods suffer from several drawbacks such as inhomogeneous surface morphology (for spray coating) and uncontrollable thickness. Moreover, insoluble compounds cannot be deposited by such solution techniques. High-vacuum evaporation (VE) overcomes solubility problems, allowing for the direct formation of films from solid materials. The lack of residual solvent during the deposition process assures the formation of high purity films: this represents a basic requirement in the gas sensing field, because of the unpredictable effects

P. Malcovati et al. (eds.), *Sensors and Microsystems: AISEM 2009 Proceedings*, Lecture Notes in Electrical Engineering 54, DOI 10.1007/978-90-481-3606-3_29,

of the retained solvent on the final response of the sensor, including occupation of adsorption sites and interference in analyte/material interactions. In this regard, recent studies have demonstrated that vacuum-evaporated thin films exhibit higher gas sensing capabilities toward alcohol vapors than chemically deposited films. In this paper, we report the employment of the VE technique for producing tetraphosphonate (**Tiiii**) and tetrathiophosphonate (**TSiiii**) sensing coatings. **Tiiii** and **TSiiii** compounds are depicted in Fig. 1.

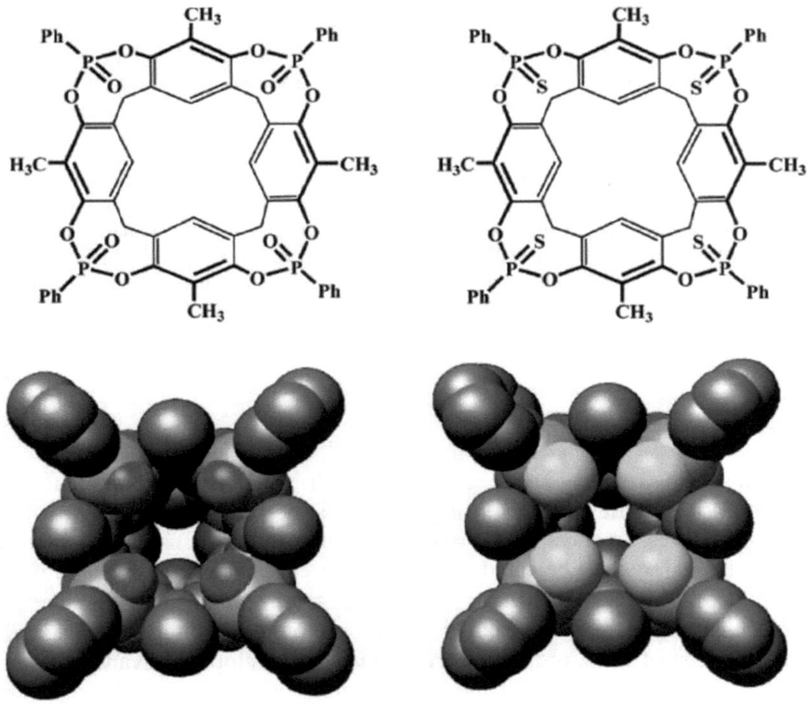

Tiiii[H, CH$_3$, Ph] **TSiiii[H, CH$_3$, Ph]**

Figure 1. Structures (*above*) and 3D CPK Models (*below*) of Tetraphosphonate Tiiii[H, CH$_3$, Ph] and Tetraphosphoante TSiiii[H, CH$_3$, Ph] Cavitands.

They present an open, conformationally rigid cavity, delimited by four inward oriented P = O/P = S bridging groups at the upper rim. Substitution of the four P = O groups with the P = S moieties completely prevents complexation by eliminating H-bonding interactions between the cavitand and the analyte.

2. Results and discussion

Sensing measurements were performed by exposing cavitand-coated QCMs to very low concentrations (ranging from 5 to 200 ppm) of ethyl alcohol (EtOH) and monitoring the shift of the QCM fundamental resonance frequency induced by the mass changes as a function of time. EtOH was chosen for this testing because its mode of interaction with **Tiiii** is well understood: Fig. 2 shows the time responses of **Tiiii** and **TSiiii** sensors exposed to 25 ppm of EtOH.

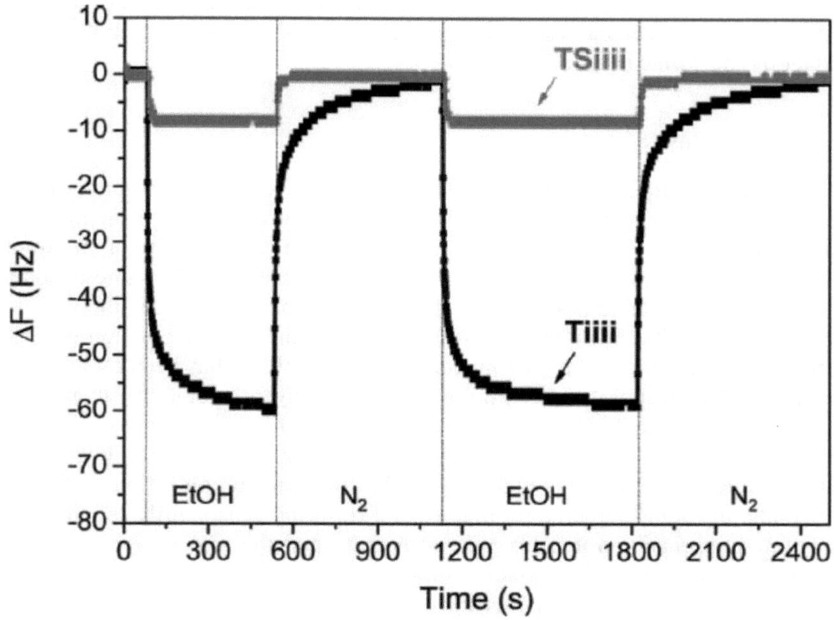

Figure 2. Change of the resonance frequency of **Tiiii** and **TSiiii** vacuumevaporated samples as a function of time during exposure to 25 ppm EtOH. Both the response and recovery phases are performed at 20°C.

Both **Tiiii** and **TSiiii** samples exhibit a fast decrease of the resonance frequency within seconds, followed by a slower decrease until saturation values are reached. As the EtOH vapor stream was switched off, a dry nitrogen flux was activated, and the original resonance frequencies were restored. The most significant result is the high difference in response intensity between the samples: **Tiiii**-coated QCM shows a resonance frequency decrease (ΔF) −60 Hz) eight times higher than that of **TSiiii** (ΔF) −7.5 Hz). Previous studies have indicated that the following key factors affect the sensing performances of phosphorus-bridged cavitands toward alcohols: (i) the preorganized cavity which provides a free volume available for the analyte, pivotal for effective H-bonding; (ii) the presence of synergistic CH-π interactions with the π-basic cavity and the energetically equivalent H-bonding options between the PdO groups at the upper rim and the analyte (specific interactions). The resulting mode of interaction between ethanol and the cavity shows how the ethanol chain fits into the cavity with its methyl residue (CH-π interactions), while the OH moiety undergoes H-bonding interactions with the P = O groups. The effective contribution of the H-bonds is absent in the **TSiiii** cavitand, because the P = S is much less polarized than the P = O one and, consequently, it is ineffective as H-bond acceptor. Therefore the observed **TSiiii** responses are due to nonspecific dispersion interactions between the organic layer and EtOH and the large difference in the sensor responses can be attributed to the specific binding of EtOH by the **Tiiii** cavitand.

3. Conclusions

Tiiii[H, CH3, Ph] cavitand represents the ultimate receptor for supramolecular mass sensing of short chain alcohols. The excision of the alkyl chains at the lower rim reduces the nonspecific interactions in the layer to a minimum, concomitantly increasing the number of cavities present for the same deposited mass. The unavoidable drawback of this optimization procedure is the very low solubility of cavitands in organic solvents, precluding their deposition on QCM transducers via either spray or spin coating. This problem has been overcome by the use of VE evaporation technique. The **TSiiii** layers operate exclusively via nonspecific extracavity adsorption, while in the **Tiiii** films, extracavity adsorption coexists with intracavity complexation. The latter dominates at low analyte concentrations (up to 50 ppm). The EtOH mode of complexation is fast and reversible at room temperature, making the resulting sensor fully reliable. This integrated approach, where both the molecular and the materials properties of the sensing layers have been finely tuned, produced supramolecular mass sensors with improved performances over the existing ones. This approach can be easily extended to many different classes of organic receptors, opening the way of sensor materials in relation to the analytes to be detected.

References

1. M. Tonezzer, M. Melegari, G. Maggioni, R. Milan, G. Della Mea, E. Dalcanale, "Vacuum-evaporated cavitand sensors: Dissecting specific from nonspecific interactions in ethanol detection", Chem. Mat., [20], 6535–6541, 2008.
2. M. Melegari, M. Suman, L. Pirondini, C. Massera, F. Ugozzoli, E. Kalenius, P. Vainiotalo, J.-C. Mulatier, J.-P Dutasta, E. Dalcanale, "Supramolecular sensing with phosphonate cavitands", Chem. Eur. J., [14], 5772–5778, 2008.
3. M. Tonezzer, A. Quaranta, G. Maggioni, S. Carturan, G. Della Mea, "Optical sensing responses of tetraphenyl porphyrins toward alcohol vapours: A comparison between vacuum evaporated and spin-coated thin films", Sens. Actuat. B [122], 620–626, 2007.
4. M. Tonezzer, "Production, characterization and testing of solid organic and metallorganic films for gas sensing applications," Ph.D. Thesis, University of Trento, Trento, Italy, 1–231, 2007.

HYDROGEN SENSING CAPABILITY OF NANOSTRUCTURED TITANIA FILMS

G. MICALI[1], A. BONAVITA[1], G. NERI[1], G. CENTI[1], S. PERATHONER[1], R. PASSALACQUA[1] AND N. DONATO[2]

[1]*Dip. Chimica Industriale e Ingegneria dei Materiali, Università di Messina, Contrada di Dio, 98166 S. Agata di Messina*
[2]*Dip. Fisica della Materia e Ingegneria Elettronica, Università di Messina, Contrada di Dio, 98166 S. Agata di Messina*

Abstract. The expected wide diffusion of hydrogen in the near future as clean fuel for PEM fuel cells and the potential danger of explosion consequent to the mixing of hydrogen with air, demands the introduction of reliable, robust and low cost hydrogen sensors. Titanium dioxide is a well known sensitive material for hydrogen sensors and the use of suitable titania nanostructures could be an efficient ways to promote the performances of these devices. Aim of this work is then aimed to synthesize and characterize different TiO_2 nanostructures and investigate their capability in hydrogen sensing. Preliminary sensing tests here reported give helpful information for the development of MOS hydrogen sensors based on TiO_2 nanostructures.

1. Introduction

Hydrogen is nowadays largely used in the chemical, petrochemical and semiconductor industry, and it is expected to gain great importance as clean fuel in automotive fuel cells in the near future [1]. On the other hand, hydrogen mixed with air in the range 4–94 vol%, is explosive. For these reasons the development of reliable, robust and low cost hydrogen sensors is strongly demanding [2]. Specifically for fuel cells applications, pre-stack sensors should be developed to monitor wide H_2 concentrations ranging from 1% to 100% in humid ambient, while hydrogen safety sensors should be designed to detect lower H_2 levels, up to 4%, in air.

Several materials and sensing technologies have been investigated in view of their capability to detect hydrogen. Among these, TiO_2 nanotubes-based MOS (metal oxide semiconductor) devices have been reported as the most promising for hydrogen sensing [3]. In recent year, it has recognized that materials in the nanostructured form exhibit peculiars sensing properties owing to their unique structural characteristics [4].Therefore, the synthesis of titania nanostructures is helpful in the design of more effective hydrogen sensors.

P. Malcovati et al. (eds.), *Sensors and Microsystems: AISEM 2009 Proceedings*, Lecture Notes in Electrical Engineering 54, DOI 10.1007/978-90-481-3606-3_30,

In this work, different nanostructures (nanotubes, nanoribbons and nanopowders) based on titanium dioxide were synthesized, characterized and used as sensing materials in MOS devices, with the final goal to develop hydrogen sensor prototypes for application in the PEM fuel cells.

2. Experimental

2.1. *TiO_2 nanostructures preparation and characterization*

Different nanostructures (nanotubes, nanoribbons and nanopowders) of titanium dioxide were prepared and characterized. The sample code used to identify them is reported in Table 1.

Table 1. TiO_2 nanostructures investigated.

Sample code	Nanostructure	Crystalline phase
TNT1	Nanotubes	Anatase
TNR1	Nanoribbons	Anatase
TNP1	Nanopowders	Anatase

TiO_2 nanopowders have been synthesized by gel-combustion starting from titanium(IV)isopropoxide ($Ti(OCH_2(CH_3)_2)_4$), isopropanol (C_3H_7OH) and hydrogen peroxide (H_2O_2) as raw materials [5]. A solution of titanium isopropoxide in isopropanol (50/50 vol) was prepared first. Then, hydrogen peroxide was added drop by drop under continuous stirring. After drying at 80°C the product obtained, a TiO_2-containing gel, was grounded in a mortar, resulting in a highly porous and softly agglomerated nanostructured powder.

TiO_2 nanotubes and nanoribbons have been prepared by anodic oxidation of Ti foils [6]. Anodic oxidation was made in a stirred electrochemical cell working at room temperature and atmospheric pressure. An Autolab PGSTAT30 potenziostat in a two electrode set-up was used to apply different anodizing voltages: 50 V for times of about 3 h 1/2 in ethylene glycol electrolyte with 0.3 wt % NH_4F and H_2O 2% vol (for nanotubes) and 20 V for times of about 1 h in H_2O with 3.5% wt HF (for nanoribbons). A platinum electrode served as cathode. The grown structures were than detached from the Ti substrates and powdered.

A detailed characterization of these TiO_2 nanostructures by SEM, TEM, XRD and BET surface area has been carried out and can be found elsewhere [5, 6].

2.2. *Sensors assembling and sensing tests*

First, the TiO_2 nanostructures were mixed with water and sonicated in order to obtain a paste with optimal rheological characteristics. Then, the paste was deposited on alumina substrates equipped by platinum interdigitated electrodes and heaters. The thick films were then thermally treated up to 600°C in order to stabilize their microstructure and make them tightly adherent to the alumina substrate. Sensing tests were performed in an home made apparatus that allows to operate at controlled temperature and to perform resistance measurements while varying the hydrogen concentration in the carrier stream.

3. Results and discussions

Figure 1 reports SEM images of the nanostructured TiO_2 materials after the high temperature treatment, showing the particular morphology of each sample. XRD analysis of the TiO_2 nanostructures as a function of the annealing temperature (not reported), showed that anatase is the main phase detected in the range of working temperature (500–600°C) of the hydrogen MOS sensor.

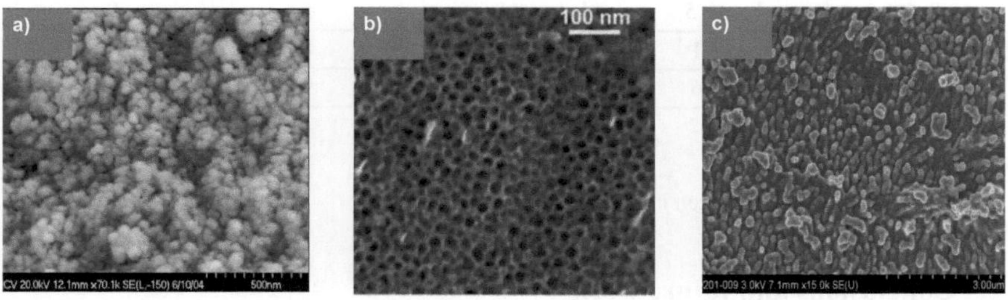

Figure 1. SEM images of nanostructured TiO_2 materials. (**a**) TNP1, nanopowders; (**b**) TNT1, nanotubes; (**c**) TNR1, nanorods.

Preliminary hydrogen sensing tests obtained with MOS sensors fabricated by using the above synthesized TiO_2 nanostructures, have shown that these devices are sensitive, at high temperature, to different hydrogen concentrations in nitrogen. As an example, in Fig. 2 is reported the calibration curve for the TNR1 sensor over a large hydrogen concentration.

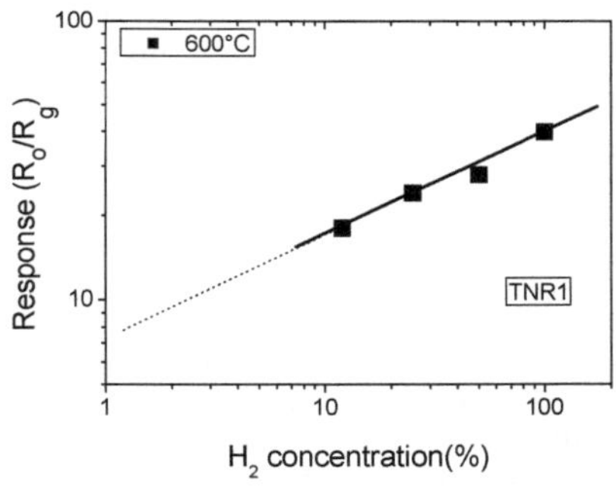

Figure 2. Calibration curves for the TNR1 sensor.

The response is linear in a wide range of concentration and the extrapolated detection level is <1%. The different sensing characteristics of the various nanostructures investigated are highlighted in Table 2. In particular, the TNT1 sensor shows a very short response/recovery time, a critical parameter for the practical application in fuel cells. It can be suggested that this behavior is related to the particular arrangement of the TiO_2 nanotubes, which favors the fast diffusion of hydrogen in the interior region of the sensing layer.

Table 2. Sensing characteristics of the TiO_2 nanostructures investigated.

Sample code	Resistance baseline[a]($M\Omega$)	Response[b] ($S = R_o/R_g$)	Response time (s)	Recovery time (s)
TNT1	13.5	5.5	5	6
TNR1	1.1	14.8	8	90
TNP1	9.6	18.8	9	45

[a]R_O, measured in pure N_2.
[b]Response to 12.5% of hydrogen in N_2.

4. Conclusions and future work

Different titania nanostructures have been synthesized, widely characterized and used to assemble MOS sensors in the thick film technology. Preliminary sensing tests have shown the applicability of these sensors based on TiO_2 nanostructures for hydrogen detection. The limit of the MOS technology in using these nanostructures as sensing layer is mainly the need to operate at high temperature. This fact could be overridden by using sensors based on different technologies as, for example, QCM or SAW operating at ambient temperature.

In this regard, studies are actually in progress and aimed to growth directly TiO_2 nanostructures (mainly nanotubes) on the electrodes of these devices. On this way some important parameters of the sensing layer (thickness, nanotubes orientation) could by carefully controlled by varying the deposition parameters.

References

1. M. Waidhas, W. Drenckhahn, W. Preidel, H. Landes "Direct-fuelled fuel cells" J. of Power Sources 61, 91–97, 1996.
2. DOE Hydrogen & Fuel Cells Annual Program Review, May 26, 2005.
3. O. Varghese, D. Gong, M. Paulose, K. Ong, C. Grimes "Hydrogen sensing using titania nanotubes" Sens. Actuators B: Chem. 93, 334–338 (2003).
4. Z. L. Wang "Nanobelts, nanowires, and nanodiskettes of semiconducting oxides – from materials to nanodevices" Adv. Mater. 15, 432–436, 2003.
5. A. Bonavita, G. Micali, G. Rizzo, G. Neri "Alternative sol-gel routes for synthesizing gas sensing nanostructured materials" ECS Transactions 3, 221, 2006.
6. C. Ampelli, R. Passalacqua, S. Perathoner, G. Centi, D. S. Su, G. Weinberg "Synthesis of TiO_2 thin films: relationship between preparation conditions and nanostructure" Topics in Catal. 50, 133, 2008.

SYNTHESIS AND GAS SENSING PROPERTIES OF ZnO QUANTUM DOTS

A. FORLEO[1], L. FRANCIOSO[1], S. CAPONE[1], P. SICILIANO[1], P. LOMMENS[2] AND Z. HENS[2]

[1]*IMM – CNR Lecce, CNR Area, c/o campus Universitario, via per Monteroni, 73100 Lecce, Italy*

[2]*Physical Chemical Laboratory, Ghent University, Krijgslaan 281-S12, B-9000 Ghent, Belgium*

Abstract. ZnO nanocrystals were prepared by a wet chemical method based on alkaline-activated hydrolysis and condensation of zinc acetate solutions. With this procedure, the mean crystallite size of the obtained powders was found in the range from 2.5 to 4.5 nm. Drop-casting of the nanocrystals onto alumina substrates allowed the fabrication of gas sensing devices. These Q-ZnO based gas sensors were tested towards NO_2, acetone and methanol and gas sensing tests showed promising results compared with the ZnO thin and/or thick films gas sensors.

1. Introduction

Several semiconducting metal oxides have attracted attention of numerous researchers and scientists taken up in gas sensors field, due to specific surface properties suitable for gas detection [1–3]. In particular, zinc oxide has been intensely studied for its versatile physical properties and characterized in literature as gas-sensing material showing good stability and gas-sensing properties to a wide range of chemical compounds. Recently, substantial efforts have been devoted to realize nanostructured materials with the aim to miniaturize the electronic and optoelectronic devices and to understand the fundamentals of nanoscale chemistry and physics [4, 5]. In fact, due to their peculiar characteristics and size effects, nanostructured materials often exhibit novel physical properties that are of great interest both for fundamental study and for potential nanodevice applications.

Recent progress in the synthesis and characterization of nanosized materials has led to realization of ZnO-nanostructured in order to improve the performances of nano-ZnO based sensor devices.

Over the past few years, one-dimensional (1D) single-crystalline ZnO nanometre materials, have been synthesized by various methods such as chemical vapour deposition (CVD), catalytic growth via the vapour–liquid–solid (VLS) mechanism, thermal evaporation, wet chemical method, and template-based growth.

P. Malcovati et al. (eds.), *Sensors and Microsystems: AISEM 2009 Proceedings*, Lecture Notes in Electrical Engineering 54, DOI 10.1007/978-90-481-3606-3_31,

In this work, we synthesized colloidal Q-ZnO particles by using a simple wet chemical procedure for the preparation of ZnO nanoparticles based on alkaline-activated hydrolysis and condensation of zinc acetate solutions. Q-ZnO based chemoresistive gas sensors were realised and gas sensing tests towards NO_2 and acetone were performed, showing promising results compared with the ZnO thin films gas sensors.

2. Experimental

Colloidal ZnO nanocrystals were prepared based on the wet chemical procedure by Schwartz et al. [6]. $(CH_3)_4NOH\cdot 5H_2O$ (TMAH, 2.5 mmol), dissolved in dry ethanol (5 ml), was added dropwise to a solution of $Zn(Ac)_2$ (0.1 M) in dimethylsulfoxide (DMSO, 15 ml), under vigorous stirring at room temperature. The ZnO nanocrystals as obtained after colloidal synthesis are washed by precipitation with a nonsolvent such as ethylacetate or heptane and resuspended in EtOH. These particles were characterised using absorption spectroscopy (Cary 500 UV–vis–NIR spectrophotometer from Varian), TEM analysis (Cs-corrected 2200FS electron microscope from Jeol) and X-ray diffraction analysis (Siemens D5000, CuKα). For the preparation of the sensors, the Q-ZnO obtained particles were deposited by dropcasting onto 2 × 2 mm alumina substrate equipped with Pt heating elements on the back side and with Pt interdigitated contact on the front size. The resulting devices were soldered onto a TO-39 support as suspended devices and placed in a stainless steel chamber for gas sensing test. The micro sensors were heated at different operating temperature by supplying a given voltage to the heating element. The sensor responses towards gases and vapours were carried out by applying a constant voltage of 4 V, between the sensor electrodes, and monitoring the current by means of a picoamperometer. They were exposed to different methanol, acetone and NO_2 concentrations. The sensors responses were calculated as $(I_{air}-I_{gas})/I_{gas}$ where I_{air} and I_{gas} are the electrical current in dry-air and in the mixture of dry-air and gas, respectively, in the case of oxidizing gases and the reverse $(I_{gas}-I_{air})/I_{gas}$ in the case of reducing gases.

3. Results and discussion

Using the colloidal synthesis procedure described in the experimental section, we can obtain charge-stabilized ZnO QDs ranging from 2.5 to 4.5 nm in diameter, thus in the confined regime. UV-Vis absorbance measurements are used to determine both the nanocrystal diameter and concentration. In this case, nanocrystals with a diameter of 3.0 nm and a concentration of 80 μM were used for dropcasting (see Fig. 1a). TEM and XRD measurements confirm that the nanocrystals are of good crystallinity with the typical hexagonal wurtzite crystal structure (Fig. 1b–c). After dropcasting, a porous ZnO QD layer is obtained.

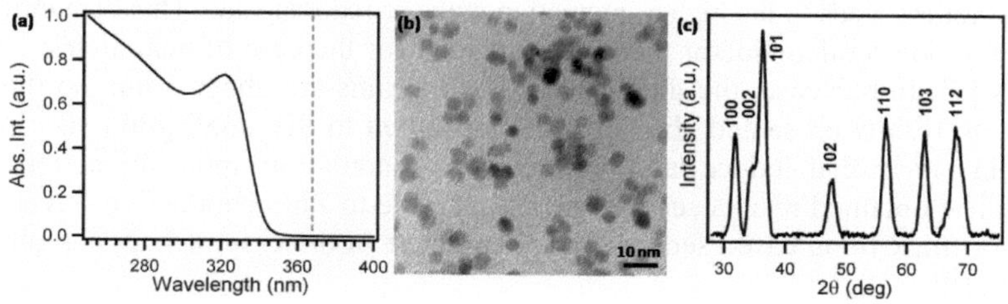

Figure 1. (**a**) UV-Vis absorbance spectrum obtained for the ZnO QD suspension used in this work. Based on this spectrum we calculated a ZnO QD diameter of 3.0 nm and a QD concentration of 80 μM. (**b**) is a typical TEM image for ZnO QDs obtained from the synthesis route used here. (**c**) Typical XRD pattern obtained for Q-ZnO QD powders after drying. These results confirm the good crystallinity (wurtzite) of the particles.

The sensing properties of nanocrystal ZnO-based sensors were evaluated by exposing them to three different gases: NO_2 (2–50 ppm), CH_3OH (10–100 ppm) and Acetone (10–100 ppm). Tests were made at different temperatures and different concentrations for each gas. Figure 2 reports the relations between working temperature and response toward different gases for a settled concentration of each target gas.

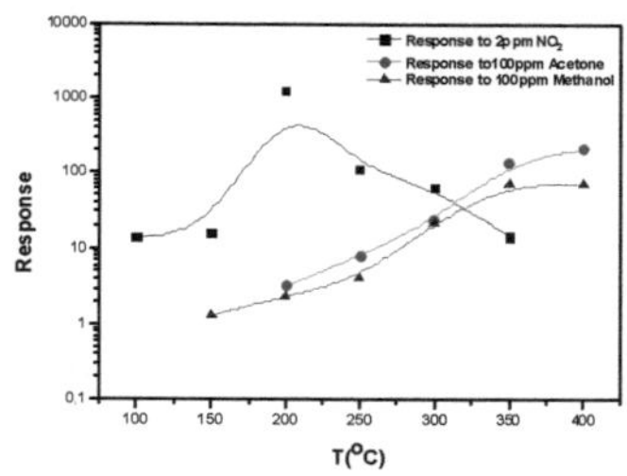

Figure 2. ZnO sensor responses versus operating temperature for 2 ppm NO_2, 100 ppm CH_3OH and 100 ppm Acetone.

As shown in Fig. 2, the response of the Q-ZnO-based sensor for NO_2 is greater than for the other gases and a good sensitivity is shown towards NO_2 at low concentration (2 ppm) and at low temperature. The best responses towards NO_2 were obtained at 200°C. At higher temperatures a significant decrease of the sensor responses takes place. In this way, at working temperature below to 300°C, the Q-ZnO based sensor is selective to nitrogen oxide, while at temperature above 350°C acetone and methanol responses further increase.

This behavior can be expected from follows the conventional metal oxide semiconductor theory and is generally observed on metal oxide – based semiconductor sensors [7, 8]. When NO_2 adsorbs on the ZnO surface creates acceptor surface states. This results in increased band bending at the grain surface and thus a rise in resistance and therefore low temperature detection is observed. With reducing gases, the gas detection mechanism is modulated by the presence of oxygen surface concentration that, with the increase of the temperature, leads to the higher interaction with the reducing gas. This conventional metal oxide semiconductor theory is extended for the case of nanometric grains sizes [9]. In our case, the dimensions of the grains are about 3 nm, so the gas sensing results presented above can be attributed to the small grain size in the sensing layer that induce large conductance modulation upon interaction with gas. The obtained results seem to be comparable to ZnO conductive – type thin and/or thick films based sensors and even better in the case of NO_2 gas sensing tests [10].

4. Conclusion

ZnO nanocrystals were synthesized at room temperature using a simple wet chemical method. The mean crystallite size of the obtained quantum dots was found in the range from 2.5 to 4.5 nm.

Drop-casting of the nanocrystals onto alumina substrates allowed the fabrication of chemoresistive gas sensing devices. The gas sensing test on resulting devices show remarkable response to NO_2 even at very low operating temperatures and good responses towards Acetone and Methanol. This result is attributed to the preparation of highly nanosized oxide particles and to their processing for the device realization. In fact, the final sensing layer results a porous and not sintered structure resulting in extensive interaction with the gases.

The results obtained are very promising and stimulating and in further developing of nano-ZnO based sensor devices.

References

1. M.J. Madouand, S.R. Morrison (Eds.), Chemical Sensing with Solid State Devices, Academic Press, New York (1989).
2. A. Vancu, R. Ionescu, N. Bârsan, in: P.Ciureany, S. Middelhoek, Thin Film Resistive Sensors, IOP Publishing Ltd., Chap. 6, 437 (1992).
3. S. Capone, P. Siciliano, Gas Sensors from Nanostructured Metal Oxides, in: H.S. Nalwa (Ed.), Encyclopedia of Nanoscience and Nanotechnology, Vol. 3, American Scientific Publishers, 769–804 (2004).
4. L.F. Dong, Z.L. Cui, Z.K. Zhang, Gas sensing properties of nano-ZnO prepared by arc plasma method, NanoStruct. Mater 8 (7) (1997) 815–823.
5. H. Gong, J.Q. H, J.H. Wang, C.H. Ong, F.R. Zhu, Nano-crystalline Cu-doped ZnO thin film gas sensor for CO, Sens. Actuators B 115 (2006) 247–251.
6. D.A. Schwartz, N.S. Norberg, Q.P. Nguyen, J.M. Parker, D.R. Gamelin, Magnetic Quantum Dots: Synthesis, Spectroscopy, and Magnetism of Co2+- and Ni2+-Doped ZnO Nanocrystals, J. Am. Chem. Soc. 125 (2003) 13205.
7. T. Wolenstein, Electronic Process on Semiconductor Surfaces during Chemisorption, Plenum Publishing Corporation, New York (1991).
8. C. Baratto, G. Sberveglieri, A. Onischuk, B. Caruso, S. di Stasio, Low temperature selsctive NO_2 sensors by nanostructured fibres of ZnO, Sens. Actuators B 100 (2004), 261–265.
9. C. Malagù, V. Guidi, M. Stefancich, M.C. Carotta, G. Martinelli, Model for Schottky barrier and surface states in nanostructured n-type semiconductors, J. Appl. Phys. 91 (2) (2002) 808–814.
10. A.A. Tomchenko, G.P. Harmer, B.T. Marquis, J.W. Allen, Smiconducting metal oxide sensors array for selective detection of combustion gases, Sens. Actuators B 93 (2003) 126–134.

OPTICAL GAS SENSING PROPERTIES OF ZNO NANOWIRES

S. TODROS[1], C. BARATTO[1], E. COMINI[1], G. FAGLIA[1], M. FERRONI[1], G. SBERVEGLIERI[1], S. LETTIERI[2], A. SETARO[2], L. SANTAMARIA[2] AND P. MADDALENA[2]

[1]*SENSOR LAB, CNR – INFM & Department of Physics and Chemistry, University of Brescia, via Valotti 9, 25133 Brescia, Italy*

[2]*CNR–INFM COHERENTIA Unity & Dipartimento di Scienze Fisiche, University of Napoli "Federico II"*

Abstract. New breakthroughs in preparation technology have given further impulse for developing chemical sensing devices based on nanostructured ZnO. Due to their peculiar characteristics and size effects, these materials often show novel physical properties. In this work, we studied the optical properties of ZnO nanowires, synthesized via evaporation-condensation (EC) process, by room temperature photoluminescence (PL) and Time Resolved PL analysis. ZnO PL spectrum shows an UV peak, attributed to free exciton, and a broad visible band, strongly depending on the preparation conditions in the range from green to yellow. UV photoluminescence of ZnO nanowires was studied in gaseous atmosphere, targeting mainly NO_2 sensing applications. Continuous wave PL shows that a reversible modification is obtained upon exposure to low concentration of NO_2. The effect of interfering species like ethanol and RH% has also been studied. Time-resolved PL shows small modifications of recombination rates due to introduction of NO_2 not proportional to simultaneous changes in PL intensity. This behaviour supports the idea that NO_2 molecules, chemisorbed on the ZnO surface, lead to suppression of radiative states.

1. Introduction

ZnO performances as chemical sensor have been known for 40 years, however new breakthroughs in preparation technology have given further impulse for developing chemical sensing devices based on ZnO. Recent deposition techniques allow finely tuning the size and the crystalline degree of this material, in order to obtain high quality single crystals and quasi one-dimensional structures. Moreover, the challenge to realize a reliable electrical contact on nanowires increases the interest in all-optical contactless gas sensors.

In the last decade, the interest in ZnO nanowires is significantly driven by its light emission properties, mainly because of its high efficiency photoluminescence (PL) properties. Due to its high-energy direct band-gap (3.37 eV energy gap) and its large exciton binding energy (60 meV), ZnO is extremely promising in view of applications such as light-emitting diodes [1] and lasers [2].

P. Malcovati et al. (eds.), *Sensors and Microsystems: AISEM 2009 Proceedings*, Lecture Notes in Electrical Engineering 54, DOI 10.1007/978-90-481-3606-3_32,

2. Experimental

2.1. *ZnO nanowires deposition*

ZnO nanowires were prepared through Evaporation-Condensation method: the source material is vaporized and transported by a gas carrier towards the substrates, where it condenses. The experimental set up consists in a furnace able to reach high temperature, a vacuum-sealed alumina tube connected to a primary pump, an automated valve and a mass flow meter to control pressure and carrier flow. First, gold particles, which act as catalyst during the deposition, were dispersed on silicon substrates by DC magnetron sputtering at a working pressure of $5 \cdot 10^{-3}$ mbar and 70 W applied power. The source material was positioned at the middle of the alumina tube, evaporated at temperature of 1,370°C and pressure of 100 mbar. The gold catalyzed substrates were placed in an alumina holder and put inside the tube in an area corresponding to a temperature range between 290°C and 370°C. The deposition conditions have been tailored in order to promote formation of one-dimensional (1D) nanostructures, through changing the evaporation temperature and the carrier gas composition and flux.

2.2. *SEM characterization*

Scanning electron microscopy (SEM) analysis was carried out by field-emission LEO 1525 microscope, equipped with In-Lens detector for secondary-electrons imaging. The SEM was operated at 10 kV accelerating voltage range to prevent the insulating substrate from electrostatic charging, thus allowing observation of uncoated specimens. The nanowires are homogeneously deposited over large area substrates; their lateral dimensions depend on the deposition parameters and their length depends on deposition time. Figure 1 report a SEM picture of nanowires grown on substrates at different temperatures: it can be observed that size and length are strongly dependent by this parameter.

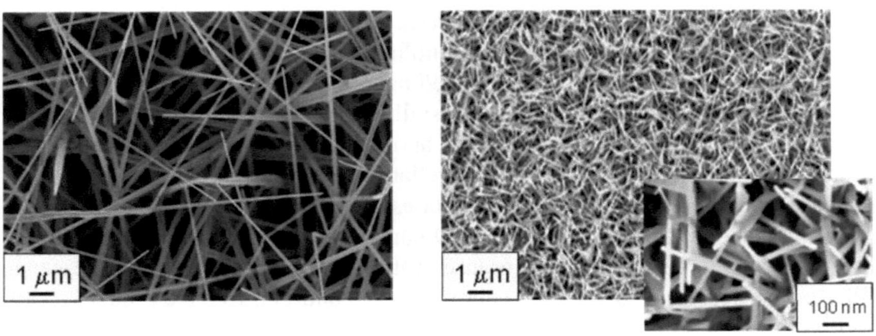

Figure 1. SEM picture of nanowires grown on substrates at different temperatures: (*left*) 260–280°C; (*right*) 340–370°C.

3. Results and discussion

3.1. *Continuous wave PL measurements in controlled environment*

The room temperature PL spectra of ZnO nanostructures typically shows an UV peak due to excitonic recombination and a broad emission band in the visible range depending on deep defect states lying inside the bandgap. The excitonic UV peak is centred at a wavelength of about 380 nm, while the spectral position of defect-related visible emission band strongly depends on the preparation conditions and may range from green to yellow [3] (see Fig. 2).

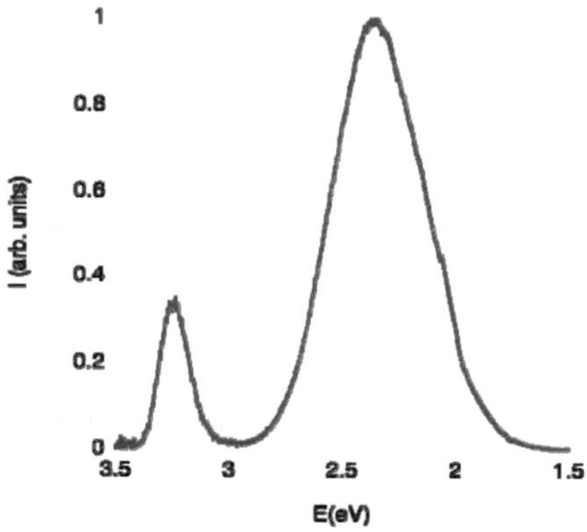

Figure 2. Photoluminescence spectra of ZnO nanowires in dry air.

We have demonstrated that PL is quenched by interaction with NO_2 at very low concentration, in a reversible way (see Fig. 3). The ability to detect attention level concentration (0.3 ppm) is very interesting in view of an optical gas sensors working at room temperature. The effect of reducing gases like humidity and ethanol has also been studied.

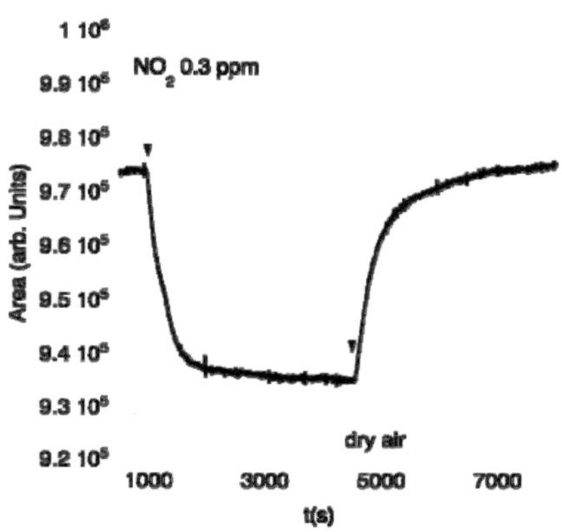

Figure 3. ZnO nanowires PL visible peak area variation at low NO_2 concentrations.

3.2. *Time-resolved PL measurements*

Further investigation on luminescence mechanisms and its quenching due to NO_2 has been performed by means of time-resolved photoluminescence (TRPL) technique, in which the sample is excited by a short time laser pulse and the resulting decaying photoluminescence emission is detected by a suitable detector with high resolution time. These measurements show small modifications of recombination rates due to introduction of NO_2, not proportional to simultaneous changes in photoluminescence intensity. No linear correlation is obtained between PL yield and decay lifetime, suggesting that interaction takes place with a static quenching mechanism, that is to say that NO_2 molecules chemisorbed on the ZnO surface lead to suppression of radiative states [4].

4. Conclusions

In this work, we studied the optical properties of ZnO nanowires by room temperature photoluminescence (PL) and Time Resolved PL analysis. Static wave PL shows a reversible modification upon exposure to low NO_2 concentration: the sensor was able to detect attention level concentration (0.1 ppm). TRPL measurements show small modifications of recombination rates due to introduction of NO_2 which are not proportional to simultaneous changes in photoluminescence intensity. This finding supports the idea that the quenching molecules act as static quenchers.

Acknowledgements This work was supported, within the EU FP6, by the ERANET project "NanoSci-ERA: NanoScience in the European Research Area".

References

1. A. Tsukazaki et al., Nature Mater., [4], pp. 42–46, 2005
2. D.M. Bagnall et al., Appl. Phys. Lett., [70], pp. 2230–2232, 1997
3. A.B. Djurisic et al., Nanotechnology, [18], pp. 095702, 2007
4. S. Lettieri et al., New. Journ. Phys., [10] pp. 043013, 2008

PORPHYRIN-PORPHYRIN DIADS AS POTENTIAL TRANSDUCERS FOR THE DETERMINATION OF CADAVERINE IN AQUEOUS SOLUTION

F. BALDINI[1], A. GIANNETTI[1], C. TRONO[1], T. CAROFIGLIO[2] AND E. LUBIAN[2]

[1]*Istituto di Fisica applicata "Nello Carrara" IFAC-CNR, via Madonna del Piano 10, 50019 Sesto Fiorentino (FI), Italy*

[2]*Università di Padova, Dip. di Scienze Chimiche and ITM-CNR, 35131 Padova, Italy*

Abstract. A Zn(II) porphyrin-porphyrin conjugated has been used as an optical transducer for the development of an optical sensor for diamines. The sensing molecule was immobilised on controlled pore glass (CPG) which was characterised with an appropriate flow cell connected to an optical spectrometer.

1. Introduction

Determination of the diamines, such as cadaverine, is an important issue both in food analysis and in biomedical applications. In food analysis, diamines can increase the toxicity of other biogenic amines (BA) content in foods, such as fishery products, cheese, wine, beer, dry sausages. BA are important markers for the food quality since their concentration increases by bacterial activity during an inappropriate storage and processing and several toxicological problems resulting from the ingestion of food containing relatively high levels of BA have been reviewed [1]. In biomedical field, elevated polyamine concentrations in human urine and serum appear to be associated with the presence of many types of cancer [2].

The present paper describes the optical characterisation of properly synthesised porphyrin(Zn)-porphyrin(Zn) dyads [3] as potential transducers for the determination of diamines and, in particular, of cadaverine.

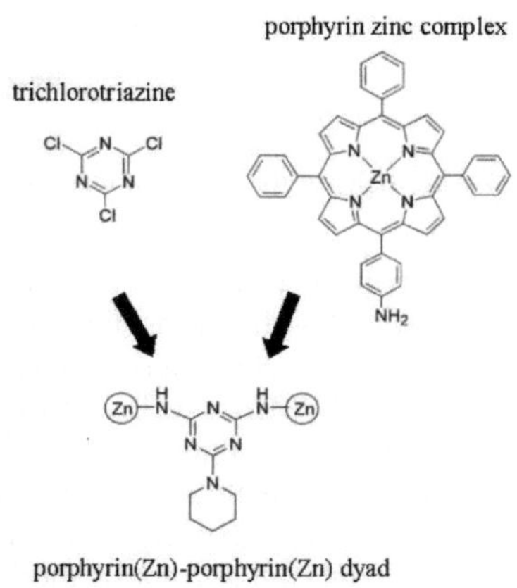

Fig. 1. The building blocks and the resulting porphyrin(Zn)-porphyrin(Zn) dyad.

P. Malcovati et al. (eds.), *Sensors and Microsystems: AISEM 2009 Proceedings*, Lecture Notes in Electrical Engineering 54, DOI 10.1007/978-90-481-3606-3_33,

2. Methodology

Figure 1 shows the building blocks used for the synthesis of porphyrin (Zn)-porphyrin(Zn) dyad according to the reference [3]. The dimer can assume different conformations due to the rotation around the C–N bonds of the cyanuric acid bridge, as shown in Fig. 2. The most relevant conformation for the molecular recognition process is the SYN-SYN conformation (Fig. 3) since it maximizes the ditopic properties of the porphyrinic receptor toward the diamines. In particular, the distance of 8.31 Å between the two zinc(II) ions, resulting from molecular modeling calculations (semiempirical AM1), justifies the partial selectivity towards cadaverine and hexane-1-6-diamine for the aliphatic diamines of general formula H_2N-$(CH_2)n$-NH_2 (n = 4–8).

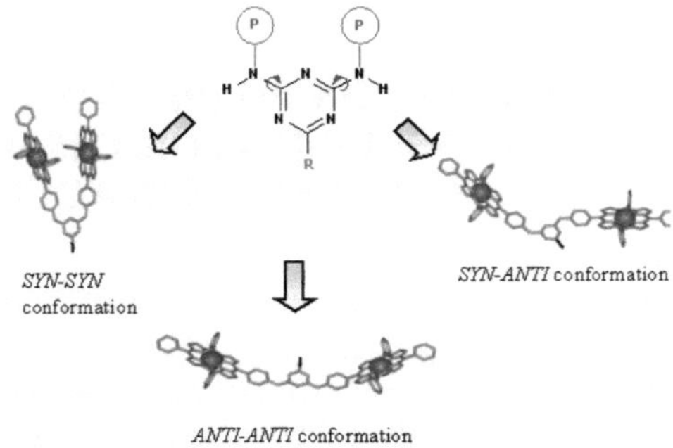

Fig. 2. Components and conformations of porphyrin (Zn)-porphyrin(Zn) dyad.

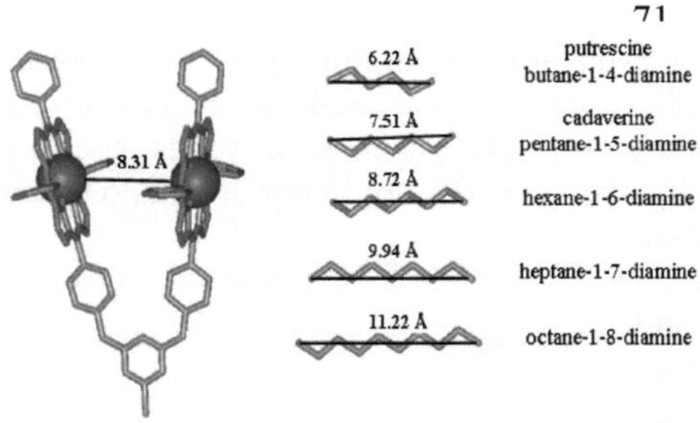

Fig. 3. Specificity of the porphyrin-porphyrin dyad to diamines.

The porphyrin(Zn)-porphyrin(Zn) dyad was covalently immobilised onto controlled pore glass (CPG) with different pore sizes (1,400 and 700 Å). To this purpose, CPG was amino-silanized according the procedure previously reported [4] then reacted with cyanuric acid bridged porphyrin-porphyrin dimer prepared as described on reference [3]. The residual amino-groups of the CPG were capped with acetic anhydride in the presence of pyridine; then the porphyrin macrocycles were metalated with zinc acetate in chloroform/methanol mixture.

The CPG were characterised with the instrumental apparatus shown in Fig. 4. A miniaturised PMMA well was filled with CPG. The well was placed over the common end of an optical fibre bifurcated bundle. The two branches of the bundle were connected to a halogen lamp (Zeiss) and to an Ocean Optics spectrophotometer, respectively. The light coming from the source was scattered by the CPG and the portion of the light, which was coupled back to the optical fibre bundle, was modulated by the absorption properties of the CPG. A peristaltic pump allowed to flow buffered solutions (phosphate buffered saline at pH 7.4) containing different concentrations of cadaverine inside the miniaturised well. The inlet and outlet were constituted by two fused-silica capillaries having an internal diameter of 300 μm.

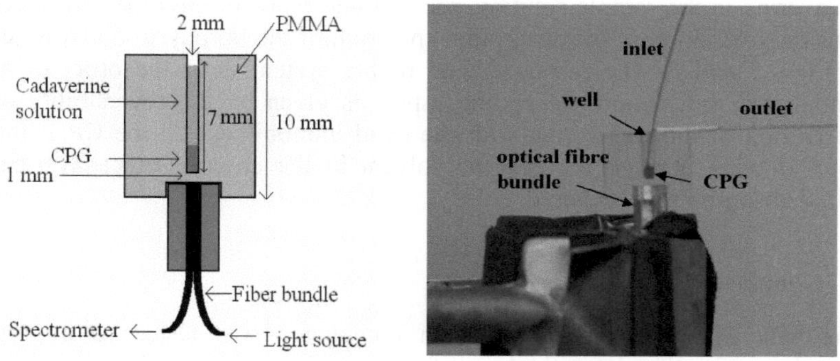

Fig. 4. *Left*: sketch of the setup for the optical characterization of the CPG. *Right*: photo of the setup with the inlet and outlet for the cadaverine solutions, constituted by two fused-silica capillaries.

3. Results and discussions

Figures 5 and 6 show the absorption spectra recorded for CPG 1,400 Å and CPG 700 Å particles when buffers at different concentrations of cadaverine (ranging from 10^{-5} to 10^{-2} M) flow through the cell. The sensitivity toward cadaverine increases upon reducing the pore size of CPG. This result was expected, based on our previous studies on similar CPG based sensors [4]. Indeed, a decreasing in the porosity corresponds to an increase of the surface available for the immobilisation.

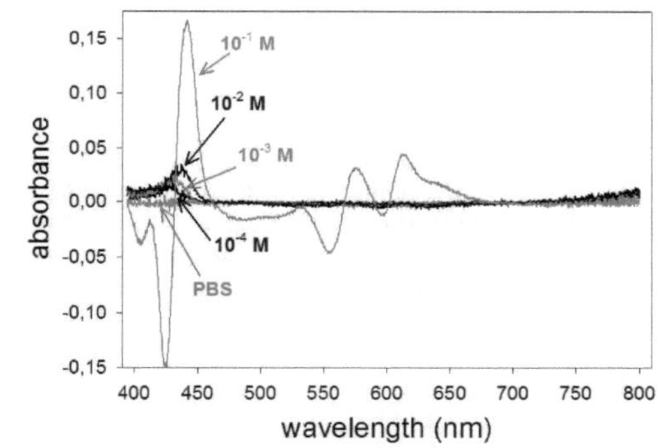

Fig. 5. Absorption spectra of CPG with pore size of 1,400 Å for different concentrations of cadaverine in buffer solution.

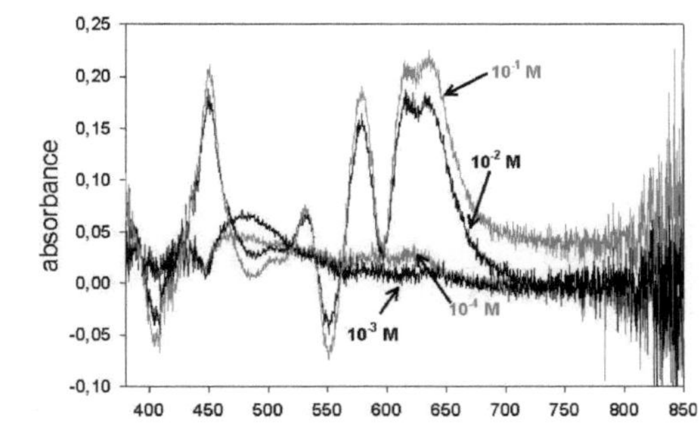

Fig. 6. Absorption spectra of CPG with pore size of 1,400 Å for different concentrations of cadaverine in buffer solution.

Since the spectrophotometer was zeroed with the porphyrin dyad in the absence of cadaverine flowing through the measuring cell, the spectra collected during the analysis are difference spectra. In particular, we examined the spectral changes that occurred in Q-band region (500 ÷ 650 nm) and the absorption in the correspondence of the porphyrin Soret band (ca. 420 nm). In the presence of cadaverine in the sample solution, the diamine binds to both the Zn(II) centers causing a variation of the absorption spectrum of the porphyrin dimer as shown in Figs. 5 and 6. The response time of the system is of the order of a few minutes. A much more interesting aspect is given by the reversibility of the interaction of the cadaverine with the dyad immobilized on the CPG. Indeed, upon flowing through the cell the solvent in the absence of cadaverine, the initial absorption is recovered.

4. Conclusions

The new Zn(II) porphyrin-porphyrin conjugated was tested as a potential transducer for the determination of cadaverine in buffer. It's important to observe that these preliminary data were collected in aqueous solution in which the water molecules could compete with cadaverine in binding to the porphyrin dyads. We are actually studying the use of less competing solvents, such as toluene or hexane, in order to achieve lowest limit of detection.

References

1. BT. Brink, C. Damink, H.J. Joosten, J. Huis in't Veld, "Occurrence and formation of biologically active amines in foods", Int. J. Food Microbiol., [11], 73–84, 1990.
2. L.J. Marton, D.H. Russell, C.C. Levy, "Measurement of putrescine, spermidine, and spermine in physiological fluids by use of an amino acid analyzer", Clin. Chem., [19], 923–926, April 1973.
3. T. Carofiglio, A. Varotto, U. Tonellato, "One-pot synthesis of cyanuric acid-bridged porphyrin-porphyrin dyads", J. Org. Chem., [69], 8121–8124, 2004.
4. F. Baldini, S. Bracci, F. Cosi, "An extended-range fibre optic pH sensor", Sens. Actuat, A, [37–38], 180–186, 1993.

ELECTROCHEMICAL CHARACTERIZATION OF PNA/DNA HYBRIDIZED LAYER USING SECM AND EIS TECHNIQUES

I. PALCHETTI, F. BERTI, S. LASCHI, G. MARRAZZA AND M. MASCINI
Dipartimento di Chimica, Università degli Studi di Firenze, Via della Lastruccia 3, 50019 Sesto Fiorentino (Fi), Italia, e-mail: ilaria.palchetti@unifi.it

Abstract. The detection of DNA hybridization is of significant scientific and technological importance as testified by the growing interest in chip-based characterization of gene expression patterns and detection of pathogens. Recently, there has been an increasing interest for DNA sensors based on the SAMs (Self Assembled Monolayers) of Peptide Nucleic Acid (PNA) modified electrodes. PNA is a DNA mimic that has a neutral peptide-like backbone with nucleobases that allows the molecule to hybridize to complementary DNA strands with high affinity and specificity. In this communication, we present hybridization studies with DNA target oligonucleotides on a mixed monolayer of PNA and MCH (mercaptohexanol) on Au electrodes using EIS (Electrochemical Impedance Spectroscopy) and SECM (Scanning Electrochemical Microscopy). Preliminary results of PNA/DNA hybridization detection in a label free approach are reported.

1. Introduction

The use of nucleic acids recognition layers in biosensor design represents an exciting area in analytical chemistry [1]. Recently, probes produced by chemical changes to the backbone of naturally occurring DNA or RNA are used more and more in NA sensing techniques [2]. Among these, Peptide Nucleic Acid (PNA) are frequently used. Owing to its neutral backbone and proper interbase spacing, PNA binds to its complementary nucleic acid sequence (DNA or RNA) with higher affinity and specificity compared to traditional oligonucleotides. The neutral backbone also implies a lack of electrostatic repulsion between the PNA and DNA strands (compared to that existing between two negatively charged DNA oligomers), and hence a higher thermal stability of PNA/DNA duplexes. Electrochemical methods can conveniently be used in NA sensing to study and detect PNA/DNA hybridization in a label free approach. The immobilized PNA probes on the electrode surface are uncharged, and hence, do not affect the charge transfer from charged redox indicator, such as $K_4Fe(CN)_6/K_3Fe(CN)_6$, to the electrode. Once DNA targets hybridize to PNA, the charge density at the sensor surface will be changed. Thus, one can use electrochemical techniques highly sensitive to the interfacial electron transfer, such as EIS or SECM, to

P. Malcovati et al. (eds.), *Sensors and Microsystems: AISEM 2009 Proceedings*,
Lecture Notes in Electrical Engineering 54, DOI 10.1007/978-90-481-3606-3_34,

conveniently monitor the PNA/DNA hybridization process. In this communication preliminary results of such studies using both the techniques are reported.

2. Experimental

Table 1 shows the thiol-terminated PNA probe and DNA target used in this study. Thiolated PNA was dissolved in 0.1% trifluoroacetic acid. Here, PNA-modified surfaces were obtained by exposing the gold substrates in a 15 μM (for SECM measurements) or 200 nM (for EIS measurements) PNA solution for about 12 h. The surfaces were then treated further with 1 mM mercaptohexanol (MCH) solution for 30 min. EIS measurements were performed with an AUTOLAB PGSTAT 30(2) digital potentiostat/galvanostat, using a Ag/AgCl pseudo reference electrode and a Pt counter electrode. SECM measurements were performed with an Uniscan instrument SECM 270 (Uniscan, UK), using Au disk microelectrodes with 10 μm diameters (CH Instruments, UK) as amperometric SECM tip and $Hg/HgSO_4$ (CH Instruments, UK) as reference electrode.

Table 1. The sequences of PNA probe and DNA targets.

Name	Sequence
Thiolated PNA	5' TAT TTA CGT GCT GCT A -$(CH)_6$-SH 3'
Fully complementary target	5' TAG CAG CAC GTA AAT A - TEG- biotin 3'
Non complementary sequence	5' AGG CCC TGC GAG CAA CAT CTC ATG–TEG-biotin 3'

3. Results and discussion

3.1. *PNA-DNA layer characterization by SECM*

The first step of the detection scheme consists of the immobilization of thiol-tethered PNA by chemisorption; then the PNA chip is postassembled with MCH to increase hybridization efficiency and to avoid unspecific adsorption of target strands. When a gold surface is exposed to thiol-containing compounds (HS-R), by varying the terminal group of R, surface properties, including chemical and electrochemical reactivity, are different. Thus, an evaluation of the behavior of the HS-PNA/MCH mixed adsorbate was performed. Approach curves obtained in 1 mM $K_4Fe(CN)_6$ solution showed a positive feedback. This behavior demonstrated that mediators are not blocked by the PNA monolayer, mainly because the backbone of PNA contains no charged phosphate groups to repeal negatively charged mediators. After DNA hybridization the feedback effect is smaller, indicating the presence of negatively charged phosphate group and of electrostatic repulsion. However this effect is not pronounced as expected. Probably this behavior is due to defects of the PNA/DNA layer; as a consequence the DNA strands are so loosely arranged that their anionic phosphate groups could not really hinders $Fe(CN)_6^{3-}$ molecules to approach the Au surface to undergo redox recycling. However, approach curves recorded in ferrocenemethanol (data not shown) showed a positive feedback demonstrating that diffusion of this uncharged mediator proceeds virtually unimpeded.

3.2. *PNA-DNA layer characterization by EIS*

EIS experiments confirmed results obtained with SECM. Since the backbone of PNA contains no charged phosphate groups, the electron transfer resistance is smaller in the case of the chip modified with thiol tethered PNA before the

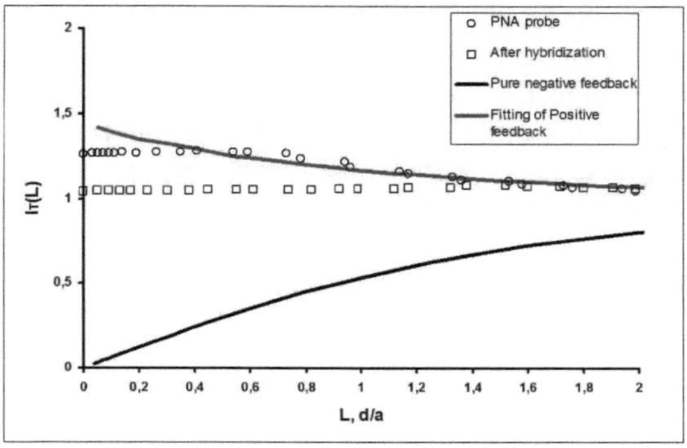

Figure 1. Approach curve experiments on a 15 μM HS-PNA modified gold substrate before and after hybridization with 30 μM DNA. The approach curves are obtained in $K_4Fe(CN)_6$ 1 mM solution in Tris-ClO_4 buffer 20 mM + $NaClO_4$ 50 mM, pH 8.6, over a modified chip, E_{tip} = + 0.100 V, E_s = OCP. For all the experiments a Au tip with a diameter of 10 μm was used. The black solid line is the SECM theoretical curve for pure negative feedback.

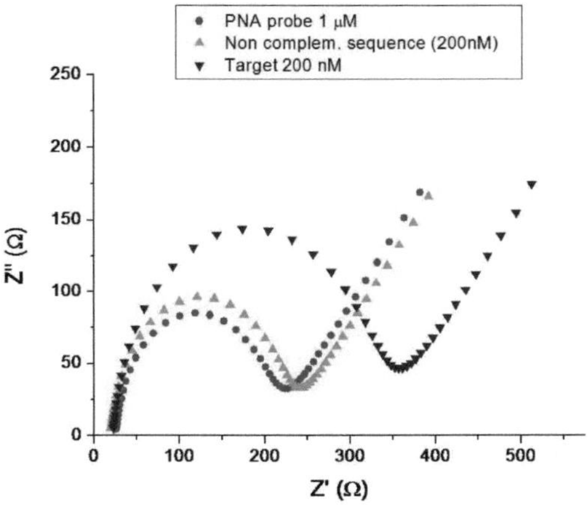

Figure 2. Nyquist plot for PNA-modified gold substrates. Experimental parameters: redox probe, 2.5 mM $[Fe(CN)_6]^{3-/4-}$ (1:1) mixture in 1 M KCl; Bias potential, + 0.2 V; frequency range, 100 kHz–100 mHz; ac voltage amplitude, 10 mV. 0.2 μM HS-PNA gold modified substrate.

hybridization with the target DNA than after the hybridization. This is consistent with the electrostatic repulsion of the redox indicator from the electrode interface by the formation of the charged PNA/DNA duplexes, thereby introducing a barrier to electron transfer. The resistance increases with the DNA target concentration (data not shown), whereas is similar to the PNA layer only when a non complementary strand was studied (Figs. 1 and 2).

4. Conclusions

The data reported in this communication demonstrated that PNA/DNA hybridization can be detected with both EIS and SECM in a label free approach by using a negatively charged redox mediator.

However, SECM detection needs the formation of highly packed probe layer, this means high probe concentration and high target concentration. EIS detection can easily discriminate specific and non specific signal at nM range of target even using low concentration of probe.

Acknowledgments Support from the Italian Government in the framework of the PRIN 2007 project.

References

1. J. Wang, "DNA biosensors based on PNA recognition layer. A review", Biosensors & Bioelectronics, [13], pp. 757–762, 1998.
2. I. Palchetti and M. Mascini "Nucleic acid biosensors for environmental pollution monitoring", Analyst, [133], pp. 825–964, 2008.

METAL-FUNCTIONALIZED AND VERTICALLY-ALIGNED MULTIWALLED CARBON NANOTUBE LAYERS FOR LOW TEMPERATURE GAS SENSING APPLICATIONS

M. PENZA[1], R. ROSSI[1], M. ALVISI[1], M. A. SIGNORE[1], G. CASSANO[1], R. PENTASSUGLIA[1], D. SURIANO[1], V. PFISTER[1] AND E. SERRA[2]

[1]*ENEA, CR Brindisi, Department of Physical Technologies and New Materials, P.O. Box 51, Postal Office Br4, 72100 Brindisi, Italy*

[2]*ENEA, CR Casaccia, Department of Physical Technologies and New Materials, Via Anguillarese 301, 00060 Rome, Italy*

Abstract. Vertically-aligned carbon nanotubes (CNTs) films have been grown by radio-frequency plasma-enhanced chemical vapor deposition system onto alumina substrates, coated with 2.5 nm thick Fe catalyst, for NO_2, H_2 and C_2H_5OH gas sensing applications, at sensor temperatures from room-temperature to 150°C. The CNTs appear in the form of *forest-like* structure. Nanoclusters of noble metals with nominal thickness of 5 nm of Pt, Ru and Ag have been sputtered on the top-surface of the vertically-aligned CNTs layers to enhance the gas sensitivity. It was demonstrated that the gas sensitivity of the metal-modified CNTs-sensors significantly improved by a factor up to an order of magnitude through a spillover effect, and a broader selectivity was provided. The gas sensing properties of the CNTs-sensors, including the metal-modified CNTs, are characterized by a change of the electrical conductivity in a model of the charge transfer with a semiconducting *p-type* character. The metal-functionalized CNTs-sensors exhibit high sensitivity, fast response, reversibility, good repeatability, sub-ppm range detection limit. A practical application of the CNTs-sensors for monitoring landfill gas is presented.

1. Introduction

Gas sensors based on carbon nanotubes (CNTs) have been largely studied in the form of networked films for highly-sensitive gas detection applications [1–9]. Due to very high surface-to-volume ratio, hollow nanostructure, high electron mobility, great surface reactivity and high capability of gas adsorption, CNTs have been investigated as building blocks for fabricating novel devices at nanoscale such as high-performance gas sensors and platforms for biosensing. These nanomaterials in the form of roped layer consisting of vertically-aligned carbon nanotubes with functionalizating metal nanoclusters can lead to an attractive strategy for very-low gas detection up to a ppb concentration level, at low temperatures of operations with minimum power consumption.

Here, vertically-aligned CNTs films in the form *forest-like* have been grown by radiofrequency plasma enhanced chemical vapor deposition (RF-PECVD)

P. Malcovati et al. (eds.), *Sensors and Microsystems: AISEM 2009 Proceedings*, Lecture Notes in Electrical Engineering 54, DOI 10.1007/978-90-481-3606-3_35,

technology onto low-cost alumina substrates, coated by nanosized Fe-catalyst for growing CNTs, to perform chemical detection of hazardous gases (NO_2, H_2, C_2H_5OH), at a working sensor temperature in the range 25–150°C. Tailored fabrication of sub-ppm level gas sensors composed of vertically-aligned CNTs films surface-modified with nanoclustered catalysts of noble metals of Pt, Ru, and Ag with 5–50 nm size has been performed for enhanced gas sensing applications. Conductometric gas sensors based on CNTs networked films with remarkable performance have been successfully reported [1–10]. The observed trend of the electrical conductance changes upon gas exposure can be modelled by a charge transfer between adsorbed gas molecules and carbon nanotubes with *p-type* semiconducting characteristics.

The possibility of monitoring the landfill gas (LFG) constituted by a multi-component gas-mixture (CO_2, CH_4, H_2, NH_3) using the metal-modified and vertically-aligned CNTs-sensors is preliminary investigated.

2. Experimental details

Vertically-aligned CNTs films [10] were grown by RF-PECVD system at a temperature of 600°C. The CNTs films were deposited onto cost-effective alumina (5 mm width × 5 mm length × 0.6 mm thickness), coated with growth-catalyst of Fe nanoclusters with a nominal thickness of 2.5 nm and sputtered at 10^{-2} Torr. The scheme of the two-pole chemiresistor is shown in Fig. 1. The Fe-catalysed alumina substrates were heated to 600°C by a rate of 10°C/min in H_2 atmosphere upon flow of 100 sccm at a total pressure of 1.5 Torr. In the gas-plasma, the flow rate ratio between C_2H_2 and H_2 was kept constant at 20/80 sccm, respectively. The CNTs deposition was performed with constant rf power, pressure and temperature of 100 W, 1.5 Torr and 600°C, respectively for 30 min by depositing a vertically-aligned CNTs film with thickness of 10 μm.

Two metal strips (1 mm width × 5 mm length) consisting of Au (300 nm) on adhesion layer of Cr (20 nm) were sputtered to serve as electrical contacts in the two-pole geometry.

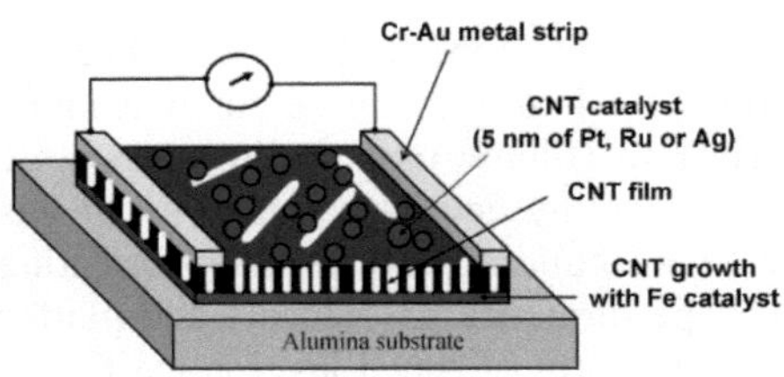

Figure 1. Scheme of the chemiresistor based on metal-modified and vertically-aligned CNTs films.

Figure 2. FE-SEM image of the vertically-aligned CNTs. The height of the CNTs film is 10 μm.

Surface-modification of the vertically-aligned CNTs films involved magnetron sputtering with a nominal thickness of 5 nm of Pt, Ru and Ag nanoclusters over the entire CNTs film. The electrical resistance, at room temperature, of the unmodified CNTs films was measured as 100–120 Ω, while for the CNTs films modified with 5 nm of Pt, Ru and Ag was measured as 70–100 Ω. Thus, the surface-modification of the CNTs layer with metal catalysts decreases the electrical resistance of the carbon nanotube films.

Before gas sensing exposure, the unmodified CNTs and metal-modified CNTs were thermally annealed at 300°C upon dry air flow of 1,500 ml/min for 2 h to purify the CNTs removing amorphous carbon and stabilize the sensing properties of catalyst nanoclusters improving their adhesion onto CNTs.

The so-fabricated CNTs sensors have been located in a test cell (500 ml volume) for gas exposure measurements. The cell case is able to host up to four chemiresitive sensors. Dry air was used as reference gas and diluting gas to air-conditioning the sensors. The gas flow rate was controlled by mass flowmeters. The total flow rate per exposure was kept constant at 1,500 ml/min. The gas sensing experiments have been performed by measuring the electrical conductance of CNTs thin films in the two-pole format upon controlled ambient of individual H_2 and C_2H_5OH reducing gases and NO_2 oxidizing gas in the range of 0.005–1%, 50–200 ppm, and 0.1–2 ppm, respectively, at sensor temperature ranging from 20 to 150°C. Moreover a four-component gas-mixture of landfill gas consisting of 800 ppm CO_2, 700 ppm CH_4, 1,000 ppm H_2 and 200–400 ppm NH_3 has been tested as well at a sensor temperature of 150°C. The dc electrical conductance of the CNTs-sensors has been measured by the volt-amperometric technique in the two-pole format by a multimeter (Agilent, 34401A). The sensors were scanned by a switch system (Keithley, 7001) equipped by a low-current scanner card (Keithley, 7158) with a multiplexed read-out.

3. Results and discussion

Figure 2 shows the FE-SEM image of the CNTs nets by revealing high abundance of the vertically-aligned CNTs at high density in the format *forest-like*. These carbon nanostructures are multiwalled, twisted each other, forming multitube ropes of uniform height of 10 μm and average diameter of the CNTs in the range of 5–35 nm. The vertical alignment is favoured by the Van-der-Waals force between the multiple tubes.

Figure 3 shows the FE-SEM image of CNTs films surface-modified by sputtered catalysts of Pt, Ag and Ru with a nominal loading of 5 nm. The size of the metal clusters varies in the range from 5 to 50 nm. These metal particles decorating the CNTs form isolated nanoclusters partially interconnected to promote gas adsorption. These features catalytically affect the gas sensing properties of the CNTs films for enhanced sensitivity and broader selectivity.

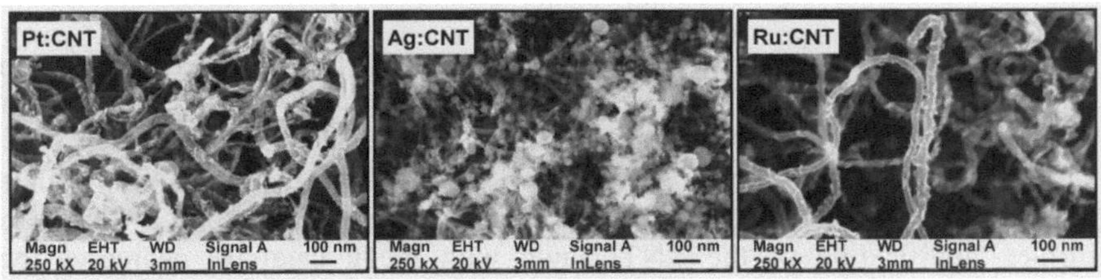

Figure 3. FE-SEM image of Pt-modified, Ag-modified and Ru-modified CNTs films, prepared by RF-PECVD on Fe-coated alumina substrates. The nominal thickness of the metal catalysts is 5 nm.

Figure 4 shows time response in terms of the percentage relative resistance change of four CNTs-chemiresistors, unmodified and functionalized with an metal-loading of 5 nm of Pt, Ru and Ag, exposed to spot-concentrations of ethanol (C_2H_5OH) reducing gas and nitrogen dioxide (NO_2) oxidizing gas in the range 78–170 ppm, and 2–1 ppm, respectively, at a sensor temperature of 25°C. As observed, the electrical resistance of unmodified and metal-modified CNTs-sensors specifically decreases upon exposure of the NO_2 oxidizing gas. This behaviour can be explained by a charge transfer between gas and CNTs-based nanomaterial according to a *p-type* conductivity electrical transport model in a semiconductor regime caused by the electron-withdrawing character of NO_2. In other words, electrons are transferred to NO_2 molecule from the valence band of the CNTs, increasing the density of holes, thereby decreasing the electrical resistance, shifting the Fermi level towards the valence band. An opposite trend in the electrical resistance is observed for all CNTs-sensors, unmodified and metal-modified, upon exposure of electron-donating gas (C_2H_5OH). Moreover, in the case of metal-modified CNTs, the test gas is likely dissociated by spillover effect at catalyst surface and the dissociated molecular fragments dissolve into the catalyst bulk. This causes the formation of interface discrete states in the band-gap of the semiconducting CNTs in proximity of the valence or conduction band, depending on character of test gases and catalyst, by enhancing the charge transfer and thus the gas response of the metal-modified CNTs-sensors.

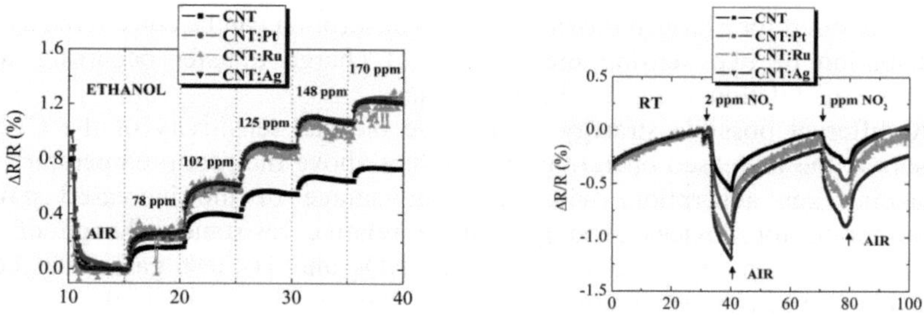

(a) (b)

Figure 4. Room-temperature time response of the percentage electrical resistance relative change for a chemiresistor based on unmodified CNTs films and metal-modified CNTs films with 5 nm loading of Pt, Ru and Ag towards (**a**) C_2H_5OH and (**b**) NO_2 gas. The CNTs films thickness is 10 μm.

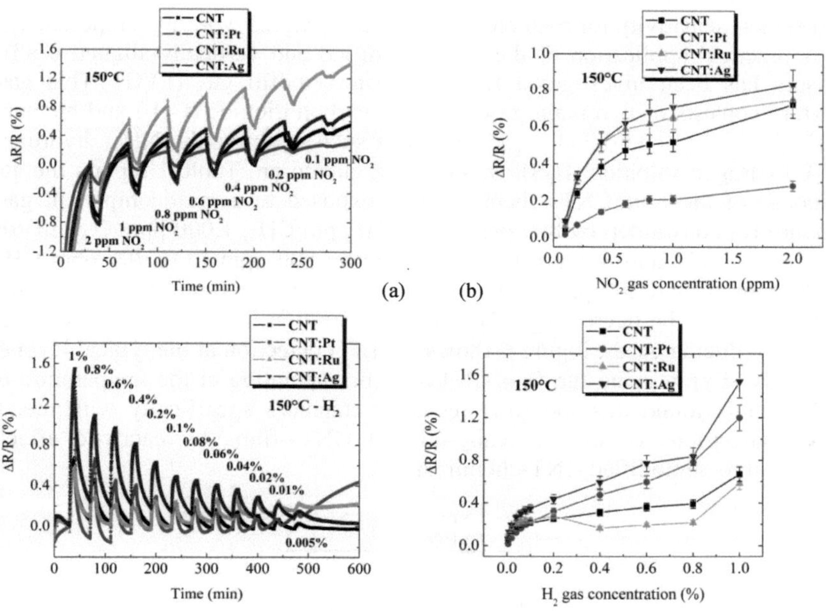

(c) (d)

Figure 5. Time responses of four chemiresistors based on unmodified CNTs (CNT), and loaded with 5 nm of Pt (CNT:Pt), 5 nm of Ru (CNT:Ru) and 5 nm of Ag (CNT:Ag), exposed to 10-min pulses of (**a**) NO_2 gas ranging from 2 ppm to 100 ppb, and (**c**) H_2 gas ranging from 1 to 0.005%, at the sensor temperature of 150°C. Calibration curves of these four CNTs-chemiresistors for (**b**) NO_2 and (**d**) H_2 gas, at the sensor temperature of 150°C. The thickness of the CNTs films is 10 μm.

The response of the metal-modified CNTs is higher than that of unmodified CNTs for both cases of ethanol and nitrogen dioxide, but no significant differentiation of the response in the metal-functionalized CNTs has been achieved for ethanol; while a tailoring of the sensor response with the metal-catalyst used has been measured for nitrogen dioxide. The gas adsorption can be considered as the contribution of two sensing mechanisms of charge transfer occurring with intratube modulation and intertube modulation.

A different possible strategy to improve the gas sensitivity of the CNTs-sensors is the increased operating temperature above the room-temperature for enhancing gas adsorption, with the disadvantage of the increased power consumption for devices. The percentage relative resistance change of the CNTs-films upon exposure to individual NO_2 and H_2 test gases has been recorded at a sensor temperature of 150°C. The results are reported in the Fig. 5a, c for NO_2 and H_2, respectively. Figure 5b, d shows the calibration curves of the unmodified and metal-modified CNTs-sensors, exposed to a broad range of gas concentrations of NO_2 (0.1–2 ppm) and H_2 (0.005–1%), at the sensor temperature of 150°C. The comparison between unmodified and metal-modified CNTs sensors demonstrates that the Ag-functionalized CNTs-sensors exhibit highest gas sensitivity for both NO_2 and H_2 gases, at this working temperature.

A practical application of the metal-modified and vertically-aligned CNTs-sensors has been investigated for monitoring landfill gas (LFG). This gas-mixture consists of two main gases such as carbon dioxide (CO_2) and methane (CH_4), including additional secondary gases such as ammonia (NH_3), hydrogen (H_2), hydrogen sulphide (H_2S), water vapor, and so on. Table 1 reports the gas response of the four CNTs-chemiresistors exposed to a multicomponent gas-mixture B1 constituted by 800 ppm CO_2, 700 ppm CH_4, 1,000 ppm H_2 and 400 ppm NH_3 in air, and mixture B2 constituted by 800 ppm CO_2, 700 ppm CH_4, 1,000 ppm H_2 and 200 ppm NH_3 in air. For a given CNT-sensor, the response decreases with the decreasing NH_3 gas concentration in the gas-mixture based on total reducing gases. Figure 6 shows the LFG detection at the typical level of hundreds of ppm using the four CNTs-sensors operating at the temperature of 150°C. It is found that the LFG response enhances specifically with metal-functionalizations of the vertically-aligned CNTs-films by catalytic effects compared to unmodified CNT-chemiresistor.

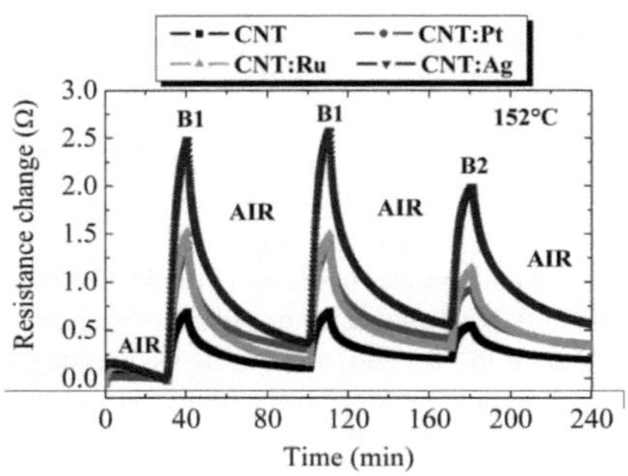

Figure 6. Time responses of four chemiresistors based on unmodified CNTs (CNT), and loaded with 5 nm of Pt (CNT:Pt), 5 nm of Ru (CNT:Ru) and 5 nm of Ag (CNT:Ag), exposed to 10-min pulses of a multicomponent gas-mixture B1 (800 ppm CO_2, 700 ppm CH_4, 1000 ppm H_2 and 400 ppm NH_3 in air) and B2 (800 ppm CO_2, 700 ppm CH_4, 1,000 ppm H_2 and 200 ppm NH_3 in air), at the sensor temperature of 150°C. The thickness of the unmodified CNTs films is 10 μm.

Table 1. Comparison of the gas response for the unmodified CNTs and Pt-, Ru- and Ag-modified CNTs sensors to a multicomponent gas-mixture B1 and B2, at 150°C.

SENSOR	Gas response, $\Delta R/R_i$ (%)		
	B1 1° Expo	B1 2° Expo	B2 1° Expo
CNT	0.86	0.65	0.39
CNT:Pt	1.45	1.17	0.50
CNT:Ru	1.03	0.89	0.55
CNT:Ag	1.53	1.36	0.88

4. Conclusions

The chemiresistors based on metal-modified and vertically-aligned CNTs-films are demonstrated to be very promising gas sensors and operating at low temperatures of 20–150°C. The metal-catalysts of Pt, Ru and Ag enhance and tune the gas sensitivity of the unmodified CNTs sensors with higher sensitivity, fast response, sub-ppm detection level, good reproducibility and reversibility. A practical application of monitoring landfill gas has been preliminary addressed. The multicomponent gas-mixture of LFG has been successfully detected by a array of four CNTs-sensors with enhanced gas sensitivity and broader selectivity for useful applications in environmental air-emission monitoring.

References

1. J. Kong, N. R. Franklin, C. Zhou, M. G. Chapline, S. Peng, K. Cho, H. Dai, *Science* **287**, 622–625 (2000).
2. L. Valentini, I. Armentano, J. M. Kenny, C. Cantalini, L. Lozzi, S. Santucci, *Appl. Phys. Lett.* **82**, 961–963 (2003).
3. T. Someya, J. Small, P. Kim, C. Nuckolls, J. T. Yardley, *Nano Lett.* **3(7)**, 877–881 (2003).
4. M. Penza, G. Cassano, R. Rossi, A. Rizzo, M. A. Signore, M. Alvisi, N. Lisi, E. Serra, R. Giorgi, *Appl. Phys. Lett.* **90**, 103101 (2007).
5. M. Penza, G. Cassano, R. Rossi, M. Alvisi, A. Rizzo, M. A. Signore, Th. Dikonimos, E. Serra, R. Giorgi, *Appl. Phys. Lett.* **90**, 173123 (2007).
6. J. Kong, M. G. Chapline, H. Dai, *Adv. Mater.* **13(18)**, 1384–1386 (2001).
7. M. Penza, R. Rossi, M. Alvisi, G. Cassano, M. A. Signore, E. Serra, R. Giorgi, *Sens. Actuators B* **135** (2008) 289–297.
8. M. Penza, R. Rossi, M. Alvisi, G. Cassano, M. A. Signore, E. Serra, R. Giorgi, *Journal of Sensors*, Vol. 2008, ID Article 107057, 8 pages.
9. M. Penza, R. Rossi, M. Alvisi, M. A. Signore, E. Serra, *Journal of Physics D: Applied Physics* **42** (2009) 072002.
10. M. Penza, R. Rossi, M. Alvisi, M. A. Signore, G. Cassano, D. Dimaio, R. Pentassuglia, E. Piscopiello, E. Serra, M. Falconieri, *Thin Solid Films* (2009). In press.

AMMONIA SENSING PROPERTIES OF ORGANIC INKS DEPOSITED ON FLEXIBLE SUBTRATES

A. ARENA[1], N. DONATO[1], G. SAITTA[1], G. RIZZO[2] AND G. NERI[2]

[1]*Dipartimento di Fisica della Materia e Ingegneria Elettronica, Università degli Studi di Messina, Italy*

[2]*Dipartimento di Chimica Industriale e Ingegneria dei Materiali, Università degli Studi di Messina, Italy*

Abstract. Highly conducting inks based on polyaniline (PANI), are obtained by doping emeraldine base with dodecylbenzene sulfonic acid (DBSA), in a mixture of chlorobenzene/water/propanol. Compact and homogeneous thin films developed by casting the inks on transparencies, are found to have sheet resistance in the 10 kΩ square range. Inexpensive devices with remarkable potentiality as disposable ammonia sensors, are developed by applying on transparencies linearly shaped carbon nanotubes electrodes, spaced by hundreds microns, coated with sensitive layers consisting of PANI-DBSA inks.

1. Introduction

In the last few years the interest towards innovative sensing systems operating at room temperature [1], and the widespread diffusion of low cost flexible electronics [2], have opened enormous perspectives in a variety of fields ranging from environmental analysis [3] to biomedical applications [4]. The idea to replace cumbersome and complex sensing systems with inexpensive and easy to handle reliable sensors has become an attractive applied research topic, and sometimes it has been at the basis of successful business, the most known example of which is the billion dollars market of disposable glucose sensors. Disposable electrochemical sensing systems allow people affected by diabetic disease to check daily the glucose level of their blood, through simple procedures. According to the market forecast, the field of flexible sensors suitable for safety, environmental, and health-care applications is expected to grow with a high rate in the next few years. In such a framework, this paper investigates about the possibility to develop inexpensive disposable ammonia sensors based on inks containing doped polyaniline, a well known conjugated polymer, applied on flexible substrates.

P. Malcovati et al. (eds.), *Sensors and Microsystems: AISEM 2009 Proceedings*,
Lecture Notes in Electrical Engineering 54, DOI 10.1007/978-90-481-3606-3_36,

2. Experimental

Emeraldine base, conducting polyaniline (2% in mixed solvents), dodecylbenzene sulfonate acid (70% 2-propanol solution), and hydroxypropylcellulose were supplied by Aldrich and used as received. MWCNTs were obtained from Bayer. Dispersions of emeraldine powder in chlorobenzene were mixed with DBSA, and water was slowly added to the mixture. After prolonged sonication at 60°, the viscous dark green solution that formed at the bottom of the flask was extracted, filtered and diluted with suitable amounts of chlorobenzene. The inks obtained following such a procedure were found to be stable, and could be drop deposited or spin-coated on a variety of substrates, in order to obtain films that change their color and their resistivity when exposed to ammonia. Flexible ammonia sensing devices were obtained using PANI-DBS inks deposited on highly conducting electrodes consisting of a MWCNTs ink. The MWCNTs ink was prepared by prolonged sonication of mixtures consisting of MWCNTs dispersed in ethanol/water solutions of hydroxyl propyl cellulose, The blends were filtrated using common filter paper, yielding dark homogenous conducting inks found to be stable for months. The inks were applied on glossy paper and transparencies by hand, in order to obtain ~2 mm wide parallel electrodes, spaced by few hundreds microns. The MWCNTs electrodes were coated with drops of PANI-DBSA inks, and, for comparison, with drops of a commercial doped polyaniline. The electrical characterization of the developed devices was performed at 24.5 ± 0.5°C, using an Agilent 34970A multimeter. The samples were contacted through a couple of point probes pressed on the electrodes with a contact area roughly estimated to be about 1 mm^2. The measurements were performed in air, under dynamic conditions, by recording the resistance of the devices, repeatedly exposed to ammonia for few seconds.

3. Results and discussion

The UV-VIS-NIR absorption spectrum of a diluted chlorobenzene solution of PANI-DBSA before and after addition of ammonia is shown in Fig. 1. The spectrum of PANI-DBSA is characterized by the presence of a main peak, positioned at about 740 nm, and two lower intensity structure centred at about 420 and 340 nm respectively. Addition of ammonia to the PANI-DBSA solution has a remarkable effect on the absorption, as the peak positioned at about 740 nm found to weaken, and to shift towards the red wavelength region as it is shown in Fig. 1. Figure 2 shows the UV-VIS-NIR absorption of a thin PANI-DBSA film deposited on a transparency, and washed with acetone to remove excess of DBSA. The spectrum agrees with those reported in literature for polyaniline in the highly conducting phase [5]. The film has a quite high transmittance, ranging between 85% and 90%, and sheet resistance of the order of 10 kΩ square.

PANI-DBSA ink is used to develop inexpensive flexible devices, as the example shown in the inset of Fig. 3, characterized by remarkable change of resistance in the presence of ammonia. The devices are deposited both on transparencies and on glossy paper substrates. They have linearly shaped electrodes, spaced by hundreds microns, obtained applying by hand the MWCNTs conducting ink prepared as described in the previous section, using standard refillable pens. The simple fabrication process is then accomplished by drop-depositing the developed sensitive ink on the top of the conducting electrodes. For comparison, flexible devices having MWCNTs electrodes have been developed also by using as sensing layer a commercially available doped polyaniline.

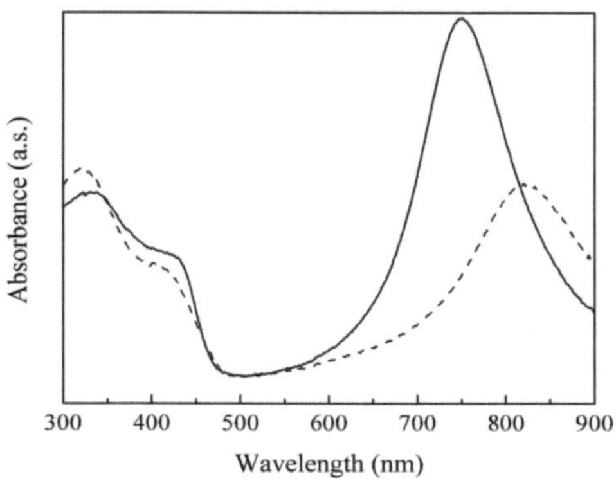

Figure 1. Absorbance of a dilute chlorobenzene solution of PANI-DBSA (solid) and of a the same solution after addition of ammonia (dashed).

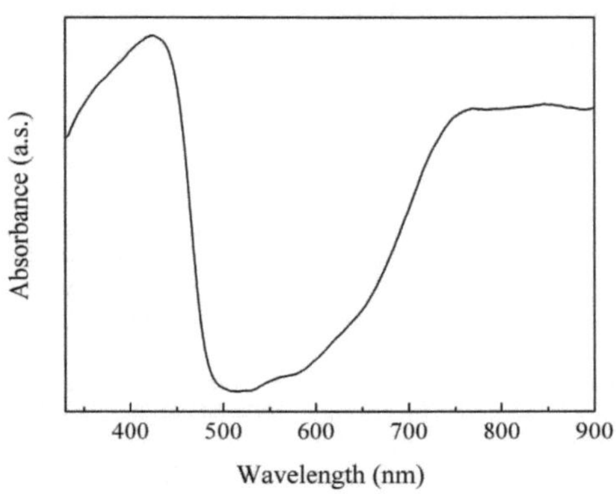

Figure 2. Absorbance of a thin PANI-DBSA film.

Preliminary measurements show that compared to commercially available conducting PANI, the PANI-DBSA developed by us has lower resistivity, higher stability, and is less affected by relative humidity and temperature changes. In addition, devices using PANI-DBSA exposed to abrupt changes in ammonia exposure have transient response time considerably shorter than devices using commercially available doped PANI, and higher sensitivity.

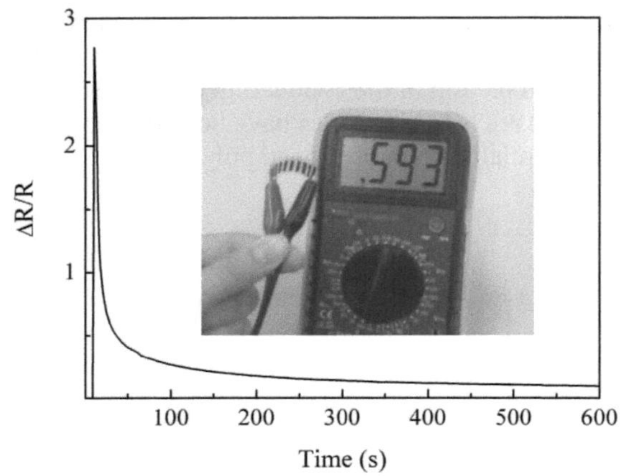

Figure 3. Normalized resistance change measured on a flexible MWCNT/PANI-DBSA/MWCNT device exposed for a few seconds to ammonia. The inset shows a sample consisting of five devices applied on a transparency.

4. Conclusion

A simple method is used to deposit highly conducting transparent polyaniline films. The material is found to be stable, and, when deposited on the top of MWCNT based electrodes has remarkable resistive response towards ammonia. These results are promising in view of obtaining inexpensive flexible ammonia sensors.

References

1. U. Yogeswaran, S.M. Chen, "A review on the electrochemical sensors and biosensors composed of nanowires as sensing material", Sensors vol. 8, pp. 290–313, 2008.
2. M.C. McAlpine , R.S. Friedman, S. Jin, K.H. Lin, W.U. Wang, C.M. Lieber "High-performance nanowire electronics and photonics on glass and plastic substrates", Nano Letters Vol. 3, pp. 1531–1535, 2003.
3. S. Dmitriev, G. Duca, I. Dementiev, A. Craciun, "Flexible substrate based gas sensors for air pollution monitoring", Flexible Electronics 2004-Materials and Device technology MRS Symposium Proceedings Vol. 814, pp. 169–174, 2004.
4. D. Feili, M. Schuettler, T. Doerge, S. Kammer, K.P. Hoffmann, P. Klaus, T. Stieglitz, "Flexible organic field effect transistors for biomedical microimplants using polyimide and parylene C as substrate and insulator layers", Journal of Micromechanics and Microengineering Vol. 16, pp. 1555–1561, 2006.
5. V. Deepak, V. Dutta, "Role of novel microstructure of polyaniline-CSA thin film in ammonia sensing at room temperature", Sensors and Actuators B: Chemical, Vol. 134 pp. 373–376, 2008.

PROSPECTIVE OF USING NANO-STRUCTURED HIGH PERFORMANCES SENSORS BASED ON POLYMER NANO-IMPRINTING TECHNOLOGY FOR CHEMICAL AND BIOMEDICAL APPLICATIONS

A. FERRARIO[1], A. DE TONI[2], L. BANDIERA[3] AND M. QUARTA[3]
[1]*Dipartimento di Scienze dei Materiali, via R. Cozzi 53, Università di Milano-Bicocca*
[2]*Dipartimento di Ingegneria dell'Informazione, via Granenigo 6/A, Università di Padova, e-mail: alessandro.detoni@dei.unipd.it*
[3]*Wetware Concepts S.r.l., via crocerossa 112, Padova*

Abstract. In recent years, the continuous development of nanotechnology manufacturing, like Nano Imprinting Lithography (N.I.L.), have permitted the design and the implementation of innovative devices based on chemical-physical properties of nano-structured surfaces. These nano-structured electronic devices have shown great ability to detect and quantify micro- and nano-molar concentrations of molecules, of great interest in biological, medical and chemical areas, by using electrochemical techniques. This paper shows the preliminary results on a new nano-structured sensor, obtained by an innovative manufacturing process based on a standard reliable disk (CD or DVD) manufacturing process.

1. Introduction

The success of biomedical devices critically depends on the ability to interact with the biological environment. The bioresponse is often determinated by the proprieties of the biointerface which requires the precise control of this interface on the micron- and nano- meter scale. Recent events in nanotechnologies have created a variety of top down approaches that can reach 100 nm, reaching a very relevant range in biology. Nowadays, typical detection approaches are micro-electronic based, and so, very expensive [1]. Our sensors are based on reliable and low cost CD/DVD technology. We have used our sensor to analyze the cow's milk toxicity, leading interesting results on the detection of the ultra-trace quantitative of Cu^{2+}. The first device we have realized is an electrochemical nano-structured sensor, implemented on a polymeric substrate by the N.I.L. technique [2, 3], realizing a perfectly organized array of gold nano dots. Such device has been characterized by electrochemical impedance spectroscopy (EIS) and cyclovoltammetry techniques. The device degradation and the analyte properties, as well as various redox systems has been studied in a wide frequency range and different analytes. Therefore, we can obtain information about the chemical-physical stability of nano-structured surface.

P. Malcovati et al. (eds.), *Sensors and Microsystems: AISEM 2009 Proceedings*, Lecture Notes in Electrical Engineering 54, DOI 10.1007/978-90-481-3606-3_37,

Finally, the results have been compared with the same obtained by conventional sensors. In particular, heavy metals and other analytes redox peaks have been measured with high precision and low noise. First studies have been focused on quantitative determination of Copper ion (Cu^{2+}) both in water and milk. The great importance of copper in biological processes (erythrocyte peroxidation, iron reserves mobilization, positive action on composition of scleroproteins and phospholipids of cellular membranes, etc), makes it the ideal analysis target for an application in biomedical area [4]. The analysis of Cu concentration is of particular importance because both high and low values can be symptomatologies of serious disturbances (high serum Cu concentrations in bovine and ovine species caused by molluschicides polluted water intoxication, low serum concentrations in calves with neuromuscular problems) [5].

2. Experimental

2.1. *The nano-structured sensor*

All the sensors used through this work are provided by Nanosims S.r.l.

These sensors are based on standard CD manufacturing device, with gold metal as outer layer. The nanostructured surface is made of a metal layer with isolated holes of about 250–500 nm dimensions; the hole to hole distance is around 600 nm.

2.2. *Preparation of the sensors*

In order to prepare the sensor, we first have defined the electrode area (Ø = 1.5 mm) with a glass cell. Then the gold surface has been treated: after cleaning with HF (2%) the surface was washed with the wash solution (280 ml of 6.25 M NaOH with 420 ml of 95% ethanol) for 1 h in an orbital shaker.

The sensors are finally rinsed by plunging them up and down in ddH_2O five times using fresh H_2O with each wash and finally dried by centrifuging for 1 min at 100 rpm.

2.3. *Electrochemical measurement*

All electrochemical measurements were performed with CHI 440A potentiostat coupled with 5.5 software (CH Instruments Inc., UK). The electrochemical cell consisted of an Ag/AgCl external reference electrode (3 M KCl), a platinum auxiliary electrode and the Au nano-structured sensor (see par. 2.1) as working electrode. The experiments were carried out at a room temperature (25°C).

Voltammetric behaviour was studied with Cow's milk, Milli-Q water and 0.5 M $CuSO_4$. A potential range from −0.6 to +0.8 V was investigated at a scan rate of 25, 50 and 100 mV/s. All the applications proposed in this paper were carried out in quiescent solution; a drop of the sample solution (300 µl) was cast in the sensor cell in contact the nano-structured area. Then the reference and counter electrodes were immersed in the cell.

3. Results

3.1. *I-V experimental data*

Preliminary measures on the nano-structured sensor were applied to detect trace of Cu^{2+} in contaminated milk.

The experiments are done in two different steps: initially, the sensor has been characterized with a water solution of Cu^{2+} ($1.52 \cdot 10^8$ ions/nl). Metal redox peaks are detected by cyclovoltammetry measures, giving us information about reduction and oxidation voltages as well as the proportionality between sensor electrical response and analyte concentration (Fig. 1, *dotted line*). Next, cow's milk has been intentionally contaminated with Cu^{2+} ($6 \cdot 10^8$ ions/nl): cyclic voltammetry experiments were performed in the same potential range (Fig. 1, *solid line*).

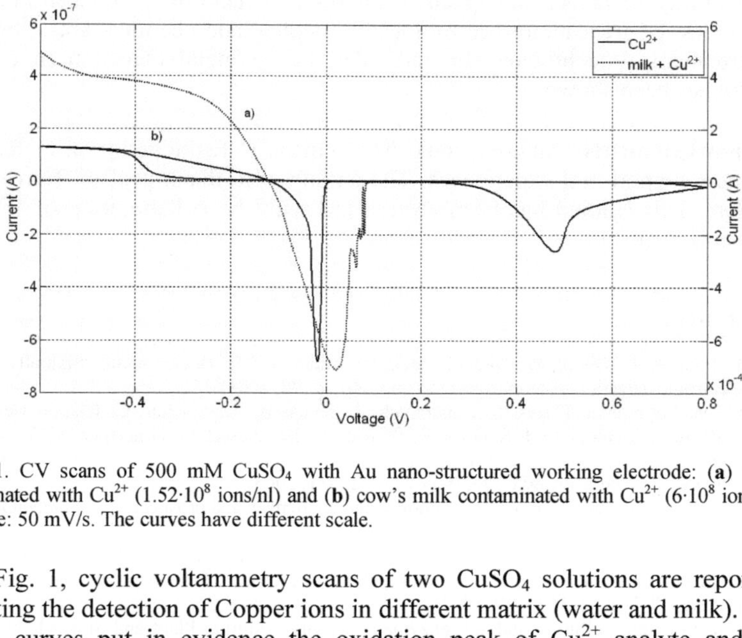

Figure 1. CV scans of 500 mM $CuSO_4$ with Au nano-structured working electrode: (**a**) water contaminated with Cu^{2+} ($1.52 \cdot 10^8$ ions/nl) and (**b**) cow's milk contaminated with Cu^{2+} ($6 \cdot 10^8$ ions/nl). Scan rate: 50 mV/s. The curves have different scale.

In Fig. 1, cyclic voltammetry scans of two $CuSO_4$ solutions are reported, exploiting the detection of Copper ions in different matrix (water and milk).

The curves put in evidence the oxidation peak of Cu^{2+} analyte and the proportionality between measured current and contaminant concentration. In particular, the dotted curve (Fig. 1a) represents the voltammogram of water contaminated with Cu^{2++} ($1.52 \cdot 10^8$ ions/nl). The negative current peak of −0.723 mA (right axis) shows the high sensitivity of the nano-structured gold surface, suggesting that this sensor enhances the reaction surface respect to the device area.

The detection of Cu^{2+} ions in milk ($6 \cdot 10^8$ ions/nl) is reported in Fig. 1b, where the oxidation peaks are compared. As can be seen, the peak potentials are nearly the same for the two matrix ($\Delta E_p \approx 32$ mV) while the two different peaks current are really different (−0.642 μA for milk matrix) indicating that the complex matrix strongly affects the redox reaction. However, the pronounced peak current of Fig. 1b exploits the great capability of the proposed nano-structured sensor to detect heavy metal species even in complex matrix, thanks to a good signal to noise ratio.

4. Conclusions

In this paper we have shown the preliminary results on a new nano-structured sensor based on Nano Imprinting Lithography (N.I.L.). This approach, obtained by an innovative manufacturing process based on a standard reliable disk (CD or DVD) manufacturing process, have permitted the design and the implementation of innovative devices based on chemical-physical properties of nano-structured surfaces. We have shown that these nano-structured electronic devices show great ability to detect and quantify micro- and nano-molar concentrations of molecules, of great interest in biological, medical and chemical areas, by using electrochemical techniques. In particular, heavy metal detection in complex matrix has been shown.

Acknowledgments Authors would like to thank F. Estini and prof. C. Reggiani for their support and experiences. The authors also thank prof. F. Naro for the support. This studied has been carried out with the collaboration of Wetware Concepts S.r.l.

References

1. F. Pouthas, C. Gentil, D. Cote, G. Zeck, B. Straub, and U. Bockelmann, "Spatially resolved electronic detection of biopolymers", *Phys. Rev. E*, 70, 2004.
2. C. M. Sotomayor Torres, S. Zankovych, J. Seekamp, A. P. Kam, C. Clavijo Cedeño, T. Hoffmann, J. Ahopelto, F Reuther, K. Pfeiffer, G. Bleidiessel, G. Gruetzner, M. V. Maximov and B. Heidari, "Nanoimprint lithography: an alternative nanofabrication approach", *Materials Science and Engineering: C*, 23, Issues 1–2, 23–31, 2003.
3. S Zankovych et al., "Nanoimprint lithography: challenges and prospects", *Nanotechnology*, 12, 91–95, 2001.
4. C. Branquinho, D. H. Brown and F. Catarino, "The cellular location of Cu in lichens and its effects on membrane integrity and chlorophyll fluorescence", *Environmental and Experimental Botany*, 38, Issue 2, 165–179, 1997.
5. "Heavy metal pollution and soil enzymatic activity", Journal Plant and Soil, 41, Number 2, 303–311, 1974.

SURFACE ACOUSTIC WAVE BIOSENSOR BASED ON A RECOMBINANT BOVINE ODORANT-BINDING PROTEIN

F. DI PIETRANTONIO[1], I. ZACCARI[1], M. BENETTI[1], D. CANNATÀ[1], E. VERONA[1], R. CRESCENZO[2], V. SCOGNAMIGLIO[2] AND S. D'AURIA[2]

[1] *"O.M.Corbino" Institute of Acoustics, CNR, via Del Fosso del Cavaliere 100-00133, Rome, Italy*

[2] *Institute of Protein Biochemistry, CNR, Via Pietro Castellino, 111-8013, Naples, Italy*

Abstract. In this work we present a Surface Acoustic Wave (SAW) biosensor system based on a recombinant bovine odorant-binding protein (bOBP) suitable to detect boletus odorant.

To optimize the bOBP's deposition method on the interdigital transducers (IDTs), the protein was deposited by means of droplet method on the surface of a 155 MHz 2-port SAW resonator on ST-cut quartz. In order to promote the protein adhesion, IDTs were made of thin gold film. The frequency shift of the resonator, after deposition, shows the presence of the protein on the IDT's surface. By using two identical 2-ports SAW resonators, operating at 392 MHz and exploiting gold/aluminum multilayered interdigital electrodes, a differential configuration was implemented. In this way, common mode parasitic effects, such as temperature variations have been reduced. The SAW resonators were used as frequency control element in a Pierce oscillator. The differential frequency shift, due to the SAW velocity changes as consequence of odorant concentration, was measured, showing a resolution of 44 ppb and a sensitivity of 314 Hz/ppm.

1. Introduction

Surface acoustic wave (SAW) devices are widely used in sensor applications in both physical and chemical fields [1]. In particular SAW biosensors [2, 3] are currently used in medicine, environmental monitoring, biotechnology, food industry and security applications. In the latter case SAW biosensors are very attractive as detectors of narcotics [4] and explosives.

In this work a SAW biosensor, able to measure small concentrations of 1-octen-3-ol, a boletus odorant, based on a recombinant bOBP [5] is presented. The 1-octen-3-ol is the natural ligand of bOBP [6].

Odorant Binding Proteins (OBPs) are a sub-class of Lipocalins, defined by their property of reversibly binding volatile chemicals that we call "odorants". They are a class of small proteins found in the nasal mucosa of several animal species. OBPs appear to play an important role in olfaction, although more than 20 years of studies were still not enough to clarify their biological role in the olfactory responses. The hypothetic roles remain transport of odorants to Olfactory Receptors, filters against high concentration of odorants in order to

P. Malcovati et al. (eds.), *Sensors and Microsystems: AISEM 2009 Proceedings*, Lecture Notes in Electrical Engineering 54, DOI 10.1007/978-90-481-3606-3_38,

avoid saturation of the OR and protectors of the nasal mucosa which are exposed to airflow and oxidative injuries, by binding cytotoxic and genotoxic molecules. The first member of this family was found in cow, but several other members were later found in other animals like rat, rabbit and pig as well as in humans [7]. OBPs are able to bind odorant molecules and, therefore, were deposited on the SAW transducer surface to implement a biosensor. The device sensing mechanism is related to change in SAW phase velocity, induced by the mass loading caused by interaction between the bOBP and the odorant molecules in air.

By using the bOBP as sensitive membrane we are able to measure the 1-octen-3-ol's concentration in air.

2. bOBP deposition

The bOBP is a protein composed of 159 aminoacids and a molecular weight of 19 kDa, it is a dimer with a symmetrical structure "up and down b-barrel" and with an elongated shape, this is in favour of his high flexibility in solution, each monomer contains a large buried cavity and there is a central pocket located at the dimer interface in communication with the solvent. This cavity might be used for odour transport or to bind different ligands.

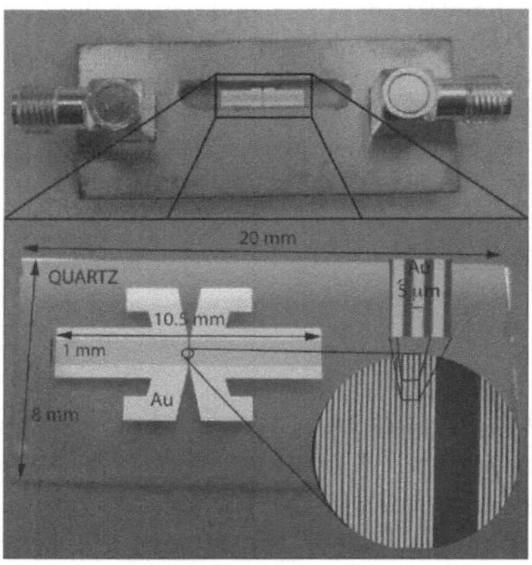

Figure 1. SAW device for deposition optimization.

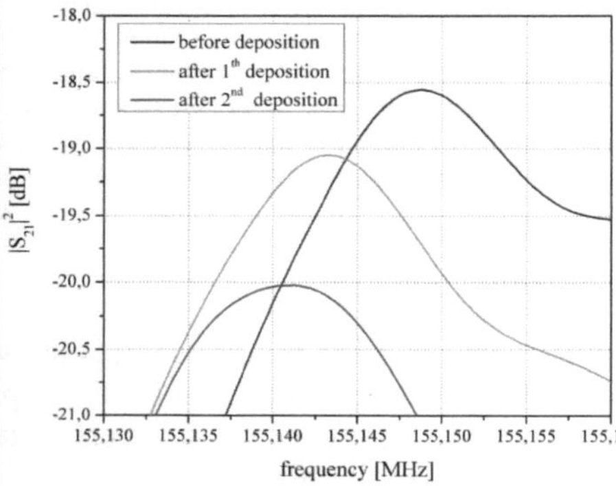

Figure 2. 2-port SAW resonator's frequency response before and after bOBP deposition.

The bOBP was deposited on the surface of the SAW device by means of droplet method. This method was optimized using a 155 MHz 2-port SAW resonator on ST-cut quartz. To promote the protein adhesion, the IDTs were made of thin gold film. SAW resonators were mounted on a PCB and connected to two SMA connectors (Fig. 1). Measurements were performed by an HP8753A Network Analyzer and data acquired by a PC, using National Instruments Labview software. The frequency shift in the resonator response (S_{21}) after deposition shows the presence of the protein on the IDTs surface as reported in Fig. 2, where the SAW resonator transfer function before and after two depositions of bOBP is shown. The frequency shift after a deposition of 20 μl of bOBP (c = 10 mM) is about 5 kHz.

3. SAW biosensor system

Odorant molecules concentration was measured by means of a differential configuration to reduce parasitic phenomena due to temperature variations. The SAW elements are two identical 392 MHz 2-ports resonators consisted of multilayered IDTs, fabricated by using a 5 nm gold film on a 100 nm aluminum film and mounted on a TO39 package. The bOBP was deposited on a SAW resonator by using the technique described previously while a second uncoated resonator was used as reference. SAW resonators were used as frequency-control element in the feedback branch of Pierce oscillator circuits. Two matching networks create the phase shift to fulfill the phase condition on the oscillator loop. A tunable capacitor allows to fine adjust the oscillation frequency to the serial resonance frequency of the SAW device [8].

4. Experimental results

The concentration of 1-octen-3-ol in a particular weight of boletus was measured by using gas chromatography technique [9].

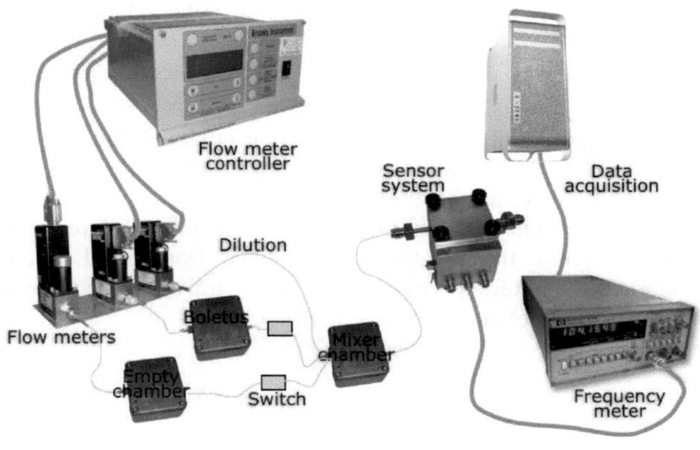

Figure 3. SAW sensor measurements setup.

To perform SAW sensor measurements, dried boletus was placed in a chamber, where pure nitrogen was fluxed. The odorant flux was diluted by a further nitrogen flux to obtain different concentrations of 1-octen-3-ol. The mixture was carried to the SAW sensor placed in a measure chamber. The differential frequency shift, due to the SAW velocity changes as consequence of odorant concentration, was measured by a frequency meter and data monitored and acquired by a PC. In Fig. 3 the experimental setup is shown.

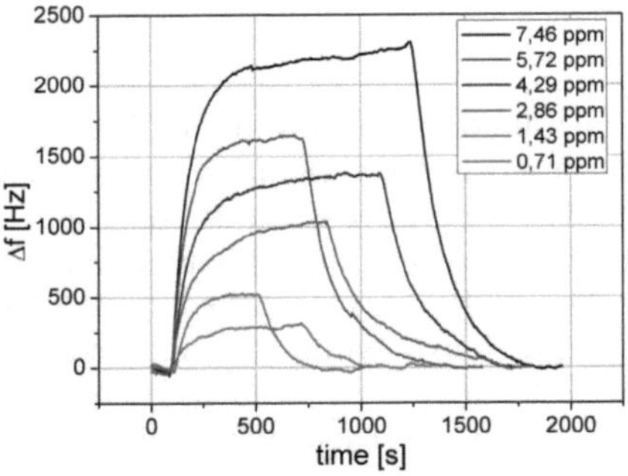

Figure 4. SAW sensor time responses for six different 1-octen-3-ol's concentrations.

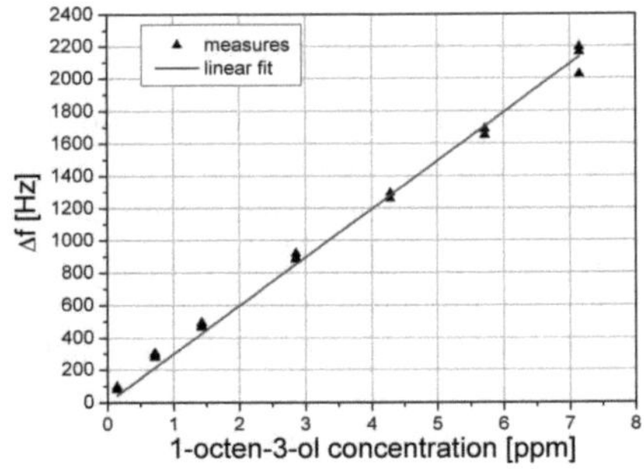

Figure 5. SAW sensor response curve.

Real time monitoring of protein-odorant interaction was performed as suggested by SAW biosensor time response. In Fig. 4 SAW sensor time responses for six different 1-octen-3-ol concentrations are reported.

To obtain the response curve, the frequency shift of SAW biosensor system at different 1-octen-3-ol concentrations was measured. In Fig. 5 the SAW sensor frequency response for 1-octen-3-ol concentrations up to 8 ppm is reported. A resolution of about 44 ppb and a sensitivity of about 314 Hz/ppm were obtained.

5. Conclusions

In conclusion the possibility to deposit bOBP on gold/aluminum multilayered IDTs on ST-cut quartz was demonstrated by using an ad hoc designed SAW device and measuring its transfer function frequency shift by Network Analyzer. A SAW biosensor system based on the chemical interaction between the bOBP and its natural ligand, the 1-octen-3-ol, obtained from dried boletus, was designed, fabricated and tested. The sensor system showed a resolution of about 44 ppb and a sensitivity of about 314 Hz/ppm.

References

1. M. Benetti, D. Cannatà, A. D'Amico, F. Di Pietrantonio, A. Macagnano, and E. Verona, IEEE Sensors Proc., Vienna, pp. 753–756, 24–27 Oct. 2004.
2. S. Shiokawa and J. Kondoh, "Surface acoustic wave sensors", Japanese Journal of Applied Physics, [Vol. 43], pp. 2799–2802, 2004.
3. K. Lange, B. E. Rapp, and M. Rapp, "Surface acoustic wave biosensors: a review", Analytical and Bioanalytical Chemistry [Vol. 391], pp. 1509–1519, 2008.
4. D. D. Stubbs', S. –H. Lee, and W. D. Hunt, "Cocaine detection Using Surface Acoustic Wave Immunoassay Sensors", 2002 IEEE International Frequency Control Symposium and PDA Exhibition.

5. S. D'Auria, M. Staiano, A. Varriale, M. Gonnelli, A. Marabotti, M. Rossi, and G. B. Strambini, "The tryptophan phosphorescence of porcine and mutant bovine odorant-binding proteins: a probe for the local protein structure and dynamics", J Proteome Res. [Vol. 7], pp. 1151–1158, 2008.
6. R. Ramoni, F. Vincent, S. Grolli, V. Conti, C. Malosse, F. D. Boyer, P. Nagnan-Le Meillour, S. Spinelli, C. Cambillau, and M. Tegoni, "The insect attractant 1-octen-3-ol is the natural ligand of bovine odorant-binding protein", The Journal of biological chemistry [Vol. 276], pp. 7150–7155, 2001.
7. A. Marabotti, M. Scirè, R. Staiano, V. Crescenzo, F. Aurilia, Tanfani and Sabato D'Auria, Journal of Proteome Research 2008 Oct 28. [Epub ahead of print].
8. M. Benetti, D. Cannata, F. Di Pietrantonio, C. Marchiori, P. Persichetti, and E. Verona, IEEE Sensors Proc., Lecce (26–29 Oct. 2008), 1024–1027.
9. Bogus law Buszewski a, Agnieszka Ulanowska, Tomasz Ligor, Marek Jackowski, Ewa K lodzinska, Jacek Szeliga. "Identification of volatile organic compounds secreted from cancer tissues and bacterial cultures", Journal of Chromatography B, [Vol. 868], 88–94, 2008.

DEVELOPMENT OF AN APTAMER-BASED ELECTROCHEMICAL SANDWICH ASSAY FOR THE DETECTION OF A CLINICAL BIOMARKER

S. CENTI, S. TOMBELLI, I. PALCHETTI AND M. MASCINI

Dipartimento di Chimica, Università degli Studi di Firenze, Via della Lastruccia 3, 50019 Sesto Fiorentino, Firenze, Italia,e-mail: sonia.centi@unifi.it

Abstract. A disposable electrochemical assay involving magnetic particles and carbon-based screen-printed electrodes (SPEs) is developed for the detection of C Reactive Protein (CRP). CRP is a plasma protein and is among the most expressed proteins in acute phase inflammation cases, being a known biomarker for inflammatory states. The assay is based on a sandwich format in which a biotinylated RNA aptamer is immobilised onto streptavidin coated magnetic beads and then coupled to the same biotinylated aptamer. A streptavidin-alkaline phosphatase solution is then added to beads and after the recognition biotin-streptavidin, the modified magnetic beads are captured by a magnet on the surface of a graphite working electrode and the electrochemical detection is thus achieved through the addition of the AP substrate (α-naphthyl-phosphate) and α-naphthol produced during the enzymatic reaction is detected using Differential Pulse Voltammetry (DPV). The performance of the assay in terms of sensitivity, reproducibility and selectivity are studied in buffer. The detection limit (LOD) is 2.0·10-2 mg/L and the average coefficient of variation (CV) results 9%. The LOD found is comparable with that reported by ELISA and it was much lower than the clinically useful borderline value (8 mg/L).

1. Introduction

Detection and quantification of C-reactive protein (CRP) in an easy, cheap, and fast way can improve clinical diagnostics in order to prevent serious inflammatory states. CRP is a protein present in plasma and is one of the most expressed proteins in acute phase inflammation cases, being a known biomarker for inflammatory states.

In this paper an electrochemical aptamer-based assay is proposed as a screening device for the detection of CRP. The proposed approach use disposable screen-printed electrodes as transducers [1–3].

2. Materials and methods

Electrochemical measurements were performed using a μAutolab type II PGSTAT with a GPES 4.9 software package (Metrohm, Italy). All measurements were carried out at room temperature by using Differential Pulse

P. Malcovati et al. (eds.), *Sensors and Microsystems: AISEM 2009 Proceedings*, Lecture Notes in Electrical Engineering 54, DOI 10.1007/978-90-481-3606-3_39,

Voltammetry (DPV) with the following parameters: range potential 0/+600 mV, step potential 7 mV, modulation amplitude 70 mV, standby potential 200 mV, interval time 0.1 s. Carbon screen-printed electrochemical sensors were printed in house using a High Performance Multi Purpose Precision Screen Printer DEK 245 (DEK, Weymouth, UK), using different thermoplastic inks.

The CRP aptamer used in this study is a 44-mer RNA aptamer with sequence 5'-GCCUGUAAGGUGGUCGGUGUGGCGAGUGUGUUAGGAGAGAUU G C-3'

The RNA aptamer biotinylated in 5' was purchased by IBA (Germany) with a spacer arm, a TEG (triethylene glycol) tail. Moreover, to enhance its stability, the aptamer was modified with F-pyrimidines. Purified Human C Reactive Protein (CRP) was purchased by Exbio (Praha, Czech Republic). Streptavidin magnetic beads (1.05 μm) were from Dynal Biotech (Milan, Italy).

A competitive assay was developed. Magnetic beads coated with streptavidin were modified by immobilisation of biotinylated aptamer. The modified beads were incubated with standard solutions of CRP and then with the same biotinylated aptamer. After molecular recognition, the extent of the affinity reaction was evaluated by addition of a enzymatic substrate, which was transformed in a electroactive product and oxidized at the electrode surface using a useful potential value.

3. Procedure

3.1. *Beads preparation and aptamer immobilisation*

All the reactions were performed at room temperature and the washing, coating and competition steps were carried out under delicate stirring in the sample mixer. The magnetic beads coated with streptavidin were 5 mM Tris-HCl, 0.5 mM EDTA, 1 M NaCl and 0.1% of Tween 20 pH 7.5 (Buffer A). A suspension of 50 μL of magnetic beads was introduced in a tube containing 250 μL of aptamer solution 0.1 μM prepared in 5 mM Tris-HCl, 0.5 mM EDTA and 1 M NaCl pH 7.5 (Buffer B). After 15 min of incubation time, the tube was positioned on a magnet holding block to allow the magnetic separation of the beads; the supernatant was then removed and the beads were washed twice with 500 μL of buffer B. Each washing step consisted of a re-suspension of the beads in the washing solution for 2 min, followed by the separation with the magnet holding block to remove the supernatant. The beads were re-suspended in 500 μL of biotin solution 800 mg/L in order to saturate the sites of streptavidin which have not bound to the biotinylated aptamer. The beads were then washed with buffer B and re-suspended in 500 μL of buffer B.

3.2. *Affinity reaction on magnetic beads and electrochemical measurement*

Addition of the analyte: 50 μL of suspension containing aptamer-coated beads mixed with 450 μL of solution in 10 mM HEPES pH 7.4 containing 2 mM $CaCl_2$ (Buffer C) containing CRP at different concentrations in the range 0–100 mg/L. After 15 min of incubation time, the beads were magnetically separated to remove the supernatant and then washed twice using buffer C.

Addition of the secondary ligand: the beads were re-suspended in 500 μL of a solution of aptamer 0.1 mg/L in buffer C for 15 min and then separated and washed twice using buffer C.

Addition of the Enzyme-conjugate: The beads carrying the affinity complex (aptamer-analyte-aptamer) were re-suspended in 500 μL of streptavidin-alkaline phosphatase conjugated (0.2 U/mL) in buffer D containing 0.5% (w/v) of casein for 10 min. After separation and washing, the beads were re-suspended in 50 μL of buffer D.

Electrochemical detection: 10 μL of the beads suspension were transferred onto the surface of the working electrode. To better localize the beads onto the electrode, the magnetic block was placed on the bottom of the electrode. Then 60 μL of a solution containing the enzymatic substrate (α-naphthyl phosphate) 1 mg/mL in buffer D were deposited on the screen-printed strip. After 5 min, the enzymatic product was determined by DPV.

4. Results and discussion

4.1. *Calibration curve*

The sandwich assay was performed using some CRP standard solutions in the concentration range 0–1,000 mg/L (Fig. 1). The assay was repeated in order to evaluate the reproducibility; at this purpose, five repetitions of each standard solution were carried out. The average coefficient of variation was 9%, calculated as mean of all the concentrations considered. The limit of detection (LOD) of the assay was evaluated as minimum detectable concentration and was calculated by the evaluation of the average response of the blank plus three times the standard deviation; in this case the recorded blank signal was 0.2 ± 0.02 μA, leading to a LOD of $2.0 \cdot 10^{-2}$ mg/L.

To test the specificity of the sandwich assay, a solution of hIgG at physiological concentration (10,000 mg/L) was used and the recorded response was compared with that measured for CRP in the same conditions. This experiment demonstrated the good specificity of the assay since a high signal was obtained only when the specific protein (CRP) was tested, whereas the signal recorded in presence of hIgG resulted as 7% of the specific signal (data not shown). Moreover, the low signals obtained with non-specific IgG showed that the modification of the aptamer with 2'-F pyrimidines to enhance its stability did not affect its specificity.

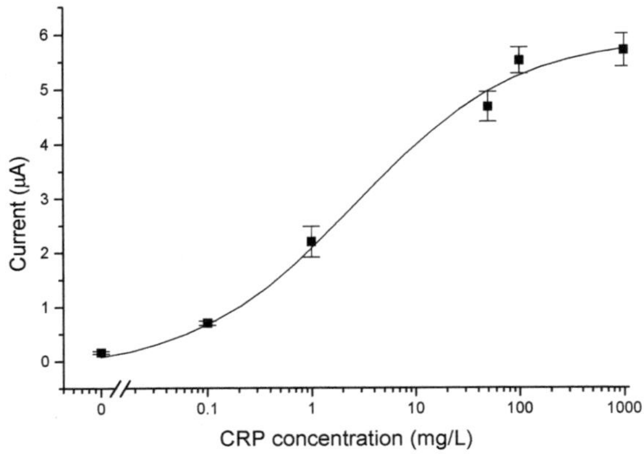

Figure 1. Dose–response curve for CRP The points correspond to the current ± SD calculated for n = 5 repetitions.

5. Conclusions

The attractive aspect of this work is the combination of the advantages taken from aptamer-based assays, magnetic beads separation and electrochemical transduction for the development of an assay for the detection of CRP. The assay exhibited a good sensitivity ($2.0 \cdot 10^{-2}$ mg/L), reproducibility (CV 9%) and selectivity.

Acknowledgments This work was partially funded by the "Healthcare by Biosensor Measurements and Networking" (CARE-MAN) (NMP4-CT-2006-017333) research project supported by the European Commission.

References

1. A. Bini, S. Centi, S. Tombelli, M. Minunni, M. Mascini, Development of an optical RNA-based aptasensor for C-reactive protein, *Analytical and Bioanalytical Chemistry*, 390, 1077–1086, 2008.
2. J. Pultar, U. Sauer, P. Domnanich, C. Preininger, Aptamer–antibody on-chip sandwich immunoassay for detection of CRP in spiked serum, *Biosensors and Bioelectronics*, 24,1456–1461, 2008.
3. C. Burtis, E. Ashwood, D. Bruns, *Textbook of Clinical Chemistry and Molecular Diagnostics*, 4th ed., Elsevier Saunders, USA, 2006.

DETERMINATION OF ETHANOL IN LEADLESS PETROLS AND BIOFUELS USING AN INNOVATIVE ORGANIC PHASE ENZYME ELECTRODE (OPEE)

L. CAMPANELLA, G. S. CAPESCIOTTI, T. GATTA AND M. TOMASSETTI
Department of Chemistry, University of Rome La Sapienza, P.le A. Moro, 5, 00185, Rome, Italy

Abstract. In the present research a new biosensor for the determination of the ethanol content in organic liquids or solvents, such as leadless petrols and biofuels was fabricated. The new biosensor is also quite peculiar: it is a 'substrate competition' OPEE (Organic Phase Enzyme Electrode), in which the enzyme, catalase, immobilized in kappa-carrageenan gel, is coupled to an amperometric gaseous diffusion Clark type oxygen electrode.

1. Introduction

An innovative biosensor, working in organic solvent, was developed for ethanol determination in leadless petrol and biofuels. This innovative biosensor is based on two parallel oxidation reactions, both of which catalysed by the same enzyme in the presence of the same hydroperoxide (tert-butylhydroperoxide or cumene hydroperoxide). In the first reaction the catalase enzyme in decane catalyses an oxidation reaction in the presence of an hydroperoxide, which produces a variation in the dissolved oxygen concentration. In this case (RH) is the used organic solvent (i.e. decane) [1] and t-buOOH the tert-butylhydroperoxide.

$$\text{t-buOOH} + O_2 + 2RH \text{ ---- } \xrightarrow{\text{catalase}} \text{t-buOH} + 2RO + H_2O \quad (1)$$

In the second reaction, catalase catalyses a reaction in which the hydroperoxide oxidizes the ethanol (the analyte to be determined) to acetaldehyde, but no change in dissolved oxygen concentration occurs.

$$\text{t-buOOH} + CH_3CH_2OH \text{ ---- } \xrightarrow{\text{catalase}} \text{t-buOH} + CH_3CHO + H_2O \quad (2)$$

As this second reaction competes with the first one for the substrate, this gives rise to a partial restoration of the initial O_2 concentration since the hydroperoxide present is now taking part in two reactions simultaneously; this therefore slows down the rate at which O_2 concentration varies under the effect of the first reaction alone. The extent of this "restoration" is measured using the Clark electrode and can be linked to the concentration of the ethanol present by constructing a suitable calibration straight line.

P. Malcovati et al. (eds.), *Sensors and Microsystems: AISEM 2009 Proceedings*, Lecture Notes in Electrical Engineering 54, DOI 10.1007/978-90-481-3606-3_40,

2. Operating conditions

The used solvent (i.e. decane) is involved in the enzymatic reaction underlying the method, the choice of solvent is of decisive importance. Decane has a low dielectric constant and is highly hydrophobic, and therefore denatures the enzymes only to a slight degree. When immersed in this solvent, the latter therefore retain most of their catalytic activity and an analysis of the experimental curves (catalase OPEE response toward hydroperoxide concentration) (Figs. 1 and 2) indicates that the highest response is obtained using a final tert-butylhydroperoxide concentration of about 0.61 mmol L^{-1} or of cumene hydroperoxide of about 5.9 mmol L^{-1} (which were thus chosen to perform all subsequent measures).

3. Results

3.1. *Analytical characteristic of the catalase biosensor to measure ethanol*

As far as the response of the bienzymatic method for measuring ethanol is concerned, the use of tert-butylhydroperoxide ensures greater sensitivity (see Figs. 1 and 2), a comparable wide linearity range, a lower limit of detection (Fig. 3), lastly much faster measurement times (see Fig. 4). In view of these conclusions it was decided to use tert-butylhydroperoxide in all subsequent experiments.

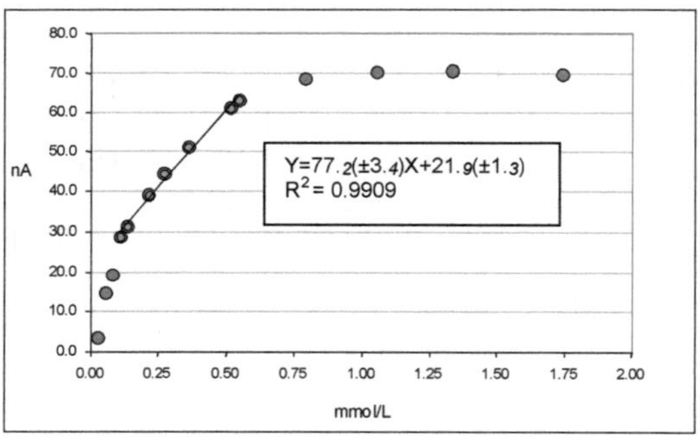

Figure 1. Response of catalase OPEE increasing tert-butylhydroperoxide concentration and linear range, using decane as solvent.

The analytical characteristics of ethanol determination are summarized in Table 1.

Also the main statistical parameters of the method involving tert-butylhydroperoxide are summarized in the same table.

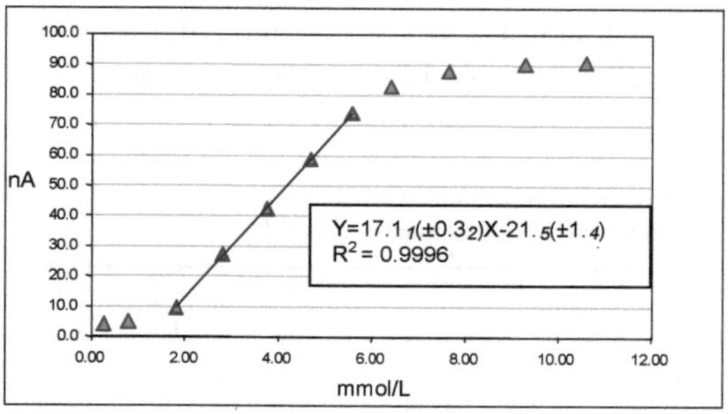

Figure 2. Response of catalase OPEE increasing cumene hydroperoxide concentration and linear range, using decane as solvent.

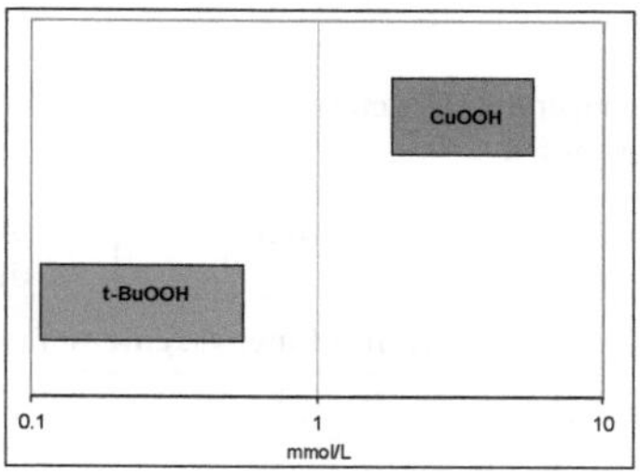

Figure 3. Comparison of linearity ranges of the two hydroperoxides using a logarithmic scale.

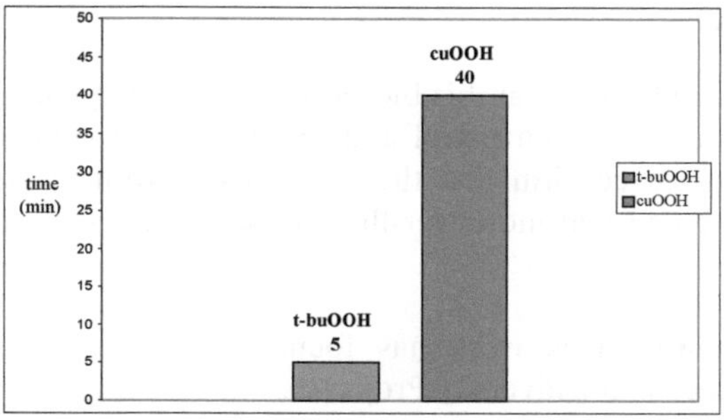

Figure 4. Comparison of response time of catalase OPEE to two different hydroperoxides using decane as solvent.

3.2. *Determination of ethanol concentration in leadless petrol samples and biofuel using the catalase biosensor*

Thus optimized and characterized, the catalase biosensor was used to determine ethanol concentration in leadless petrol samples purchased in random fashion from normal service stations belonging to two different distributors: IP and ESSO and in a biofuel sample (see Table 2).

The concentration values obtained were very low in leadless samples and much higher in biofuel sample, in keeping with the fact that the presence of ethanol in current leadless petrols is very low, at "trace" level as stated in the literature.

Table 1. Main analytical data and repeatability of the biosensor method applied to standard ethanol solutions, using tert-butylhydroperoxide and decane as solvent.

Linearity range	4.2×10^{-2}–3.7×10^{-1} (mmol L^{-1})
LOD	0.02 (mmol L^{-1})
Standard deviation	0.0013
RSD%	1.17%

Table 2. Results of ethanol concentration determined in two different leadless petrol and biofuel samples using the catalase biosensors.

Analyzed sample using catalase biosensor (n = 5)	**Found ethanol**	**Standard deviation**	**RSD%**
	(unit of measurements) (v/v)%		
IP	0.0033	±0.0003	7.8
ESSO	0.0025	±0.0005	18.8
BIOFUEL	0.057	±0.0030	5.3

4. Conclusions

The results obtained show that this biosensor method may be used to determine the ethanol content of green petrol and biofuel. The analytical data obtained using these samples confirm that the measures performed are comparatively rapid and precise and that moreover the real samples tested do not need to be pretreated.

Acknowledgements This work has founded by University of Rome "La Sapienza", "Ateneo and University Projects".

References

1. L. Campanella, G.S. Capesciotti, M.V. Russo, M. Tomassetti, "Study of the catalytic mechanism of the enzyme catalase on organic hydroperoxides in non-polar organic solvent", Current Enzyme Inhibition, [4], pp. 86–92, 2008.

IMMUNOSENSORS FOR THE DIRECT DETERMINATION OF PROTEINS: LACTOFERRIN AND HIgG

L. CAMPANELLA, E. MARTINI AND M. TOMASSETTI

Department of Chemistry, University of Rome "La Sapienza" P.le Aldo Moro 5, 00185, Rome, Italy

Abstract. New immunosensors for the direct determination of proteins, i.e. lactoferrin and human immunoglobulins G (HIgG), were developed. Direct measurement does not need any competition procedure; nevertheless, an enzymatic marker was used. For all the immunosensors developed, a systematic study was made to ensure their complete analytical characterisation using standard solutions of lactoferrin or HIgG, respectively. A comparison by the classical "competition" ELISA type immunosensors, studied in previous works, is also suggested.

1. Introduction

The immunosensors for direct measurement already described in the literature, that is, which do not involve any "competition" procedure and generally do not make use of a marker, and in which the signal is often obtained directly as a result of immunocomplex formation (which may, for example, be a cause of potential variation in the membrane) have generally not been very successful for various reasons even though they afford a considerable reduction in analysis time. Nevertheless, we recently resumed studies on this type of immunosensor as part of research conducted on new electrochemical immunosensors developed for the analysis of several proteins contained in serum (Human immunoglobulin G) or milk (lactoferrin) [1–3], designed to perform measurements using different operating schemes. In the latter case, however, unlike what is usually reported in the literature for this type of immunosensor, an enzymatic marker was again used to perform the electrochemical measurement whether the transducer used was of the amperometric or of the potentiometric type.

2. Methods

For the purpose of direct measurement of a real sample, completely innovative procedures were therefore developed. A one-off experimental measure was performed of the concentration of the labelled antigen or antibody required to form the complex conjugated with the respective antibody or antigen, which was immobilized on an ad hoc membrane (Immobilon). The other measurements were

P. Malcovati et al. (eds.), *Sensors and Microsystems: AISEM 2009 Proceedings*, Lecture Notes in Electrical Engineering 54, DOI 10.1007/978-90-481-3606-3_41,

then performed by conjugating the antigen to be measured, contained in the real sample, with the antibody (immobilized or in solution). Lastly, the conjugation between immobilized antibody and a fixed quantity of labelled antigen, or else between the antigen immobilized on the membrane and a fixed quantity of labelled antibody, was "completed." Each final enzymatic measurement was then performed by adding the specific substrate after probe separation from the solution and washing. Lastly, the signal was obtained which would be used to enter the calibration curve by taking the difference between the signal recorded during the one-off measurement and that obtained each time by the procedure described above. In particular, in the case of immunosensors used for lactoferrin determination by this procedure, two different types of amperometric electrochemical transducers were used alternately, either the hydrogen peroxide electrode or the Clark type electrode, while the enzyme marker used was always peroxidase (HRP); the conjugation with lactoferrin was obtained via a biotinylation method. On the other hand, a gaseous diffusion potentiometric transducer for NH_3 was always used to assemble the immunosensors for HIgG (immunoglobulins), while in this case the enzyme marker used was always urease, the conjugate of which with the antigen is available on the market. Lastly, the direct procedure for the measurement of HIgG may be considered a slight variation of the one described above.

3. Results

The two direct procedures used for the measurement of lactoferrin and HIgG in real samples are shown respectively in Figs. 1 and 2, while the relative calibration curves for the lactoferrin and HIgG determinations are shown in Figs. 3–5, respectively.

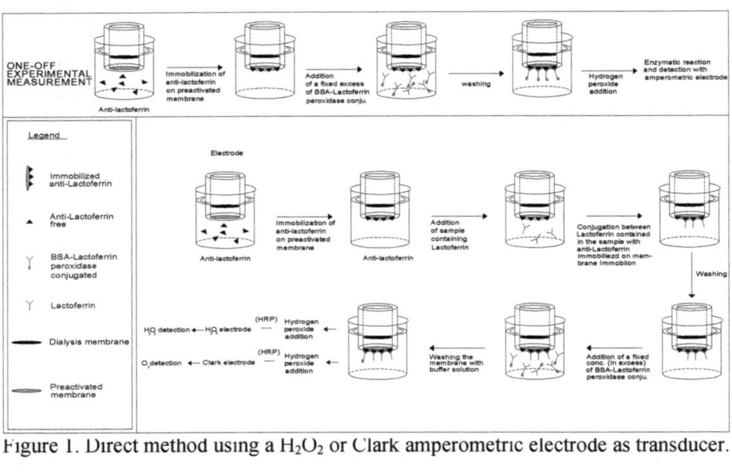

Figure 1. Direct method using a H_2O_2 or Clark amperometric electrode as transducer.

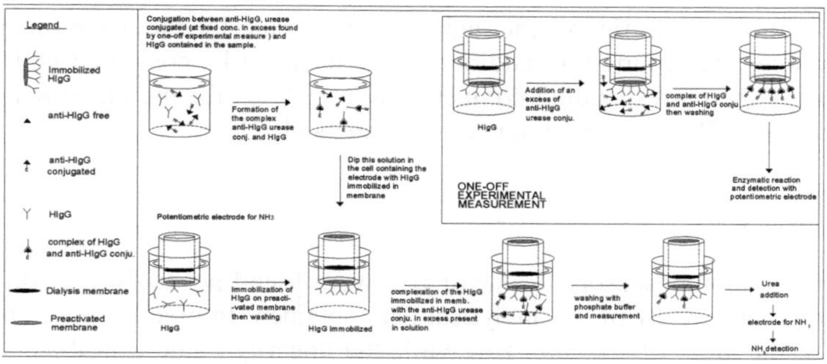

Figure 2. Direct method using a NH_3 potentiometric electrode as transducer.

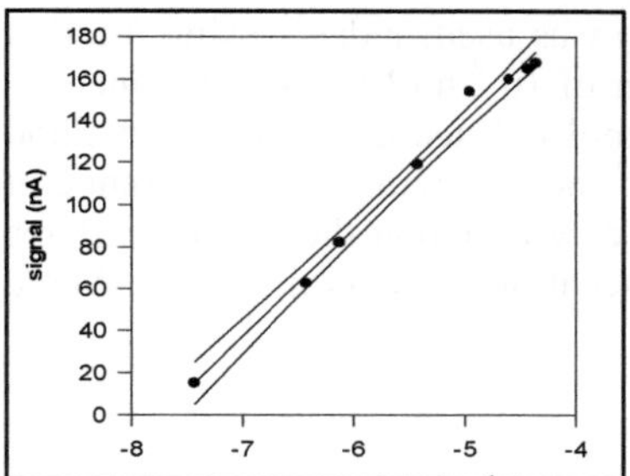

Figure 3. Direct method: calibration curve and confidence interval for the lactoferrin determination, plotted in a semilogarithmic scale, obtained using an immunosensor assembled using an Immobilon membrane and an amperometric electrode for H_2O_2 as transducer.

$Y = +51.4\ (\pm 5.24) \log X + 396.1\ (\pm 38.2)$
Linear range = $4.0 \times 10^{-5} - 3.5 \times 10^{-8}$ mol L^{-1}
LOD = 1.8×10^{-8} mol L^{-1}
pooled SD% = 6.5

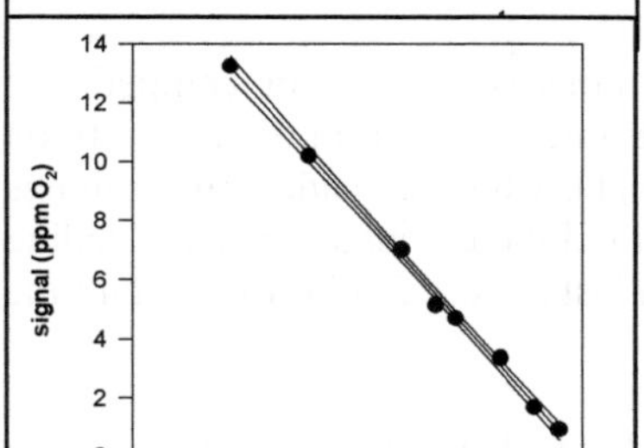

Figure 4. Direct method: Calibration curve and confidence interval for the lactoferrin determination, plotted in a semilogarithmic scale, obtained using an immunosensor assembled using an Immobilon membrane and a Clark electrode as transducer.

$Y = -4.22\ (\pm 0.98) \log X - 16.9\ (\pm 2.2)$
Linear range = $4.0 \times 10^{-5} - 7.0 \times 10^{-8}$ mol L^{-1}
LOD = 3.8×10^{-8} mol L^{-1}
pooled SD% = 6.1

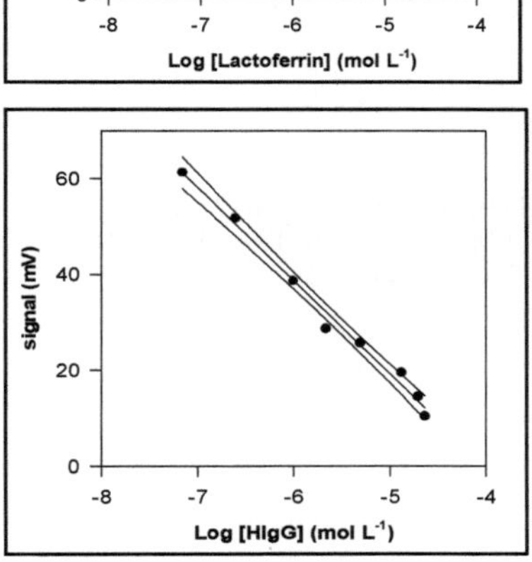

Figure 5. Direct method: Calibration curve and confidence interval for HIgG determination plotted in a semilogarithmic scale, obtained using an immunosensor, assembled using an Immobilion membrane and a potentiometric gas-diffusion sensor for NH_3 as transducer.

$Y = -19.4\ (\pm 3.1)\ X - 77.9\ (\pm 13.4)$
Linear range = $2 \times 10^{-5} - 7.0 \times 10^{-8}$ mol L^{-1}
LOD = 3.7×10^{-8} mol L^{-1}
pooled SD% = 6.8

Lastly, in the same figures the main analytical data for the determination of these proteins are reported. In addition, these data can be compared with those found using the corresponding competition procedures for the measurement of lactoferrin and HIgG, reported in previous papers [1, 2].

4. Discussion

The various immunosensors for lactoferrin were found not to be very different from the point of view of the analytical data obtained. In practice the LOD was always of the order of 10^{-8} mol L^{-1}, while the linearity range was about 10^{-5}–5×10^{-8} mol L^{-1}, whatever the transducer or the method (direct or competition procedure, see [1]) used. Conversely, in the case of immunosensors for the HIgG, the analytical results obtained varied considerably according to whether the direct or the competition method (see [2]) was used: indeed, using the first method, the LOD was found to be of the order of 10^{-8} mol L^{-1} and the linearity range was between about 10^{-5} and 5×10^{-8} mol L^{-1}; using the second method the LOD was found to be of the order of 10^{-12} mol L^{-1} with a linearity range of about 5×10^{-11}–5×10^{-12} mol L^{-1}. In HIgG determination the two procedures thus allowed two comparatively different concentration intervals to be covered in which to perform the measurements.

5. Conclusion

Of course the two different operating procedures (i.e. direct or competition) involve above all very different measurement times. Indeed, in the competition method, the competition step lasts about 1 h [1], which considerably prolongs the total analysis time, while in the direct method the conjugation steps leading to the formation of the immuno-complex each last about 15 min, for a total time of about 30 min.

Acknowledgments This work was funded by University of Rome "La Sapienza", "Ateneo and University Projects."

References

1. Campanella L, Martini E, Tomassetti M, J. Pharm. Biomed. Anal., 2008, 48(2), 278–287.
2. Campanella L, Attioli R, Colapicchioni C, Tomassetti M, Sens. Act. B, 1999, 55, 23–31.
3. Campanella L, Martini E, Tomassetti M, Anal. Lett. 2007, 40, 113–125.

A METHOD BASED ON SCATTERING PARAMETERS FOR MODEL IDENTIFICATION OF PIEZOACTUATORS WITH APPLICATIONS IN COLLOIDAL SUSPENSION MONITORING

R. P. PAGANELLI[1], A. GOLFARELLI[2], A. ROMANI[2], M. MAGI[2] AND M. TARTAGNI[2]

[1]*CNR-IEIIT, a National Research Council Institute c/o DEIS, University of Bologna, Viale Pepoli, 3/2 – 40123 Bologna, Italy*

[2]*School of Engineering II, Campus of Cesena, University of Bologna, Via Venezia, 52 – 47023 Cesena, Italy*

Abstract. Liquid colloids are mixtures where one substance is dispersed evenly throughout another. Piezoactuators can be reliably used to sense physical properties of colloids. However, to better understand and forecast the behavior of the system, a suitable and reliable model should be used to fit experimental data. In this work we will show how the scattering parameters, introduced for the first time in this field, can greatly improve model identification accuracy and give deeper insight into the resonant response of the system, especially in the case of viscous systems, whenever resonance peaks tend to shift or to vanish. Any change on the density and/or viscosity of the interacting fluid causes a well-predictable modification on each possible frequency response computable from measured data: the best choice in terms of sensitivity and orthogonal correlation has been investigated by means of parametric simulations with variable loads. Experimental results demonstrate that the conceived sensor is able to reliably discriminate different bentonite clay suspensions.

1. Introduction and principle of operation

Screening the physical characteristics of colloids is very important in order to quantify single components of multi-phase mixtures widely used in drilling engineering as well as in food and manufacturing industry. Piezoactuators can be reliably used to sense colloids properties when proper waterproof is provided.

The adopted device is a millimeter cantilever beam whose dimensions are roughly 3 × 6 cm and whose first natural frequency in air is approximately 65 Hz (Fig. 1). The beam is firmly mounted on the trough containing the mixture to be sensed and well dipped into the fluid. If the mixture changes, the device input impedance varies, according to the different mechanical loading condition. The device simultaneously acts both as actuator and sensor, definitely enabling a lower system parts count. The piezoelectric cantilever beam is energized by means of a low frequency AC voltage, sweeping in a range starting from practical quasi-DC values to few KHz, in order to excite the lower order modes of mechanical vibration [1].

P. Malcovati et al. (eds.), *Sensors and Microsystems: AISEM 2009 Proceedings*,
Lecture Notes in Electrical Engineering 54, DOI 10.1007/978-90-481-3606-3_42,

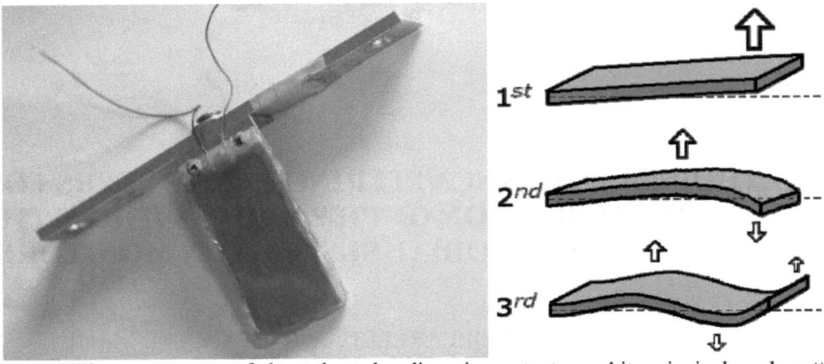

Figure 1. Millimeter waterproofed two-layer bending piezoactuator and its principal mode patterns.

2. Lumped element model for the description of the system

A reliable system model should be used in order to better understand and forecast the frequency response for different loading fluids. In the band of the lowest resonant frequencies the lumped model of Fig. 2 is proposed [2].

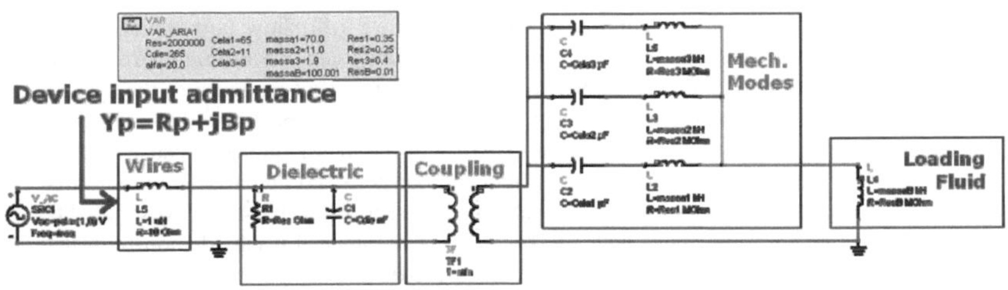

Figure 2. Lumped-element electro-mechanic circuit modeling the piezoelectric device: only the first three natural frequencies of the system are modeled.

The core of the lumped element equivalent circuit is the transformer: it couples the electrical domain at the primary winding and the mechanical domain at the secondary winding. The shunt capacitor on the left side models the dielectric properties of the device. The three series resonator on the right side respectively model the first three mechanical resonances associated with the modes of vibration; only three resonators have been considered because only three resonant frequencies have been spotted in the band of interest. For the unloaded device a short circuit is applied at the load side. Rather, when the fluid is loading the beam, a simple LR series is assumed to work properly, respectively modeling the fluid mass density and its viscosity. Since the LR termination affects the input port impedance, information on the loading fluid are simply attained from input voltage and current measurements [3]: then, by means of a simple bilinear transformation, we can calculate the S-parameter as well [4].

3. Scattering parameters and advantages connected to their use

We have seen that the use of scattering parameters, introduced for the first time in this field, improves both measurement interpretation and model identification accuracy, especially in the case of high viscous systems, whenever the resonance peaks tend to shift or even to vanish. The main advantage coming from the use of S-parameters is that any passive impedance is displayed on the complex plane plainly inside the unitary circle centered on the origin. Therefore the resonant responses cannot go out of scale but, rather, are transformed in clearly visible loops (Fig. 3), whose amplitude is associated to the resonator Q: the higher the resonator Q, the wider the loop of the complex S-parameter.

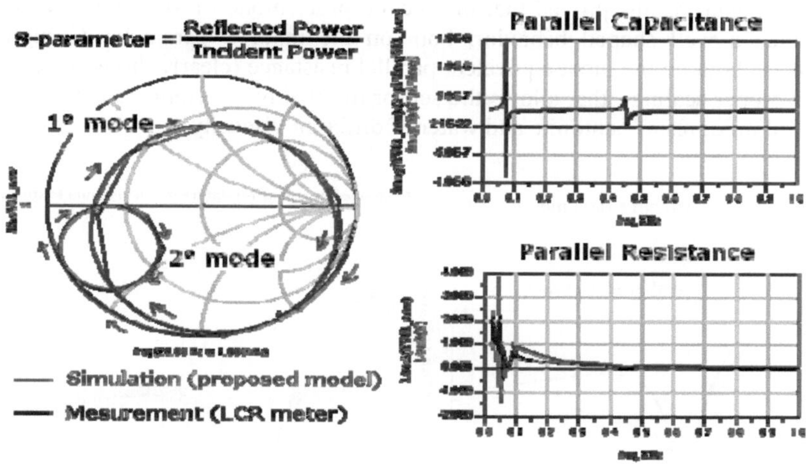

Figure 3. Input characteristic of the unloaded device, both simulated and measured: on the left the complex S-parameter is pictured (clock-wise oriented trace for increasing frequencies) whereas, on the right, the corresponding imaginary and real parts of admittances are depicted.

The first step is the identification of the model of Fig. 2 for the unloaded device. During this phase the use of S-parameter loops allow a better identification of the mode resonator Qs, as can be seen on Fig. 3. Therefore, the extraction of the lumped component values is clearly improved by an S-parameter fitting.

When a loading fluid is present, the more viscous the fluid, the smaller the S-parameter loop is, and eventually it also disappears. Anyway, since all the resonant loops of the modes are simultaneously displayed, it is easy to spot the mode whose resonance is yet not completely damped and available for load discrimination; using usual impedance/admittance parameters the Y-scale does not allow to display all the modes in sufficient detail to properly compare them.

Moreover, on the S-parameter trace the effect of the impedance transformation caused by the input shunt capacitor (dielectric of the piezo-device) is clearly visible, because causes a clock-wise rotation of the loops that changes the antiresonant frequencies of the modes of vibration: therefore, the more suited input parameter to scrutinize is the real part of the admittance, because it is less affected from impedance transformation, and much more adherent to the pure mechanic resonators, on which the loading fluid has the most direct influence.

4. Simulation results and experimental verifications

Parametric simulations with variable loading condition show that increasing the viscosity of the load dampens the input resonant peaks, as expected; besides, the damping effect is less efficient on higher order resonant modes. Therefore, depending on the viscosity range involved, the best suited resonant mode able to discriminate between the different viscous loads is found to be the second (Fig. 4). Experimental results proved that the second order mode of vibration is able to produce a well shaped behavior, both on the input S parameter (straightly spotted), both on the input equivalent parallel resistance (clearly discernable only after proper scaling), that allows the sensor to reliably discriminate between the several mixtures of bentonite and water of different viscosity.

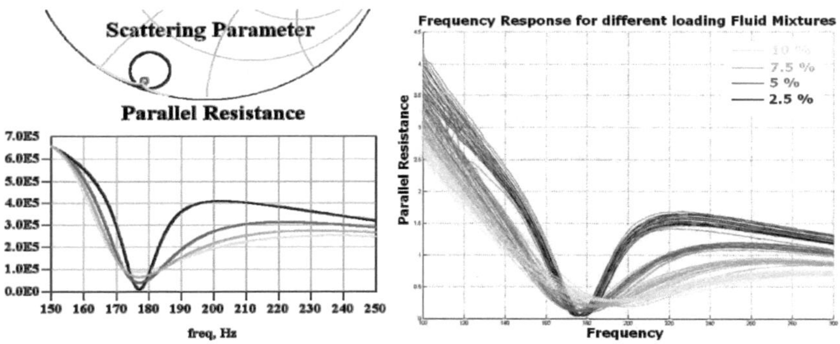

Figure 4. On the *left*: CAD simulation results for increased load viscosity (density unchanged). On the *right*: experimental results for different compounds of bentonite clay suspensions.

4. Conclusions

Experimental results obtained with variable bentonite clay suspensions confirmed the ability of the sensor to plainly discriminate between the different mixtures. The S-parameters allowed a better comprehension of the resonance phenomena and aided both the model identification procedure, both the singular mixture discrimination process.

References

1. M.P. Norton, Fundamentals of noise and vibration analysis for engineers, Cambridge University Press, 1989, pp. 81–86
2. IEEE Standard on Piezoelectricity, ANSI/IEEE Std 176-1987, p. 50

3. T.L. Wilson, G.A. Campbell, R. Mutharasan, Viscosity and density values from excitation level response of piezoelectric-excited cantilever sensors, Sensors and Actuators A, 138 (2007), pp. 44–51
4. G. Gonzales, Microwave Transistor Amplifier Analysis and Design, Prentice-Hall, 2nd edition (1996), pp. 45–60

MEMS TILT SENSOR WITH IMPROVED RESOLUTION AND LOW THERMAL DRIFT

DAMIANO CRESCINI, MARCO BAÙ, AND VITTORIO FERRARI

Dipartimento di Elettronica per l'Automazione,
Università degli Studi di Brescia, Via Branze 38, 25123 Brescia

Abstract. The purpose of this paper is to investigate the design methodology, test results, and potential applications of a high resolution and low thermal drift calibrated digital inclinometer based on a MEMS capacitive sensor. The inclinometer provides ±10° of measurement range in a rotational plane that is perpendicular to the earth's gravity and it uses a primary sensing element based on high aspect ratio microstructures, processed by air gap insulated micro structure (AIM) technology. A signal conditioning electronics, processing the MEMS signals, enables flexibility in terms of linearization, thermal compensation and signal calibration. Resolution of ±0.002° has been found around 0° together with a thermal zero drift lower than ±0.002°/°C from 0°C to +50°C. The overall system consumes 15 mA dc current from a 5V supply. Examples of application fields include: platform stabilization, drill orientation, event detection and telematics/GPS.

1. Design and manufacturing of the tilt sensor

The inclinometer provides a ±10° of measurement range in a rotational plane that is perpendicular to the earth's gravity and it uses a single-axis accelerometer that provides the base-sensing function resolving the earth's gravity into two orthogonal vectors.

The MEMS primary sensing element is based on high aspect ratio micro-structures, processed by air gap insulated micro structure (AIM) technology [1–4]. The MEMS element is zero level packaged by glass frit bonding of a silicon cap wafer to ensure hermetic sealing of the accelerometer sensor cavity. In Fig. 1a and b SEM images of the tilt sensor are reported.

A signal conditioning electronics, processing the MEMS signals, enables flexibility in terms of linearitazion, thermal compensation and signal calibration. In Fig. 2 a block scheme of the MEMS sensor and the signal conditioning electronics is reported. The main advantages of this solid-state tilt sensor are the long term bias stability, the high resolution, the low temperature coefficient even without compensation, and the low power consumption. A power-efficient approach to a common trigonometric identity converts the orthogonal vectors obtained by the accelerometer sensor into an inclination-angle measurement.

The digital post processing circuit digitises the sensor outputs and applies sensitivity/offset calibration coefficients prior to angle calculations.

P. Malcovati et al. (eds.), *Sensors and Microsystems: AISEM 2009 Proceedings*,
Lecture Notes in Electrical Engineering 54, DOI 10.1007/978-90-481-3606-3_43,

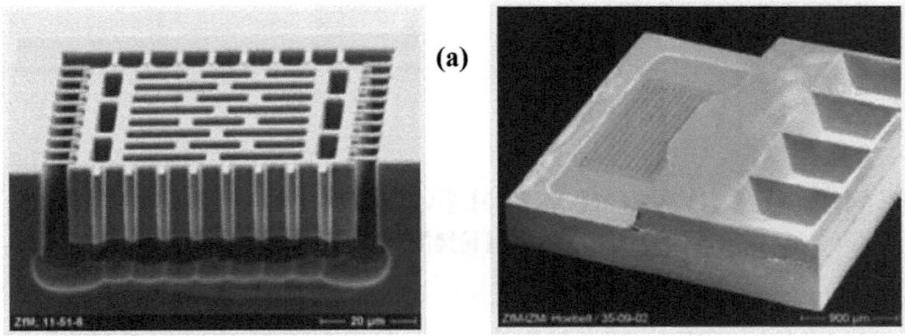

Figure 1. **(a)** Air gap surrounded single crystal silicon structure (courtesy of MEMSFAB). **(b)** SEM image of a glass sealed AIM partially opened (courtesy of MEMSFAB).

A factory calibration produces these coefficients using a +/– 10° mechanical rotational apparatus. This eliminates the need for system-level calibration in many cases. In addition to calibrating the sensor elements, the digital inclinometer corrects for power-supply dependent parameters, providing a more robust calibration. The accuracy of the inclination-angle measurements relies on three important factors: the absence of external (aside from gravity) acceleration, managing offset errors introduced during system-level configuration, and maintaining a proper axis of rotation (rotation plane perpendicular with earth's gravity).

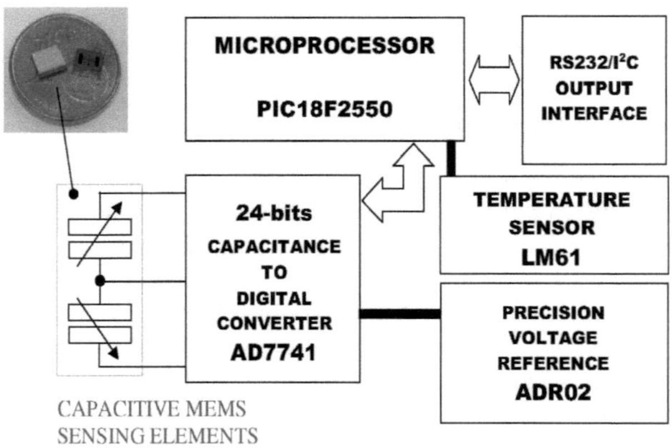

Figure 2. Block scheme of the tilt sensor.

All of these factors can influence the acceleration measurements and introduce errors. An internal temperature sensor monitors the accelerometer temperature and it provides convenient temperature measurement for system-level characterization and thermal drift compensation. In Fig. 3a a photograph of the sensor prototype is reported where the MEMS element and the signal conditioning are well distinguishable.

2. Experimental results

Performance testing is accomplished in an inclinometer test facility which includes an high precision rotation stage (see Fig. 3b), tilt, and cross-axis sensitivity measurements; temperature chambers for performance and life cycle tests; and shock machines for testing sensor robustness.

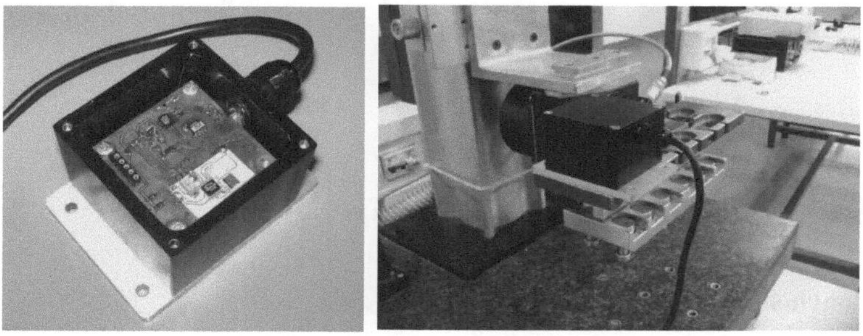

Figure 3. **(a)** Photograph of the tilt sensor in the aluminium housing. **(b)** The tilt sensor during calibration.

Figure 4 reports the calibration curve of the inclinometer. A linearity error of about +/–0.04% FSO has been measured at room temperature in the range +/–10°.

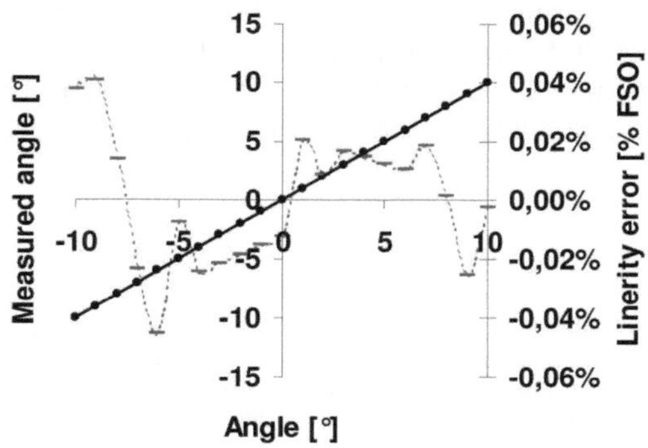

Figure 4. Calibration curve plot and linearity errors.

Resolution of ±0.002° has been found around 0° together with a thermal zero drift lower than ±0.002°/°C from 0°C to +50°C (see Fig. 5).

Finally, output drift analysis at constant room temperature has been carried out with the aim to determine the long term stability. A bias stability of about ±0.01° has been found over 3 months.

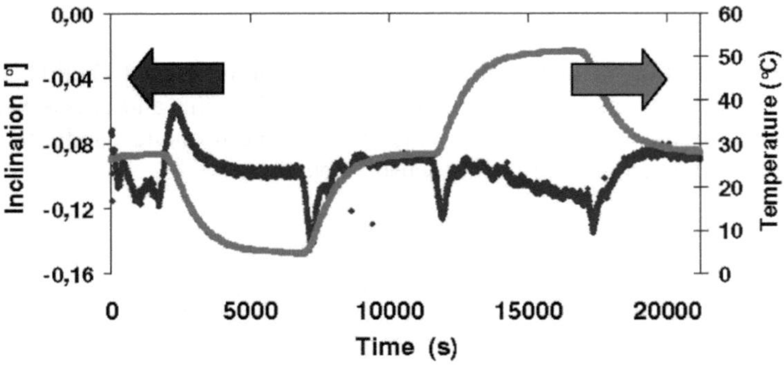

Figure 5. Sensor output vs temperature.

3. Conclusions

Investigations, design methodology, test results, and potential applications of a high resolution and low thermal drift calibrated digital inclinometer based on a MEMS capacitive sensor have been presented. The inclinometer provides ±10° as full scale and it uses a primary sensing element based on high aspect ratio microstructures, processed by air gap insulated micro structure (AIM) technology. A signal conditioning electronics, processing the MEMS signals, enables flexibility in terms of linearization, thermal compensation and signal calibration. Resolution of ±0.002° has been found around 0° together with a thermal zero drift lower than ±0.002°/°C from 0°C to +50°C. Examples of applications fields include: platform stabilization, drill orientation, event detection and telematics/ GPS.

References

1. A. Bertz, M. Küchler, R. Knöfler, T. Gessner: A novel high aspect ratio technology for MEMS fabrication using standard silicon wafers, *Sensors and Actuators A 97–98, (2002) 691–701.*
2. M. Lemkin et al.: A low-noise digital accelerometer using integrated SOI-MEMS technology, *Proceedings of the Conference on Tranducers '99, Sendai, Japan, June 1999, pp. 1294–1297.*
3. T. Gessner, G. Ebest: Subproject C2, A novel high aspect ratio technology for MEMS fabrication using standard silicon wafers, *Fraunhofer-IZM Chemnitz Report.*
4. C. Lohmann, D. Reuter, A. Bertz, T. Geßner: Chapter "High aspect ratio micromachining using the AIM technology", in the book: *The World of Electronic Packaging and System Integration, pp. 544–548.*

AN OFFSET COMPENSATION METHOD FOR INTEGRATED THERMAL FLOW SENSORS

P. BRUSCHI[1], M. DEI[1] AND M. PIOTTO[2]

[1]*Dipartimento di Ingegneria dell'Informazione, Università di Pisa, Via G. Caruso, 16 56122 Pisa*

[2]*IEIIT – Pisa, CNR, Via G. Caruso, 16 56122 Pisa*

Abstract. The effectiveness of a new method for the offset compensation of integrated thermal flow sensors is demonstrated. The method is based on flow sensors with a double heater configuration, designed with a commercial CMOS process and fabricated by means of a post-processing technique. The power unbalance between the two heaters is used to compensate the unavoidable sensor offset. Measurements in nitrogen at room temperature confirm an offset decrease by nearly one order of magnitude.

1. Introduction

Integrated thermal flow meters, fabricated by means of micromachining techniques, are characterized by lower power consumption, shorter response times and greater resolution than their macroscopic counterparts [1]. Nevertheless, they are usually affected by an intrinsic offset that limits the possibility of using them in very low flow applications. Their operation is based on a thermal symmetry at rest condition which is modified by the fluid flow and unavoidable asymmetries of the sensing structure and package deform the heat distribution causing an offset which is usually greater than the sensor resolution [2]. It should be noted that this sensor offset cannot be reduced with the traditional methods (e.g. chopper amplifier) used to compensate the offset of the electronic read out circuits. Furthermore, a significant offset temperature drift can be expected due to the dependence of the sensor signal on the thermal gas properties. This strongly reduces the effectiveness of standard offset compensation techniques, based on software or hardware subtraction of a constant term.

In this work, we demonstrate the effectiveness of a compensation method proposed in [3] and based on a differential temperature calorimeter equipped with a double heater. The power unbalance between the two heaters is used to compensate the intrinsic offset as described in the next section.

P. Malcovati et al. (eds.), *Sensors and Microsystems: AISEM 2009 Proceedings*, Lecture Notes in Electrical Engineering 54, DOI 10.1007/978-90-481-3606-3_44,

2. Principle of offset compensation

The sensor structure, schematically shown in Fig. 1, is made up of two polysilicon heaters (H_1 and H_2), placed between two thermopiles (TP_1 and TP_2). The device was designed with the BCD3s process of STMicroelectronics and a post-processing technique was used in order to thermally insulate the heaters and the thermopiles from the substrate. A small amount of silicon was removed from the front-side of the chip by means of an anisotropic silicon etching, as described in detail in [3].

Supposing a linear relationship between the heater powers (P_1 and P_2) and the thermopile overheating, the voltages, V_{T1} and V_{T2}, produced by the two thermopiles (supposed linear) are given by:

$$\begin{aligned} V_{T1} &= s_1\left(P_1 a_{1,1} - P_2 a_{1,2}\right) \\ V_{T2} &= s_2\left(P_2 a_{2,2} - P_1 a_{2,1}\right) \end{aligned} \tag{1}$$

where s_1 and s_2 are the respective Seebeck coefficients of the thermopiles while a_{ij} are flow dependent coefficients (thermal resistances) taking account of the heat exchange mechanisms. The output signal of the device (V_{out}) is the difference between the voltages V_{T1} and V_{T2}:

$$V_{out} = P_2\left(s_2 a_{2,2} - s_1 a_{1,2}\right) - P_1\left(s_1 a_{1,1} - s_2 a_{2,1}\right) \tag{2}$$

In an ideal symmetrical structure at zero flow, $s_1 = s_2$ and $a_{22} = a_{11}$, $a_{12} = a_{21}$. In a real device, unavoidable structure and package asymmetries generate an offset voltage. Nevertheless, Eq. (2) indicates that a condition of null offset can still be obtained if:

$$\frac{P_2}{P_1} = \frac{\left(s_1 a_{1,1} - s_2 a_{2,1}\right)}{\left(s_2 a_{2,2} - s_1 a_{1,2}\right)} \tag{3}$$

Thus, using a double heater structure with a proper power unbalance $\left(P_1 \neq P_2\right)$ it is possible to compensate for the offset.

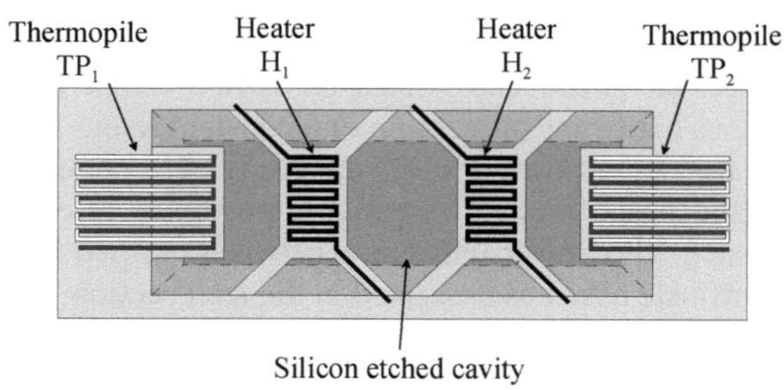

Figure 1. Schematic view of the device structure.

3. Experimental results

The effectiveness of the proposed method has been proven by means of the experimental set-up schematically shown in Fig. 2. The chip was glued to ceramic DIP cases by means of epoxy resin and wedge bonding was used to connect selected chip pads to the case pins. A PMMA package was designed in order to connect the sensor to a reference gas line, equipped with a precision flow controller (MKS 1179B) with a 10 sccm full scale range. The package was glued to the chip following the procedure described in detail in [4].

The packaged device was placed inside a hollow aluminium cylinder whose temperature was varied using a Peltier cell driven by an electronic temperature controller. The gas reaches the sensor after passing through pipes drilled through the cylinder walls, in order to get isothermal with the latter.

The read out electronics has been developed on two printed circuit boards (PCBs). The first one, connected to a PC via USB, includes a MicroConverter® ADuC842 (Analog Devices) with two 12-bit voltage output DACs used to drive the device heaters. The two DACs have been calibrated to obtain an accuracy of ± 1 mV.

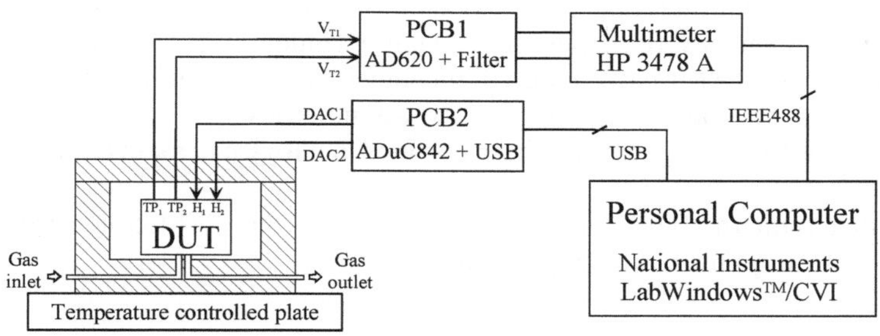

Figure 2. Schematic view of the experimental set-up: DUT indicates the device under test.

The second one, connected to a digital multimeter (HP 3478 A), includes a low noise instrumentation amplifier (AD 620) in cascade with a second order 10 Hz low pass filter used to read the sensor output signal. An application in the National Instruments LabWindows™/CVI environment has been developed in order to drive both the ADuC842 and the multimeter. The offset of the readout electronics was cancelled by subtracting the output voltage measured with both the heaters turned off.

The sensor response in nitrogen at room temperature with and without the offset correction is shown in Fig. 3a: an enlarged view around the origin of the coordinates is also reported for clarity in Fig. 3b. It can be noted that an intrinsic offset of about 2.47×10^{-5} V is present when the heaters are biased with the same power ($P_1 = P_2$). The power unbalance is capable to reduce the offset to a residual value of -1.2×10^{-6} V. Therefore, the proposed method allows the reduction of the sensor offset by more than one order of magnitude.

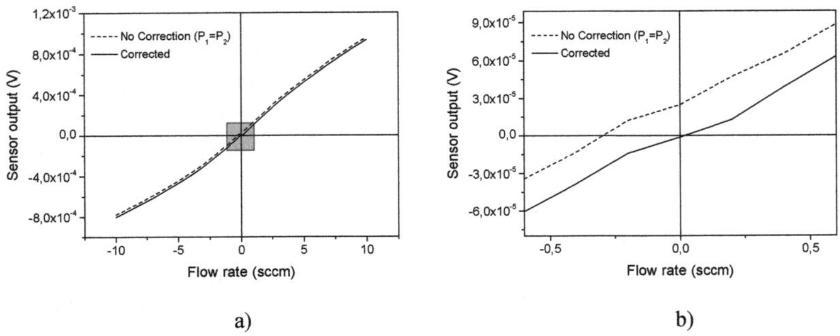

Figure 3. (**a**) Sensor response at room temperature with ($P_1 \neq P_2$) and without ($P_1 = P_2$) the offset correction; (**b**) enlarged view around the origin of the coordinates.

4. Conclusions

A new method for the compensation of the intrinsic offset of integrated thermal flow sensors has been described. The method is based on flow sensors with a double heater configuration designed with a commercial CMOS process and fabricated by means of a post-processing technique. The power unbalance between the two heater has been used to compensate the intrinsic offset generated by structural mismatches and packaging asymmetries. An offset reduction by nearly one order of magnitude has been obtained in nitrogen at room temperature.

Acknowledgments The authors wish to thank the STMicroelectronics of Cornaredo (Italy) for fabricating the chips.

References

1. B. W. van Oudheusden, "Silicon thermal flow sensors", Sensors and Actuators A, [vol. 30], pp. 5–26, 1992.
2. S. P. Matova, K. A. Makinwa, and J. H. Huijsing, "Compensation of packaging asymmetry in a 2-D wind sensor" IEEE Sensors Journal, [vol. 3], pp. 761–765, 2003.
3. P. Bruschi, M. Piotto, and N. Bacci, "Postprocessing technologies, interface circuits and packaging strategies for CMOS compatible gas flow sensors", proc. of IWASI 2007, Bari 26-27 June 2007, pp. 117–122, 2007.
4. P. Bruschi, M. Dei, and M. Piotto, "Single chip sensing of multiple gas flows", proc. of DTIP 2008, Nice 9-11 April 2008, pp. 242–246, 2008.

A NEW PRINCIPLE FOR ENVIRONMENT RESISTANT INTEGRATED ANEMOMETERS

P. BRUSCHI[1], M. DEI[1] AND M. PIOTTO[2]
[1]*Dipartimento di Ingegneria dell'Informazione, Università di Pisa, via G. Caruso, 16, 56122 Pisa, Italy*
[2]*IEIIT – Pisa, CNR, via G. Caruso, 16, 56122 Pisa, Italy*

Abstract. Development of compact, inexpensive and low power sensors is the key step towards the application of wireless sensor networks in real scenarios. In this work we propose a new approach for monitoring wind direction and speed using integrated flow meters, included into a protecting cylinder. An original channel geometry drilled through the cylinder produces a flow that depends on the wind direction according to a cosine law, facilitating wind direction estimation. The effectiveness of the approach is proven by means of measurements performed on a prototype.

1. Introduction

The development of compact anemometers, capable of being interfaced with wireless sensor network (WSN) nodes, has several potential applications, ranging from pollution monitoring, airport wind surveillance, plantation disease spreading prediction and optimization of air circulation in food storage and drying rooms. The possibility of deploying low cost networks in a short time for capillary detection of the wind distribution can be a key element for predicting the evolution of pollution accidents. Traditional mechanical anemometers are not easily scalable, since, similarly to micro-motors [1], reduced size devices present unfavorable drag-to-force ratios, resulting in low sensitivity and irregular rotation. Ultrasonic anemometers are being proposed as a relatively robust and durable alternative to mechanical wind sensors. However, the requirement of proper spacing between the acoustical transducers prevented miniaturization below dimensions of several centimeters [2]. Macroscopic thermal sensors are very sensitive devices but their power consumption rules out their application in battery powered WSN nodes. Recently we have presented [3] a directional anemometer based on an arrangement of multi-functional chips on a ceramic membrane. This solution was proven very sensitive and suitable for both velocity and wind direction measurements. However, due to the necessity of heating the whole volume of the chips, the power dissipation is still too high for portable devices. Use of integrated MEMS thermal anemometers is often claimed as a

P. Malcovati et al. (eds.), *Sensors and Microsystems: AISEM 2009 Proceedings*, Lecture Notes in Electrical Engineering 54, DOI 10.1007/978-90-481-3606-3_45,

method for significantly reducing power consumption. Unfortunately, their fragility does not allow direct exposure to the wind in real situations.

In this work we propose a novel approach for protecting MEMS thermal anemometers while providing the necessary interaction with the air flow.

2. Device description

The structure of the proposed device is shown in Fig. 1a. The wind stream creates a pressure distribution around the external surface of the cylinder that depends on both the wind speed and direction. Only wind velocities lying on a plane perpendicular to the cylinder axis are considered (2D anemometer). The pressure differences are sampled with apertures in the cylinder surface, connected to the channel structure depicted in Fig. 1b [4]. The reference axis of the channel structure is indicated with x. Besides the apertures along x, other four apertures indicated with $A_{1\text{–}4}$ are introduced. An integrated flow sensor (MFS), similar to that described in Ref. [5], is connected across H_1 and H_2, thus measuring a flow proportional to the pressure difference across these points.

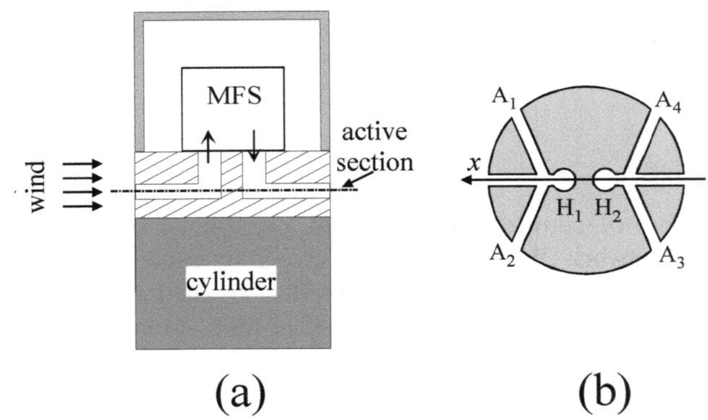

Figure 1. Structure of the proposed anemometer showing the placement of the integrated mass flow sensor (MFS) (**a**). Configuration of the active section (**b**). The MFS communicates with the active section through holes H_1 and H_2.

The channel configuration is optimized by means of fluid dynamic simulations in order to produce a flow through the MFS proportional to the cosine of the angle formed by the wind with *x*-axis. Simulations have been performed using the COMSOL™ finite element method program. The best cosine approximation was obtained with apertures A_1, A_3 and A_2, A_4 placed across diameters forming angles of 40° and −40° with *x*, respectively. The dependence of the flow rate on the wind direction, obtained with the optimized configuration, is shown in Fig. 2 (left). The maximum difference from the ideal cosine curve is about 2.5%. Using this characteristic, a 2D wind sensor capable of detecting the wind direction can be easily obtained by combining two devices such that of Fig. 1a mounted with perpendicular reference axes. In this way two flows $Q_X = f(u)\cos(\theta)$ and $Q_Y = f(u)\sin(\theta)$, where $f(u)$ is a function of the wind velocity u

and θ the angle formed by the wind with *x,* are produced. The wind direction θ and *f*(*u*) can be estimated by simple trigonometric operations, while the wind velocity can be extracted from *f*(*u*) after calibration. Note that with simpler structures proposed in existing patents [6], where sensing of only one diametric pressure difference is proposed, the non monotonic behaviour of Fig. 2 (right) can be expected.

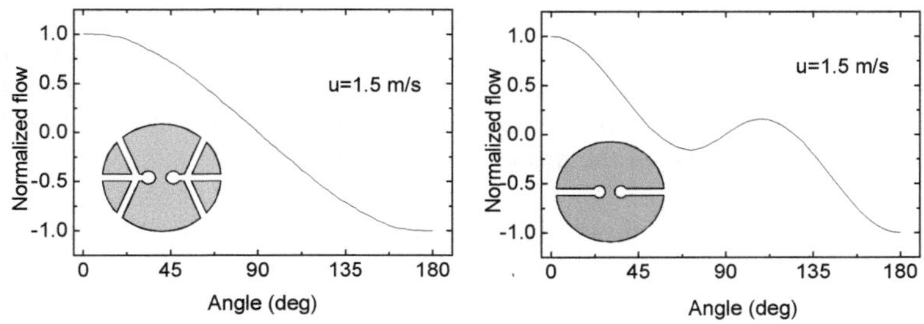

Figure 2. Simulated flow through the MFS in the case of the optimal structure (*left*) and a simpler active section with only the two aperture along the reference diameter x (*right*).

3. Results

To demonstrate the feasibility of this approach, a prototype, consisting of a high sensitivity integrated flow meter, has been built. Fabrication and packaging of the flow meter have been described elsewhere [5]. A PMMA (PolyMethyl-Methacrilate) cylinder with a diameter of 3 cm and a height of 14 cm was used. The sensor has been characterized within a small wind tunnel for a velocity range from 1 to 8 m/s. Angle variation was obtained by rotation of the cylinder. An example of curves obtained at a wind velocity of 2.7 and 5 m/s is shown in Fig. 3 where the ideal cosine curves fitting the data are shown for reference. A monotonic behavior in good approximation with the target cosine curve can be observed.

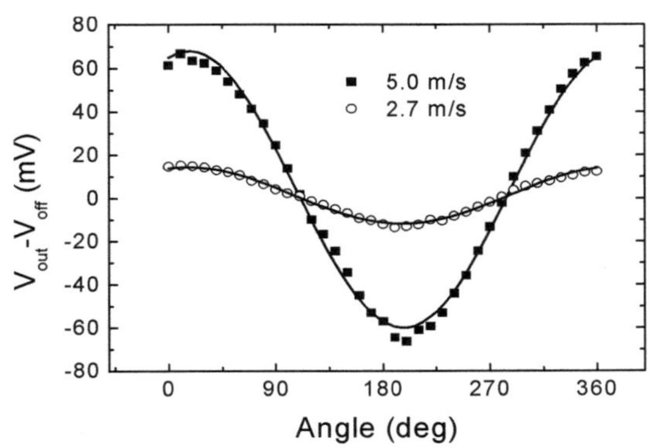

Figure 3. Sensor output signal as a function of the wind direction (angle θ) measured at two different wind velocities. The solid lines are the best cosinusoidal fittings.

The curves are slightly shifted by about 10° with respect to the zero angle reference possibly due to mechanical misalignment between the actual active section symmetry axis and sample holder used in these preliminary experiments.

4. Conclusions

A solid state anemometer, based on an integrated thermal flow sensor inserted into a PMMA cylinder, has been described. An original channel configuration connects the outer surface of the cylindrical body to the flow sensor in such a way that the measured flow is proportional to the cosine of the angle formed by the wind and a reference axis of the structure. Preliminary measurements have confirmed the effectiveness of the proposed approach.

Acknowledgments The authors wish to thank the STMicroelectronics of Cornaredo (Italy) for fabricating the chips.

References

1. Z. Rymuza, "Control tribological and mechanical properties of MEMS surfaces. Part 1: critical review" Microsystem Technologies, [vol. 5], pp. 173–180, 1999.
2. D. Han, S. Kim, S. Park, "Two-dimensional ultrasonic anemometer using the directivity angle of an ultrasonic sensor" Microelectronics Journal, [vol. 39], pp. 1195–1199, 2008.
3. P. Bruschi, N. Nizza, M. Piotto, M. Schipani, "Solid state directional anemometer for harsh environments", Sensors and Microsystems proceedings of the 11th Italian Conference, pp. 117–121, World Scientific, 2008.
4. P. Bruschi, M. Piotto, "Sensore a basso consumo per il rilevamento della velocità e della direzione del vento", Italian patent, appl. n. PI2007A000071, 8/6/2007.
5. P. Bruschi, V. Nurra, M. Piotto, "A compact package for integrated silicon thermal gas flow meters" Microsystem Technologies, [vol. 14] pp. 943–949, 2008.
6. R. Klause, D. Klaus, F. Uwe, "Thermal anemometer", UE patent n. EP1629287, 1/3/2006.

DISTRIBUTED DYNAMIC STRAIN MEASUREMENT USING A TIME-DOMAIN BRILLOUIN SENSING SYSTEM

R. BERNINI[1], A. MINARDO[2] AND L. ZENI[2]

[1]*Istituto per il Rilevamento Elettromagnetico dell'Ambiente – Consiglio Nazionale delle Ricerche, Via Diocleziano, 328 – 80124 Napoli, Italy*

[2]*Dipartimento di Ingegneria per l'Informazione, Seconda Università di Napoli, Via Roma 29 – 81031 Aversa, Italy*

Abstract. We present a novel technique for dynamic strain measurements in optical fibers, based on the stimulated Brillouin scattering interaction between two counter-propagating optical pulses. The technique permits to address dynamically and randomly the position at which vibration is measured. Preliminary experimental results demonstrate the validity of the proposed technique.

1. Introduction

In recent years, there has been a growing interest in distributed fiber-optics sensors, especially thanks to their unparallel ability to carry out continuous measurements of the quantity of interest over the whole fiber length. In particular, distributed sensors based on stimulated Brillouin scattering (SBS) permit to perform distributed strain measurements with high accuracy and spatial resolution, making them ideally suited in the field of structural health monitoring [1, 2]. The typical detection technique, referred to as Brillouin optical time-domain technique (BOTDA), is based on the use of a cw probe beam and a counter-propagating optical pulse. Such a technique requires an acquisition time in the order of a few minutes, so that it is only suitable for static measurements, On the other hand, the possibility to extend the use of distributed optical fiber sensors to dynamic measurements would be of special interest in some application fields, such as measurement of vibrations in civil or aeronautic structures.

In this paper, we propose a novel technique capable of dynamic strain measurements. The proposed method exploits the SBS interaction between two counter-propagating optical pulses, and permits very high sampling frequencies (in the order of hundreds of kHz) to be achieved at the sensing location, while keeping a reasonable spatial resolution (in the order of a few meters). Moreover, the technique permits to carry out quantitative measurements, i.e. the exact value of strain at the sensing location can be retrieved as a function of time, as long as strain excursion is within certain limits. Finally, the position of the sensing

P. Malcovati et al. (eds.), *Sensors and Microsystems: AISEM 2009 Proceedings*, Lecture Notes in Electrical Engineering 54, DOI 10.1007/978-90-481-3606-3_46,

location can be dynamically addressed along the sensing fiber, so as to monitor different fiber regions by the same set-up.

2. Principle of operation

One of the main factors limiting the acquisition rate in standard BOTDA schemes is the need for sweeping the pump-probe frequency difference over a suitable range. When dynamic strains are to be detected, it is possible to overcome this limitation by using a slightly different detection scheme. The proposed technique is applied by using two counter-propagating optical pulses with a fixed optical frequency difference. In particular, the latter is set to a spectral distance from the local Brillouin frequency shift approximately equal to half the BGS linewidth (see Fig. 1). In this way, any vibration-induced modulation of the local Brillouin frequency shift will be easily measured as an intensity variation of the Stokes pulse peak intensity, as it emerges from the sensing fiber.

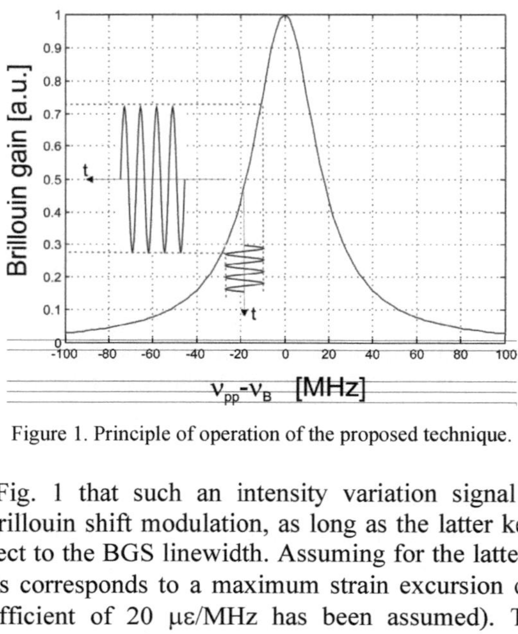

Figure 1. Principle of operation of the proposed technique.

Note from Fig. 1 that such an intensity variation signal will reproduce faithfully the Brillouin shift modulation, as long as the latter keeps sufficiently small with respect to the BGS linewidth. Assuming for the latter a typical value of 35 MHz, this corresponds to a maximum strain excursion of about 700 $\mu\varepsilon$ (a transfer coefficient of 20 $\mu\varepsilon$/MHz has been assumed). The use of two counter-propagating pulses permits to restrict the SBS interaction over a fiber region whose extension is equal to half the pulse length. The position of the sensing region along the fiber is chosen by properly setting the time delay between the two optical pulses. Moreover, the sampling rate of the dynamic signals is strictly related to the pulse repetition rate, the latter being only limited by the time-of-flight of pulses, and hence by the length of the sensing fiber. As an example, a sensing length of 100 m would allow for a sampling rate up to 1 MHz.

3. Experimental results

The experimental set-up used for testing the proposed technique is illustrated in Fig. 2. Pump and probe signals were obtained from the same distributed feedback laser diode (DFB-LD), by means of the sideband technique described in Ref. [3]. Optical pulses were formed by two electro-optic intensity modulators (IM), driven by the same electrical delay/pulse generator (Stanford Research Systems, model DG535). The pulse generator provided 30-ns-long pulses (3-m spatial resolution). Pulse repetition rate was set to 50 kHz. The relative delay between pump and probe pulses was set in order to obtain a spatial overlap of the two optical pulses in correspondence of the dynamically strained fiber section. In particular, dynamic sensing at the generic position z was obtained by delaying the probe (Stokes) pulse of the quantity $\tau = 2n/c(z - L/2)$ with respect to the pump pulse, n being the fiber effective group index, c the velocity of light in vacuum, and L the total fiber length.

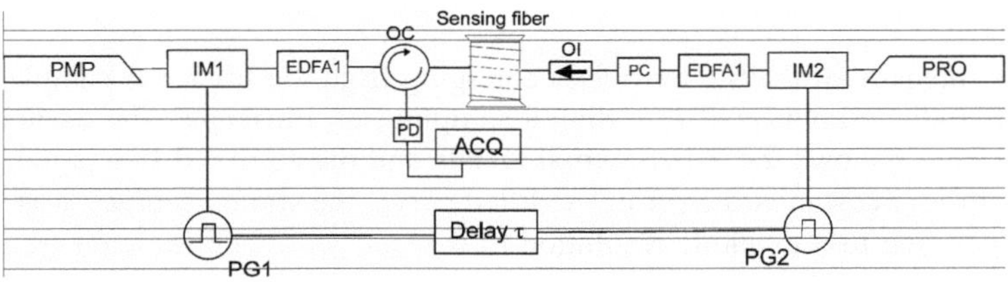

Figure 2. Experimental setup for distributed dynamic strain measurements. PMP = pump laser. PRO = probe laser. EDFA = erbium-doped fiber amplifier. PC = polarization controller. IM = intensity modulator. OI = optical isolator. OC = optical circulator. PD = photodiode. PG = pulse generator.

The first 3 m of a 30 m-long sensing fiber were fixed at one end, with the other end connected to a DC motor via a connecting rod, in order to provide periodic strain to the fiber. Dynamic strain measurements were carried out by setting the pump-probe frequency shift to 10,908 MHz.

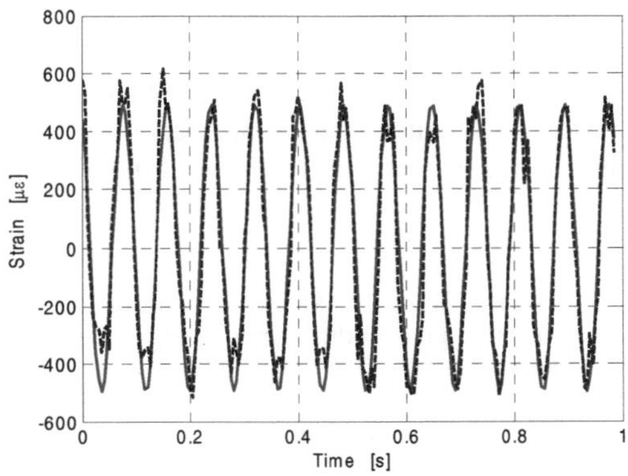

Figure 3. Dynamic strain measured as a function of time at a sampling rate of 200 Hz (dashed blue line), when a 12.3-Hz vibration is applied to the 3-m sensing section. Solid red line is the fitting curve.

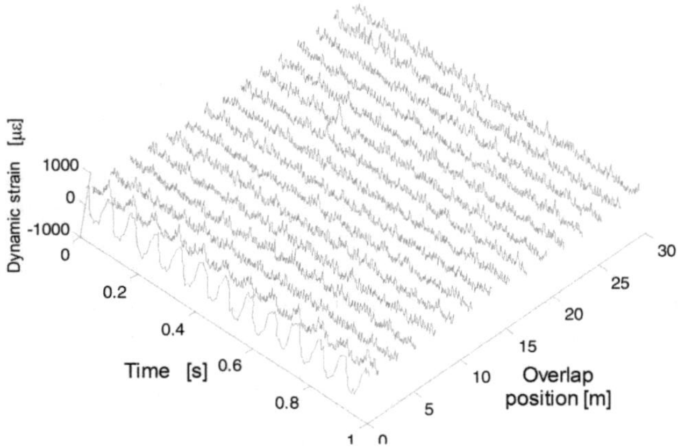

Figure 4. Dynamic strain measured as a function of time and the overlap position of the optical pulses. The pump-probe frequency difference was set to 10,908 MHz.

The intensity of the Stokes pulses emerging from the fiber was sampled by a fast photodiode connected to a digital sampling oscilloscope. The dashed blue line in Fig. 3 shows the experimental results and the solid red line is the fitting curve. From the standard error of the fitted curve, the strain accuracy was 95 με, corresponding to a dynamic resolution of $\approx 27\,\mu\varepsilon/\sqrt{Hz}$. The resonant frequency of the vibration is estimated to be 12.3 Hz.

A further demonstration of the validity of the technique was carried out by repeating the measurement for a pump-probe frequency offset fixed to 10,908 MHz, while varying the position where the two optical pulses overlap. The results, shown in Fig. 4, demonstrate that the technique has spatial resolution capabilities. Actually, vibration is only detected when the location at which the two pulses cross corresponds to the dynamically strained section.

4. Conclusions

We presented a technique for SBS-based distributed dynamic strain measurements. The technique can provide dynamic sensing at a randomly addressed position, at a sampling rate only limited by the time-of-flight of the pulses.

References

1. H. Ohno, H. Naruse, M. Kihara, and A. Shimada, "Industrial applications of the BOTDR optical fiber strain sensor," Opt. Fiber Technol., [vol. 7], pp. 45–64, January, 2001.
2. R. Bernini, A. Minardo, and L. Zeni, "Accurate high-resolution fiber-optic distributed strain measurements for structural health monitoring", Sensors and Actuators A, vol. 134, pp. 389–395, March, 2007.
3. M. Niklès, L. Thévenaz, and P. A. Robert, "Brillouin gain spectrum characterization in single-mode optical fibers", Journal of Lightw. Technol., vol. 15, no. 10, pp. 1842–1851, October, 1997.

EPOXY/MWCNT COMPOSITE BASED TEMPERATURE SENSOR WITH LINEAR CHARACTERISTICS

H. C. NEITZERT[1], A. SORRENTINO[2] AND L. VERTUCCIO[2]

[1]*Dipartimento di Ingegneria dell'Ínformazione e Ingegneria Elettrica, Università degli Studi di Salerno, Via Ponte Don Melillo,1 84084 Fisciano (SA), Italy*

[2]*Dipartimento di Ingegneria Chimica e Alimentare, Università degli Studi di Salerno, Via Ponte Don Melillo,1 84084 Fisciano (SA), Italy*

Abstract. An ultra-low cost temperature sensor has been developed that uses as active sensing material an epoxy layer, mixed with a small concentration of multi-walled carbon nanotubes. A stable NTC characteristics has been measured during temperature cycling. Under Joule heating conditions, a linear current-temperature characteristics has been obtained.

1. Introduction

Electronic temperature sensors are based on a great variety of materials and operating principles. Important properties are: High sensitivity, wide operating temperature range, stability against other ambient influences such as for example humidity, linearity and low production costs. Most sensors fulfil a limited subset of these conditions. Just some examples: Silicon based sensors are linear over a wide temperature range, but rather insensitive and need a stable interface electronics. NTC-resistors have a high sensitivity, but a highly nonlinear resistance-temperature characteristics, which has to be corrected for by using polynominal fits. Platinum resistor based sensors, like the PT100 elements, are linear and sensitive over a large temperature range, but rather costly. Epoxy material is often employed in temperature sensors as encapsulation material, because of the good stability against other ambient influences. Carbon nanotubes alone have been proposed as temperature sensors for example, deposited on patterned ITO substrates and exhibited a nonlinear temperature-resistance characteristics with PTC behaviour [1]. In the case of polyethylene/CNT nanocomposites, PTC and NTC regions have been observed, but temperature cycling resulted in a continuous change of the resistivity due to the rearrangement of the CNT- and polyethylene networks and a rather high CNT concentration was needed [2].

P. Malcovati et al. (eds.), *Sensors and Microsystems: AISEM 2009 Proceedings*, Lecture Notes in Electrical Engineering 54, DOI 10.1007/978-90-481-3606-3_47,

2. Experimental

2.1. *Material*

In this work, a diglycidyl ether of bisphenol A-epoxy resin (DGEBA) and 4,4' diamine-dibenzyl-sulfone hardener (DDS) both supplied by Sigma-Aldrich Chemicals, were used. The Multiwalled Carbon Nanotubes (MWNTs) were obtained from Nanocyl S.A., synthesized by catalytic carbon vapor deposition (CCVD) process and consisted of 95 vol% carbon.

Regarding the production of the composites, 0.5 wt% of carbon nanotubes were sonically dispersed for 20 min in the liquid epoxy resin before curing with DDS. The composite was then used to completely fill a rectangular mold cavity of 100 × 30 × 2 mm in dimensions.

Two metal electrodes were than applied at the end of the samples and cured within the polymeric matrix. The cure reaction was carried out in an oven with a standard temperature profile. The typical procedure consists of a ramp from room temperature to 130°C at the heating rate of 10°C/min, then a 60 min of isothermal stage at this temperature followed by a second heating ramp up to 180°C and finally an isothermal stage at 180°C for 180 min.

The specific composite sample, used for the here presented temperature sensor, had been cured in the presence of a voltage of 200 V, applied to the copper electrodes. This resulted in a dissipated power of about 4 W and substantial Joule heating of the sample.

2.2. *Electrical characterization*

The curing process and subsequent temperature cycles have been done in an electronically controlled Heraeus oven and the temperature during these processes has been monitored using a thermoelement, positioned in the vicinity of the epoxy/CNT composite sample.

Electrical characterization has been done, using a Keithley 2400 source-measurement unit. In the specific case presented here, we used copper electrodes and a two terminal measurement configuration. It should, however, be mentioned that we checked on samples with the same concentration of 0.5 wt% CNTs and evaporated contacts, that four-point and two-point measurements gave very similar results and that for composite samples with this CNT concentration, the contact resistance is sufficiently low.

3. Results and discussions

The development of the oven temperature and of the sample current as a function of time, during the cooling of the sample, when maintaining the same applied voltage of 200 V as during the curing process, is shown in Fig. 1 for the temperature range between 160°C and 45°C. A monotonic and smooth decrease of the current with decreasing oven temperature is observed and plotting the measured current as a function of the oven temperature (see Fig. 2) a strictly linear relation between both entities has been found. In the inset of Fig. 2, the linear fitting parameters are given. A room temperature resistance value of about 11 kΩ has been calculated and we obtained a roughly 4.5% increase of the conductivity when increasing the temperature by 100°C.

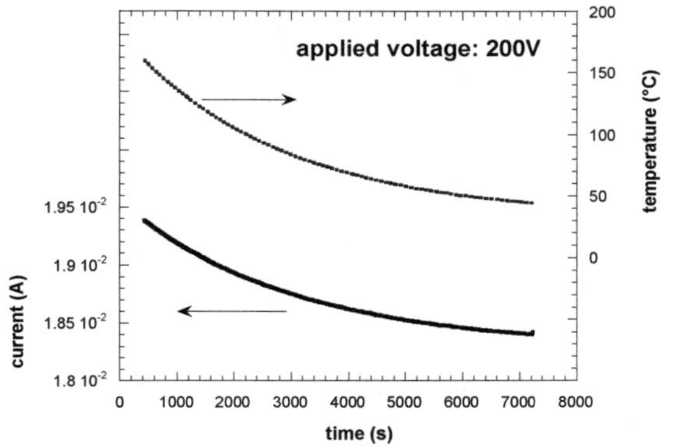

Figure 1. Temperature and current, monitored during cooling when applying a voltage of 200 V to the expoxy/nanotube composite.

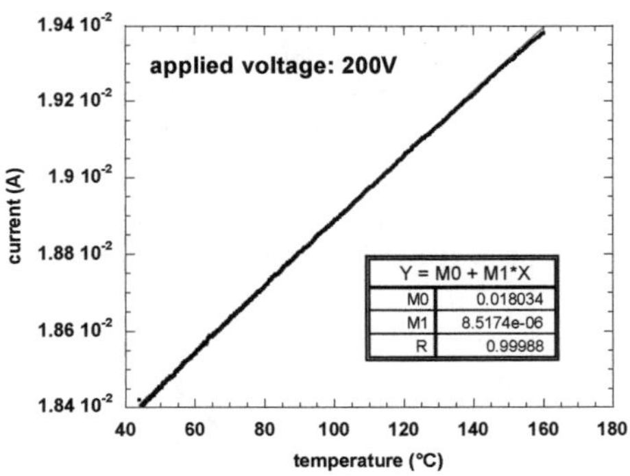

Figure 2. Temperature dependence of the current, measured during cooling when applying a voltage of 200 V to the expoxy/nanotube composite and linear fit of the characteristics (inset).

Subsequently we performed measurements during temperature cycling on the same composite sample, but applying a smaller bias voltage of 10 V, that did not result in Joule heating of the sensor. The results are shown in Fig. 3 (monitoring of temperature and current during two subsequent temperature cyles) and Fig. 4 (current–temperature characteristics during the cooling of the first temperature cycle). Again we observe a very smooth characteristics without any instabilities during temperature cycling.

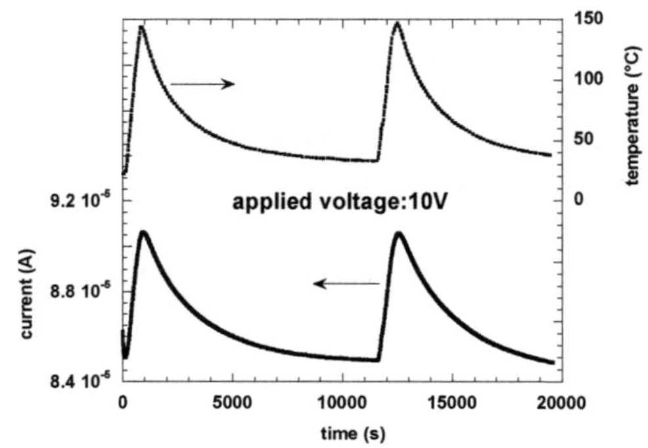

Figure 3. Temperature and current, monitored during cooling when applying a voltage of 10 V to the expoxy/nanotube composite.

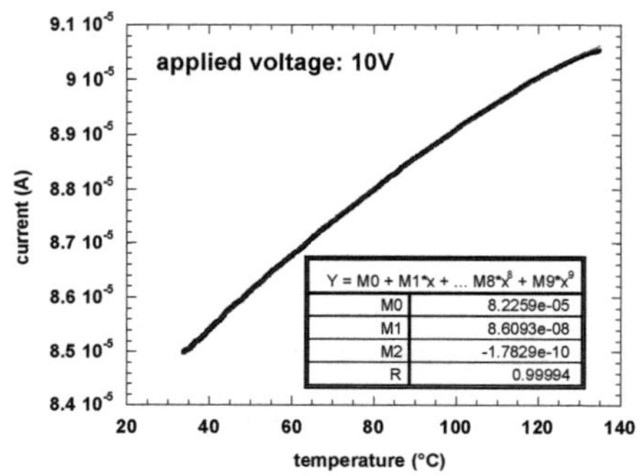

Figure 4. Temperature dependence of the current, measured during first cooling when applying a voltage of 10 V to the expoxy/nanotube composite and polynomial fit of the characteristics (inset).

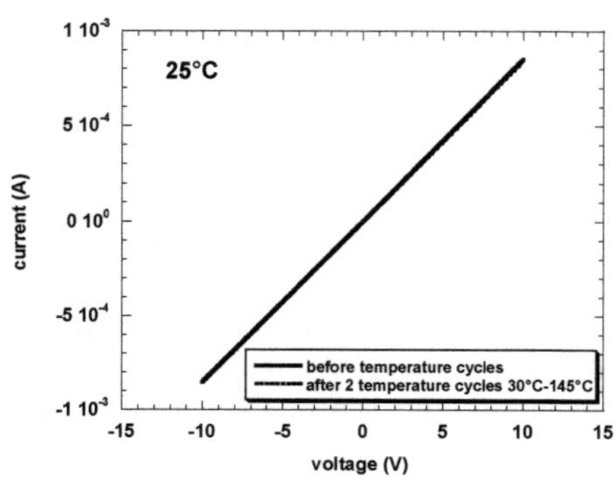

Figure 5. Room temperature current–voltage characteristics of an epoxy/CNT composite before and after two temperature cycles between 30°C and 145°C.

The current–temperature characteristics, shows again the negative temperature coefficient (NTC) behavior with a decreasing sample resistance with increasing temperature. Without Joule heating in the investigated temperature range the characteristics can now be fitted by adding a quadratic term to the polynomial fitting algorithm (see inset of Fig. 4).

The current–voltage characteristics (see Fig. 5) shows good ohmic behavior and no difference between the measurements, performed before and after these two temperature cycles, can be detected. This confirms the good stability of the sensor, when remaining below the glass transition temperature of the composite, that has a value of about 230°C.

4. Conclusions

A low-cost temperature sensor, based on the temperature increase of the conductivity of an epoxy loaded with multi-walled carbon nanotubes has been developed. Stable operation under thermal cycling has been demonstrated.

Acknowledgments The authors would like to sincerely thank Prof. Vittoria Vittoria and Dr. Liberata Guadagno for their thoughtful and helpful comments.

References

1. Y. Arai, C. Ng, P. Liu, L. Dong, Y. Imaizumi, K. Maeda, H. Maruyama, A. Ichikawa and T. Kukuda, "Ultra-small site temperature sensing by carbon nanotube thermal probes", Proceedings of the 4th IEEE Conference on Nanotechnology, pp. 146–148, Munich, 16–19. August 2004
2. H.C. Neitzert, N.G. Rainone, O. Valentino, M.R. Nobile, M. Sarno and P. Ciambella, "Monitoring of the sample electrical conductivity during temperature cycling of polyethylene/ CNT composites", Proceedings of the 24th Meeting of the Polymer Processing Society, S13–783, p. 286, Salerno, 15–19. June 2008

THERMOELECTRIC SENSOR FOR DETECTION OF CHEMICAL RADIATION HEAT

A. CATINI[1], E. GIOIA[1], E. GIROLAMI[1], L. SPAGNOLO[1], C. DI NATALE[1], A. D'AMICO[1], M. BARI[2] AND L. AHMAD[3]

[1]*Department of Electronic Engineering, University of Tor Vergata, Rome*
[2]*Department of Molecular Biology, University of Tor Vergata, Rome*
[3]*Invoy Technologies, L.L.C.*

Abstract. This work provides a method for the evaluation of the radiation heat emitted during a chemical reaction between two reagent substances. This method allows the measure of the temperature variations through the use of a thermoelectric sensor [1]. The procedure has led to satisfactory results and does represent a promising tool for medical applications.

1. Introduction

In recent years the strong development in biotechnology has led to a continuous research on new methodologies for accurate measurements at the limits of the real possibilities of the existing instrumentations.

Radiation based heat transfer can be fruitfully used to estimate the heat developed by a biochemical binding reaction relevant in the field of biomedicine.

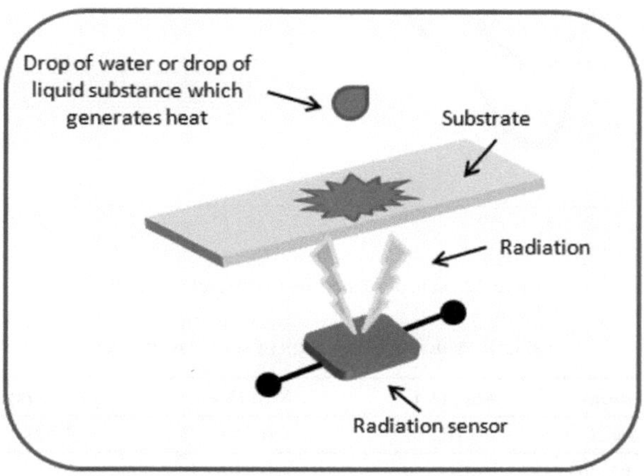

Figure 1. Schematic of the sensor system used for the detection of a chemical reaction.

P. Malcovati et al. (eds.), *Sensors and Microsystems: AISEM 2009 Proceedings*, Lecture Notes in Electrical Engineering 54, DOI 10.1007/978-90-481-3606-3_48,

This work considers the possibility of measuring small heat quantities by using highly sensitive temperature transducers taking advantage of the radiation heat which is almost always present in a given biochemical reaction.

Figure 1 shows the principal on which the measurement is based. Onto a given biocompatible substrate a biochemical reaction is left to develop itself giving it the possibility of emitting infrared radiation in the region of wavelength compatible with the temperature distribution involved. This radiation is detected by a very low noise integrated thermopile followed by a suitable preamplifier.

2. Experimental results

Several tests were performed to estimate the radiation heat using a thermopile as thermal sensor. The reaction takes place because of the contact between the substrate filter and a drop of water solution containing the reagent. The preparation method of the substrate concerns a chemical protocol [2] in which successive washing and fixing phases are made with appropriate substances. The interactions of these chosen substances (antigen-antibody) give rise to a chemical reaction with a theoretical enthalpy of about 120 kJ/mol. Through quantitative analysis of the substances involved in the test, it was calculated that the amount of heat released by these kinds of reactions are in the range of 0.2–0.8°C. Figure 2 shows, as an example, experimental results obtained in presence of a given reaction and in the case of the simple interaction between the functionalized substrate and water.

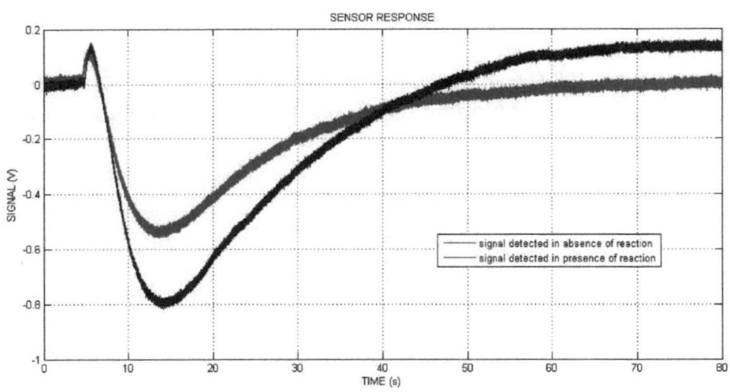

Figure 2. Sensor responses to different stimuli.

Table 1. Principal parameters of sensor responses.

Condition	Max (V)	Min (V)	Time (s)
Reaction	0.165	−0.569	47.21
No reaction	0.11	−0.819	61.12

The signals depicted in Fig. 2 show, except an initial peak (Table 1), a first decrease due to a cooling effect of the sensor surrounding in presence of a water drop, followed by an exothermal effect induced by the chosen chemical reaction.

The steady state is reached more quickly in presence of the chemical reaction, so the time taken to reach the equilibrium can be a useful sign for the discrimination of the two conditions: presence of water drop or presence of thermal reaction.

The experimental setup is shown in Fig. 3. The most relevant part concerns the sensor chamber visible on the left where the hot junction of the thermopile is in contact with air while the cold junction are embedded in an aluminium case which has the duty of keeping about constant the reference junction temperature. The output of the thermopile goes to a low noise amplifier characterized by a selectable low frequency filter and then to a data acquisition system able to collect about 10,000 sample/s. A standard PC completes the measuring chain.

In practice, first a drop of water is placed on the sensitive part of the substrate and a signal is recorded due to the radiation induced by the drop itself; second a drop of water containing antigen is located in the same substrate and during the bioreaction the thermopile again detects, by radiation changes, the reaction temperature function. Numerical evaluation follows for the rational estimation of the differences among the two signals. From the thermal point of view the measurement has to consider different kind of heat sources rates all connected each other by the general equation:

$$\dot{Q}_1 + \dot{Q}_2 + \dot{Q}_3 + \dot{Q}_4 = 0$$

where $\dot{Q}_1$ is related to the mass of the antigen and substrate and to their thermal capacities, $\dot{Q}_2$ represents the heat rate due to the drop water evaporation, $\dot{Q}_3$ represents the emission of heat developed by the reaction and $\dot{Q}_4$ represents the heat of reaction proportional to the total mass of the reagents, to the reaction velocity according to Arrhenius and finally to the degree of the biochemical reaction.

All these leads to the estimation of the reaction temperature profile which is compared with the expected heat theoretically determined. The satisfactory agreement between the theoretical model and the experimental findings can be estimated to be below 10%.

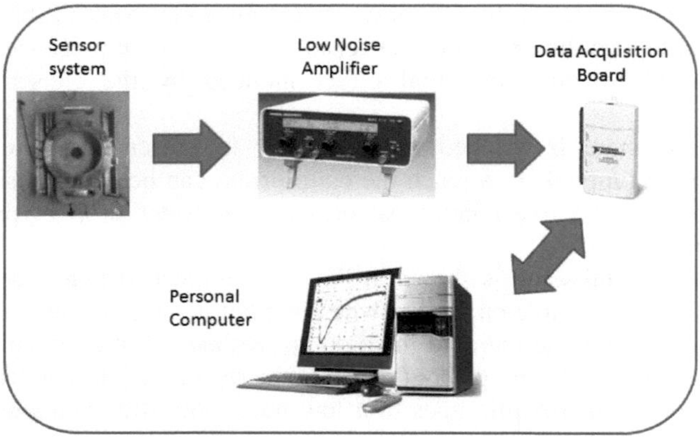

Figure 3. Experimental setup.

3. Conclusions

Many reactions at biological level between antigen and antibody or among different kinds of proteins develop a certain amount of heat which is not completely random but it seems to contain useful information. In this work we have designed a measuring chain which can allow temperature changes of the order of 0.1°C to be detected during biochemical reaction. The study of the time evolution of the heat developed in circumstances where a certain degree of pathology is present may help to understand intrinsic biomechanisms responsible of the most important diseases. This work has demonstrated that small amount of heat can be estimated by detecting the radiation emitted by a given biochemical reaction occurring on a suitable substrate. Further improvements are possible along the line of reducing lateral heat diffusion on the substrate and orient the radiation, as much as possible, on the sensitive part of the thermopile.

References

1. Jürgen Schilz – THERMOELECTRIC INFRARED SENSORS (THERMOPILES) FOR REMOTE TEMPERATURE MEASUREMENTS. PYROMETRY – EG & G Heimann. Optoelectronics: thermoelectric IR sensors (13 Aug. 1999).
2. Suggested Chemical Interacting Materials for Immobilization on Pyroelectric Detector, Invoy Technologies, L.L.C.

SQUID SENSORS FOR HIGH SPATIAL RESOLUTION MAGNETIC IMAGING AND FOR NANOSCALE APPLICATIONS

A. VETTOLIERE, C. GRANATA*, P. WALKE, E. ESPOSITO, B. RUGGIERO AND M. RUSSO

Istituto di Cibernetica "E. Caianiello" del CNR, Via Campi Flegrei, 80078 Pozzuoli (NA), Italy

Abstract. We present the improved Superconducting Quantum Interference Devices (SQUID) for magnetic microscopy and for nanoscale investigations. Low critical Temperature SQUIDs with integrated micro pick-up coils (with inner diameter ranging from 5 to 50 μm) which can be used as magnetic field sensors in magnetic microscopy have been developed. The level of flux noise spectral density, measured in flux locked loop configuration and using a direct coupled scheme, is about 3 $\mu\Phi_0/\sqrt{Hz}$ in the white region at T = 4.2 K.

A high sensitive dc-SQUID based on niobium Dayem bridges for nanomagnetism is presented. The sensor has a flux capture area as low as 0.04 μm^2, allowing the study of nano-object magnetic properties. The authors report the main design rules of the devices, the fabrication processes and their characterization including also the supercurrent decay measurements at T = 4.2 K.

*E-mail: c.granata@cib.na.cnr.it

1. Micro-SQUID for magnetic microscopy

SQUID (Superconducting Quantum Interference Device) sensors are successfully employed for magnetic microscopy (SSM) which is a powerful technique for imaging magnetic field distribution.

A SQUID sensor for magnetic microscopy should have a high spatial resolution to detect very small magnetic sources, and a high ratio between the measured magnetic flux Φ_r and that one applied to pick-up loop Φ_p. Moreover, it should be insensitive to any background magnetic field [1].

In this framework, SQUID sensors in both magnetometer and gradiometer configuration have been designed, fabricated and characterized.

In such configurations a circular pick-up loop is a part of the SQUID's self inductance and it is connected to Josephson junctions by a stripline consisting of two superconducting layers separated by a double insulation layer. In order to modulate the sensor and to operate in flux locked loop mode (FLL) [2] a washer structure has been inserted to 1.4 mm away from pick-up loop, between the end of the stripline and the Josephson junctions. The sample fabrication process,

P. Malcovati et al. (eds.), *Sensors and Microsystems: AISEM 2009 Proceedings*, Lecture Notes in Electrical Engineering 54, DOI 10.1007/978-90-481-3606-3_49,

capable to routinely produce high quality window-type junctions, has been well described elsewhere [3]. A picture of the device is showed in Fig. 1. For both configurations, SQUIDs with four pick-coil sizes of inner diameter ranging from 5 to 50 μm have been fabricated. In order to reduce the flux noise contribution of the preamplifier to a tolerable value, the dc SQUID has been properly designed to obtain a high intrinsic responsivity which, for a low critical temperature SQUID, is given by $V_\Phi = 4I_0R/\Phi_0(1+\beta)$; here R is the shunt resistance of the Josephson junctions, I_0 is the critical current, Φ_0 is flux quantum and $\beta = 2LI_0/\Phi_0$ is the inductance parameter where L is the SQUID inductance.

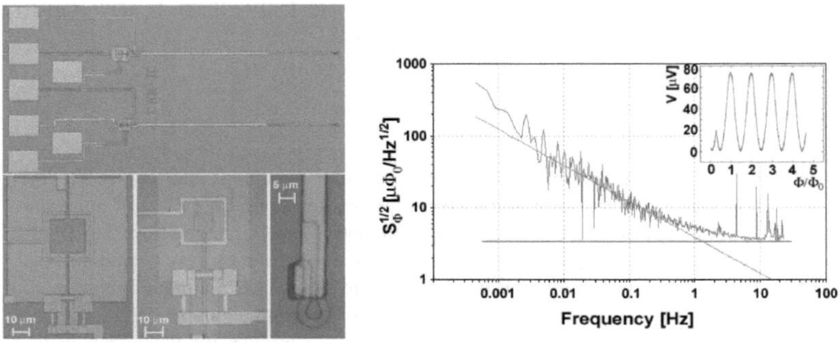

Figure 1. A picture of a microSQUID is shown in the left figure in both magnetometer and gradiometer configuration; a 5 μm inner diameter micro-coil is also shown. In the right figure the spectral density of low frequency flux noise of a micro-SQUID is reported. The inset shows the voltage-flux characteristic of the device.

An increase of V_Φ has been obtained increasing the critical current rather than the shunt resistors value. In fact, a higher resistance could lead to a more significant increase of McCumber parameter ($\beta_c = 2\pi I_0CR^2/\Phi_0$, C is the junction capacitance) value, due to its quadratic dependence from the resistance. In such a way a V_Φ value up to 1 mV/Φ_0 has been obtained which is about four to five times greater than the typical values of micro-SQUIDs reported in the literature, allowing to reduce the equivalent flux noise of the readout electronics to a value of about 0.5 μΦ_0/Hz$^{1/2}$, being the voltage noise of the readout amplifier as low as 0.5 nV/Hz$^{1/2}$. The magnetic flux noise spectral density of the SQUID sensors exhibit, typically, a white magnetic flux noise level less than 3 μΦ_0/Hz$^{1/2}$ measured in FLL at T = 4.2 K (Fig. 1). In order to verify the 1/f behavior and to determine the corner frequency, we have extended the measurement of the noise spectrum to low frequencies. The spectral noise density reported in Fig. 1 showing good 1/f behavior over more than 2 decades below 1.5 Hz. Such devices have been successfully employed to study the spontaneous current in superconducting arrays of π-junctions [4].

2. Nano-SQUID based on niobium Dayem bridges

Recently, great efforts have been devoted to the development of nanoscale SQUIDs to measure small spin populations and the magnetization of nanoparticles [5].

An improved nano-SQUID based on niobium Dayem bridge shunted by an aluminum film is presented. The device includes two micrometric integrated niobium coils located very close to the nano-sensor in order to modulate, tune and operate the SQUID in FLL mode. With respect to external coils, they show a higher mutual inductance allowing to effectively measure the voltage-flux characteristics (V–Φ) and an easy flux biasing of the SQUID at the point corresponding to the maximum value of the responsivity (V_Φ).

The device consists of a niobium loop in a washer shape, to enhance the heat dissipation, with a hole of 0.04 μm^2 having two nano-constriction with a length and a width of 100 and 80 nm, respectively (Fig. 2).

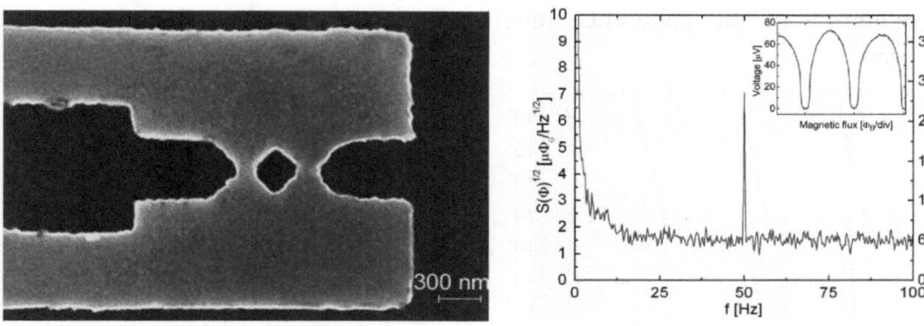

Figure 2. Scanning electron micrograph showing the SQUID loop with the nano-bridge Josephson junctions (*left*) and spectral density of magnetic flux noise (*right*). The inset shows the flux to voltage characteristic of a nanoSQUID.

The fabrication process of thin films is based on patterning by both electron beam lithography (EBL) and optical lithography as well as lift off and reactive ion etching (RIE) processes [6]. The V–Φ characteristic is reported in the inset of Fig. 2; an intrinsic voltage swing and a maximum responsivity of 75 μV and 1 mV/Φ_0 respectively are measured from the characteristic. The sensor exhibits a magnetic flux noise level of 1.5 μΦ_0/Hz$^{1/2}$ in the white region, measured in open loop mode (small signal mode). It corresponds to a spin noise in Bohr magneton unit of $S_n^{1/2} = 2aS_\Phi^{1/2}/\mu_0\mu_B \approx 60\,spin/Hz^{1/2}$ where *a* is the radius of the SQUID loop, μ_0 is the magnetic vacuum permeability and μ_B is the Bohr magneton [7].

The Fig. 3 shows the measurements of the switching current distributions (P) from the zero voltage state and of the related escape rates (Γ) as function of the bias current (I), performed by standard fly time technique at different temperature. Theoretical curves are obtained in the framework of the thermal theory in the moderate damping limit [8] with Ic = 506 μA, R = 20 Ω and considering an estimated capacitance of 1 fF, due to the edges of both the current leads and banks of the washer, approximated by a parallel plate configuration. However, in the range of few fF and tens of Ohm the theoretical predictions do not show evident variations [9]. We can define a magnetic flux resolution as the minimum external magnetic flux required to yields a switching from the metastable zero voltage state to the normal one and it is related to the current sensitivity and to the switching current distribution width (σ) by:

$$\Delta\Phi = \frac{1}{2}\frac{\sigma}{(\partial I_c/\partial\Phi)} \quad ; \qquad \sigma = \sqrt{\langle I^2\rangle - \langle I\rangle^2}$$

In Fig. 3 the magnetic flux resolution ΔΦ as function of the temperature T is also reported, showing that the magnetic flux resolution does not improves significantly (less than 10%) by decreasing the temperature. In fact, the decreasing of σ is compensated by a decreasing of the current responsivity due to a reduction of the coherence length which occurs when the temperature decreases.

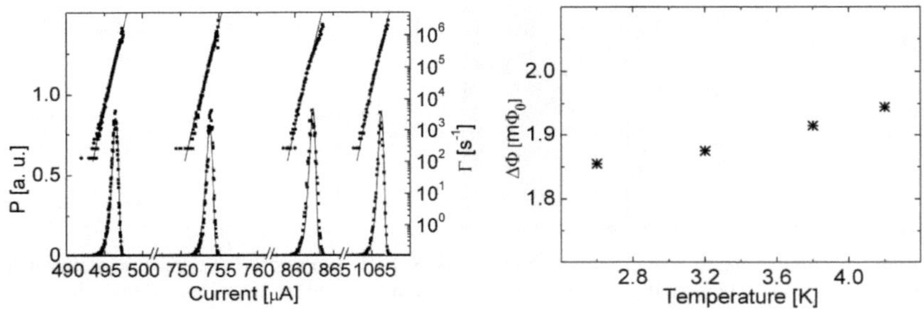

Figure 3. Experimental data (dots) and theoretical predictions in the moderately damping limit (solid lines) of the switching current distribution (P) and the escape rates (Γ) vs the bias current I of a nano-SQUID measured at different temperatures. Magnetic flux resolution ΔΦ as function of the temperature of a nano-SQUID.

3. Conclusions

Low noise SQUID sensors for high spatial resolution magnetic imaging and nanoscale applications have been presented. The micro-SQUIDs have been successfully employed as magnetic field sensors in a scanning magnetic microscopy [4]. Improved nano-SQUIDs, including an integrated calibration micrometric superconducting circuitry has been fabricated and characterized. Possible applications for such nano-devices are in the study of magnetic of nano-particles, in the measurement of nano-object magnetization, in quantum computing and in the memory devices.

References

1. C. Granata, A. Vettoliere, R. Vaccarone, and M. Russo, IEEE Trans. on Appl. Supercond., [vol. 17], 796, 2007.
2. R.K. Fagaly, Rev. Sci. Instrum., [vol. 77], 101101, 2006.
3. C. Granata, A. Vettoliere, and M. Russo, Appl. Phys. Lett., [Vol. 88], p. 212506, 2006.
4. S.M. Frolov, M.J.A. Stoutimore, T.A. Crane, D.J. Van Harlingen, V.A. Oboznov, V.V. Ryazanov, A. Ruosi, C. Granata, and M. Russo, Nat. Phys. 4, [vol. 32], 2008.
5. A.G.P. Troeman, H. Derking, B. Borger, G. Pleikies, B. Veldhuis, and H. Hilgenkamp, Nano Lett., [vol. 7], p. 2152, 2007.
6. C. Granata, E. Esposito, A. Vettoliere, L. Petti, and M. Russo, Nanotechnology [vol. 19], p. 275501, 2008.
7. S.K.H. Lam and D.L. Tilbrook, Appl. Phys. Lett. [vol. 82], p.1078, 2003.
8. H. Grabert, P. Olschowski, and U. Weiss, Phys. Rev. B, [vol. 36], p. 1931, 1987.
9. C. Granata, A. Vettoliere, R. Russo, E. Esposito, M. Russo, and B. Ruggiero, Appl. Phys. Lett. [vol. 94], p. 062503, 2009.

PERMING EFFECT IN RESIDENCE TIMES DIFFERENCE FLUXGATE MAGNETOMETERS

B. ANDO'[1], S. BAGLIO[1], S. LA MALFA[1], C. TRIGONA[1] AND A.R. BULSARA[2]
[1]*DIEES – University of Catania V.le A. Doria 6, 95125, Catania, Italy*
[2]*Space and Naval Warfare System Center, Code 2363, 49590 Lassing Road, San Diego, CA 92152-6147, USA*

Abstract. The aim of this paper is to present experimental results on the estimation of perming effect in two different experimental prototypes of fluxgate magnetometer based on the Residence Times Difference (RTD) readout strategy. The perming effect can affect the output signal of the device (by an offset) after a magnetic shock. It is similar to hysteresis, but the applied field must be much higher than the full-scale range. All the sensors containing ferromagnetic material are susceptible to perming. The obtained experimental results allow for the definition of guidelines for the optimal re-magnetization strategy of the ferromagnetic core in the particular case of RTD fluxgate magnetometers.

1. Fluxgate magnetometers

Fluxgate magnetometers are suitable to sense low magnetic fields or magnetic field perturbation, at room temperature. In particular, The Residence Times Difference (RTD) Fluxgate magnetometer represents an innovative solution to detect quasi static magnetic field. The authors have proposed RTD fluxgate as competitive devices to the traditional second harmonic architectures [1–5]. Low cost, small dimension, high sensitivity, low noise floor, low power consumption and an intrinsic digital form of the output signal are the main advantages of this innovative strategy. The RTD fluxgates find applicability in fields such as space, biomedical, vehicle navigation, security, military, geomagnetic field measurement and security applications.

1.1. *The μWire RTD fluxgate prototype*

The μWire fluxgate considered through the rest of the paper is based on a filiform ferromagnetic core that exhibits very interesting magnetic properties such as a sharp hysteresis loop and low coercitive fields value (about 38 A/m, experimentally estimated). The "wire core" material has a 100 μm magnetic structure diameter (FeSiB Amorphous magnetic microwire). The RTD fluxgate exploiting the "wire core" can be used to measure spot and directional magnetic fields allowing to point at very small targets. A simplified construction description can be summarized

P. Malcovati et al. (eds.), *Sensors and Microsystems: AISEM 2009 Proceedings*, Lecture Notes in Electrical Engineering 54, DOI 10.1007/978-90-481-3606-3_50,

by the following steps: a two coils structure (primary coil and secondary coil) is wound around a (glass or plastic) support, where a 100 μm ferromagnetic amorphous wire is hosted (Fig. 1). The main features of the prototype are summarized in Table 1.

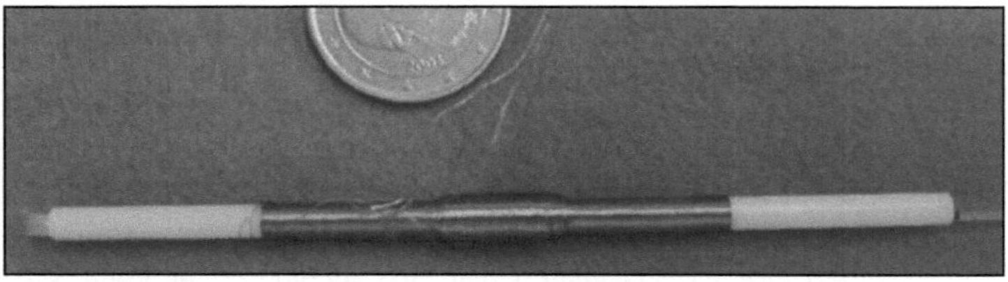

Figure 1. The μWire RTD fluxgate prototype.

Table 1. Main specifications of the μWire RTD Fluxgate prototype. Values refers to the operating condition defined by an excitation current of 20 mA @ 80 Hz.

Parameter	**Value**
Sensitivity	0.05 us/nT
Resolution	0.1 nT
Power consumption (peak)	12 mW

2. Perming effect

The perming effect can affect the output signal of the device (by an offset) after a magnetic shock. It is similar to hysteresis, but the applied field must be much higher than the full-scale range. All the sensors containing ferromagnetic material are susceptible to perming. This procedure ensures a defined magnetic state and is called re-magnetization. In the case of fluxgate sensors, the re-magnetization is performed by forcing a current through the primary coil. Anyway, it must be considered that no significant perming is observed if the device is polarized by a high current value [6].

To suppress the offset, it might be useful to employ the same technique used by the AMR sensor [7], consisting in the application of a high current pulse to the core to stabilize its the magnetic status before the normal operation of the sensor.

2.1. *The experimental set-up to investigate the perming offset*

The experimental set-up consists of a Residence Times Difference fluxgate magnetometer (μWire prototype) operated with a triangular bias current @ 80 Hz and a four-layers solenoid in order to produce the external reference target magnetic field necessary for the sensor characterization. The experimental characterization procedure is schematized in Fig. 2 and consists of six consecutive steps:

1. In the first step the amplitude of the driving current is linearly decreased to zero.
2. A triangular shaped magnetic field is forced through the external solenoid to infer the core re-magnetization. The amplitude of the applied field must be high enough to ensure a sufficient saturation of the core (in our case the peak-to-peak amplitude is 20 mT). During this phase the magnetometer is not driven by any excitation current as it could vary its magnetic state.
3. The driving current is restored to its nominal value.
4. 30 s of the output signal (RTD) with nominal zero applied field are acquired.
5. A high DC magnetic field is applied through the external solenoid for 15 s.
6. The applied field is turned off and a RTD shot of 30 s is acquired. As it can observed an offset appears in the output signal as respect to the same signal analyzed in phase (D). This quantity, divided by the sensor sensitivity, can be considered as the perming offset. It must be noted that the output signal of the sensor is not acquired during phases A, B, C and E (grayed areas in Fig. 2).

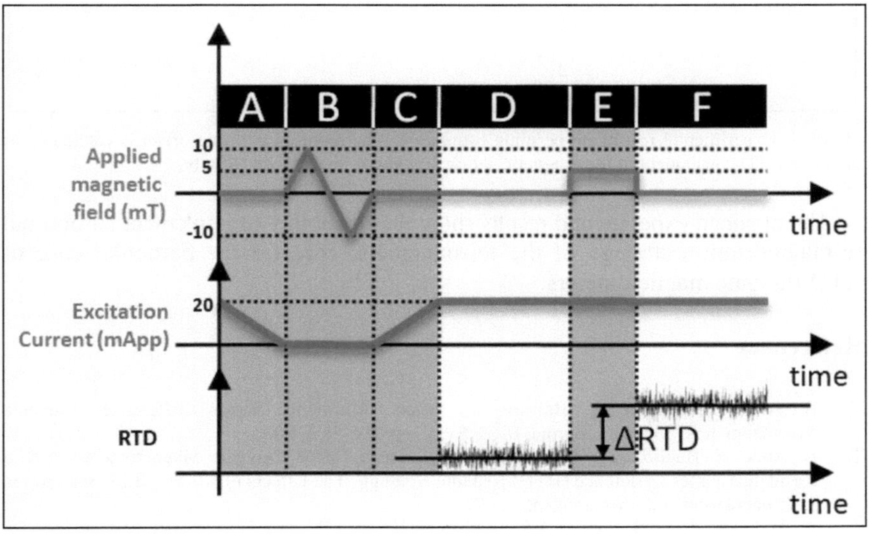

Figure 2. Qualitative schematization of the experimental procedure.

3. Results and conclusion

Experimental results concerning the estimation of the perming offset in the μWire RTD fluxgate are summarized in Fig. 3. For each excitation current amplitude (2, 5, 10, 15, 20 mA) the variation of the RTD measured before and after the application of the magnetic shock, ΔRTD, was estimated. Results obtained show that the perming effect increases quite linearly with the field intensity. Moreover, this effect is less evident when excitation current amplitude increases. This is quite reasonable because an higher driving current produces a higher saturation of the ferromagnetic core that hence becomes less susceptible to perming.

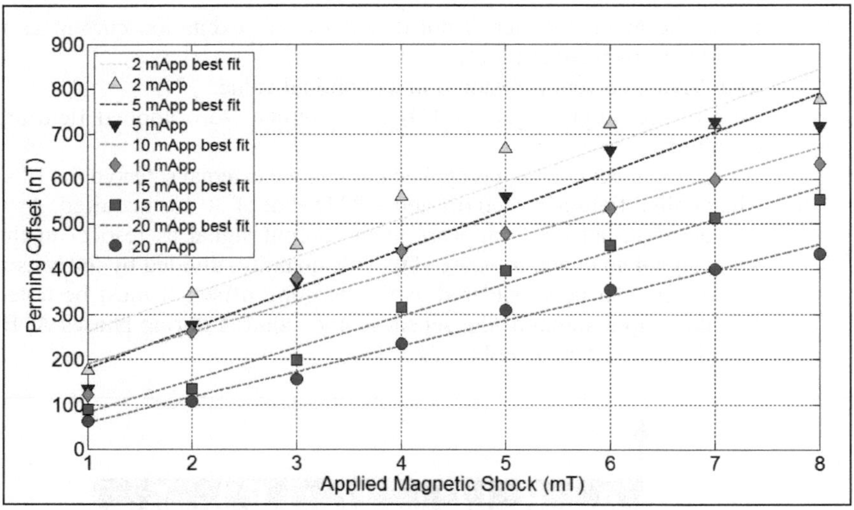

Figure 3. Experimental results on perming estimation. The reported perming offset is calculated as the delta-RTD consequent to the magnetic shock divided by sensor's sensitivity.

The obtained experimental results show the possibility to implement an optimal re-magnetization strategy of the ferromagnetic core for the particular case of RTD fluxgate magnetometers.

References

1. B. Andò, S. Baglio, A. Bulsara, V. Sacco, "Residence Times Difference" Fluxgate Magnetometers", Sensors Journal, IEEE 5 (5). pp. 895–904, 2005.
2. B. Ando, S. Baglio, A.R. Bulsara, V. In V. Sacco, "PCB Fluxgate Magnetometers with a Residence Times Difference (RTD) Readout Strategy: The Effects of Noise", IEEE transaction Instrumentation and Measurements, 2007.
3. B. Ando, A. Ascia, S. Baglio, A.R. Bulsara, C. Trigona, V. In RTD Fluxgate performance for application in magnetic label-based bioassay: preliminary results, proceeding of IEEE – EMBC, 2006.
4. B. Ando, A. Ascia, S. Baglio, A. R. Bulsara, V. In N. Pitrone, C. Trigona, Residence Times Difference (RTD) Fluxgate Magnetometer for Magnetic Biosensing, AIP American Institute of Physics, 2007.
5. B. Ando, S. Baglio, V. Caruso, V. Sacco, A. Bulsara, Multilayer based technology to build rtd fluxgate magnetometer, IFSA, Sensors & Transducers Magazine, Vol. 65, issue 3, pp. 509–514, 2006.
6. P. Ripka, Magnetic Sensors and Magnetometers. Boston, MA: Artech House, 2001.
7. Honeywell, "1- and 2-Axis Magnetic Sensors, HMC 1001/1002 and HMC1021/1022," Datasheet 900248 Rev. B.

DIFFUSE-LIGHT ABSORPTION SPECTROSCOPY BY MEANS OF A FIBER OPTIC *SUPERCONTINUUM* SOURCE – AN INNOVATIVE TECHNIQUE

A.G. MIGNANI[1], L. CIACCHERI[1], I. CACCIARI[1], H. OTTEVAERE[2], H. THIENPONT[2], O. PARRIAUX[3] AND M. JOHNSON[4]

[1]*CNR-IFAC, Sesto Fiorentino (FI), Italy*
[2]*Vrije Universiteit Brussel, Dept. Applied Physics and Photonics – Brussels, Belgium*
[3]*Laboratoire Hubert Curien CNRS, Université Jean Monnet, Saint-Etienne, France*
[4]*Sagentia Ltd., Harston, Cambridge, UK*

Abstract. An innovative fiber optic setup for the scattering-free absorption spectroscopy of liquids is presented. It makes use of an integrating sphere that contains the sample under test, coupled to a fiber optic *supercontinuum* source and to a fiber optic spectrometer. A collection of turbid lubricant oils was considered as a test case for verifying and validating the innovative scheme of diffuse-light absorption spectroscopy. Scattering-free spectra were successfully measured and processed as product fingerprints for the prediction of turbine types.

1. Motivation: the drawbacks of absorption spectroscopy

When carrying out absorption spectroscopy in liquids, the measurements are found to be influenced by dispersed scattering particles that cause turbidity. Scattering dependence is the main drawback of absorption spectroscopy, especially when particles settle down, or change their position in time. For this reason, liquids must be filtered prior to measurements, with a consequent loss of time and increased costs. In addition, unless a maintenance program that contrasts with automatic operation is applied, filtering is not suitable for industrial online applications.

Diffuse-light absorption spectroscopy, that is, spectroscopy carried out by means of an integrating sphere, is a valid alternative for overcoming scattering-dependence aspects [1, 2]. Unfortunately, because it requires high-intensity sources, diffuse-light absorption spectroscopy provides low efficiency when carried out by means of optical fibers for illumination and detection.

This paper presents an innovative and efficient setup for diffuse-light absorption spectroscopy measurements. It is fully assisted by optical fibers, and, consequently, is a precursor of online measurements. A collection of turbid lubricant oils was considered as a test case for verifying and validating the innovative measuring

P. Malcovati et al. (eds.), *Sensors and Microsystems: AISEM 2009 Proceedings*, Lecture Notes in Electrical Engineering 54, DOI 10.1007/978-90-481-3606-3_51,

setup. Scattering-free spectra were successfully measured and processed as product fingerprints for the prediction of turbine types.

2. Diffuse-light absorption spectroscopy: concept and setup

Diffuse-light absorption spectroscopy makes use of an integrating sphere that contains the sample under test. The source and the detector are butt-coupled to the sphere, as shown in Fig. 1-top. Almost all the light shining on the sphere surface is diffusely reflected, and the detector can be placed anywhere in the sphere in order to gather the average flux [3]. By inserting an absorbing medium in the cavity, a reduction of the radiance in the sphere occurs. The reduction is related to the absorption of the medium and to its volume, and is independent of non-absorbing objects within it, such as suspended scattering particles.

Efficient diffuse-light measurements need bright sources. Recently, the revolutionary advent of compact, high brightness *supercontinuum* fiber optic sources has changed the perspectives of optical spectroscopy [4, 5]. The bright light generated by the holey fiber over a wide spectral range can easily be coupled to a conventional optical fiber and guided to the input port of an integrating sphere. Another port of the sphere can accommodate an optical fiber coupled to a spectrometric detector, so as to achieve an efficient setup for diffuse-light absorption spectroscopy, as shown in Fig. 1-bottom.

Commercially-available components were used for the practical implementation of the experimental setup [6]. The Fianium-SC400 fiber optic *supercontinuum* source was used for illumination: it emits 4 W throughout the entire 415–1,800 nm spectral range. The Instrument System-Spectro 320 fiber optic spectrometer was used as detector, and scanned the wide 400–1,700 nm spectral range with a resolution of 1.37 nm. The Labsphere LMS100 cavity was used as a diffusing sphere, the ports of which were equipped by means of fiber optic connectors for coupling to both the source and the detector.

3. Results

A collection of 27 lubricant oils was considered as a test case. These samples came from three different types of turbines: six from an aeronautical-type gas turbine (TGAD) – six from a steam turbine (TV) – 15 from a gas turbine (TG). TGAD oils were of polyester-synthetic type, while the others were mineral-paraffin oils. These oils were previously analyzed by means of several fiber optic spectroscopy schemes. These showed the possibility of correlating the optical spectral signature to important quality and functional indicators [7]. None of the previous spectroscopy measurements was considered scattering-free, while suspended particles were still considered a problem to be coped with.

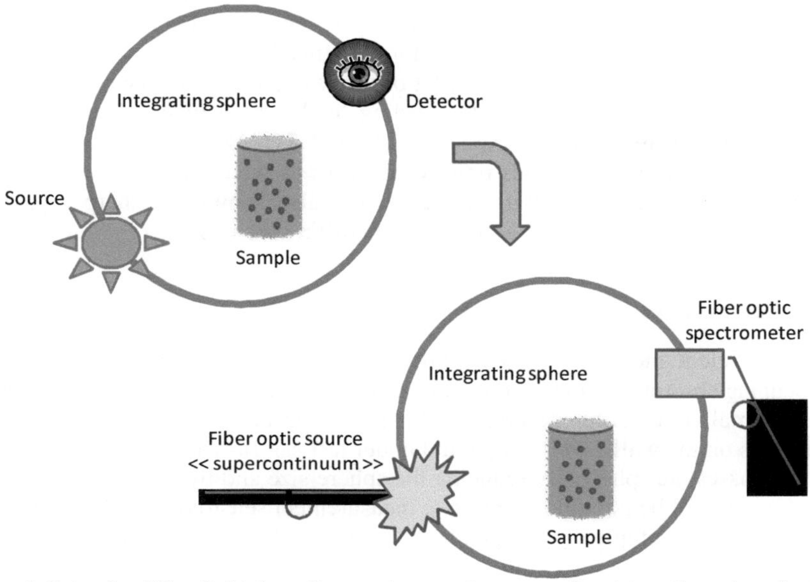

Figure 1. Setup for diffuse-light absorption spectroscopy by means of an integrating sphere (*top*), and its implementation with the use of a fiber optic *supercontinuum* source and a fiber optic spectrometer (*bottom*).

Figure 2-left shows the measured scattering-free diffuse-light absorption spectra. These spectra are considered to be product signatures. As previously performed by means of other non scattering-free spectroscopy techniques, an attempt was made to correlate the spectral signatures to oil indicators such as the type of the oil and the turbine of origin.

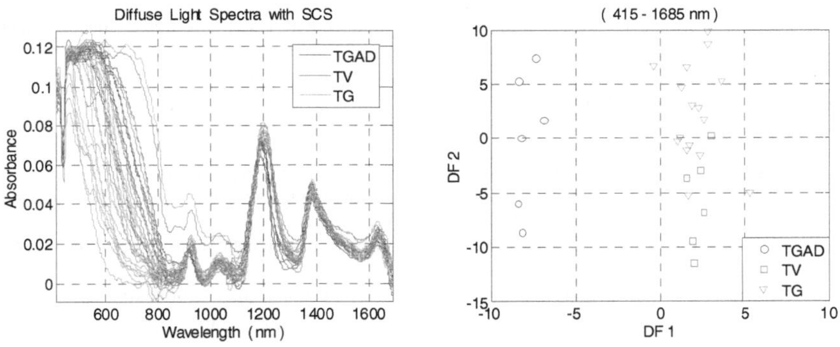

Figure 2. Scattering-free diffuse-light absorption spectra of lubricant oils as a test case (*left*), and clustering of oil types according to the turbine of origin (*right*).

The lubricant oil spectra were classified using a two step procedure involving Principal Component Analysis (PCA) for reducing data dimensionality, and Linear Discriminant Analysis (LDA) for classification.

Figure 2-right shows the score plot of the two main discriminating functions. The oils were clearly clustered according to the three types of machinery in which they were used. As expected, TV and TGAD oils were more tightly grouped, being of two brands only, while TG oils were widely spread being of several brands. As expected, the score plot also shows the sharp separation between polyester-synthetic and mineral-paraffin oil types.

4. Perspectives

The efficient measuring scheme presented is the precursor of a scheme for obtaining absorption spectroscopy information online. The probe can be configured as a small integrating sphere, similar to a ping-pong ball, with fiber optics fixtures on the walls, while the liquid under test can flow inside a glass capillary that crosses the sphere. By reducing the sphere size and by increasing the liquid volume it will be possible to improve measurement sensitivity while keeping the measurement completely independent of any turbidity of the liquid.

Acknowledgements The European Network of Excellence on Micro-Optics, and the ICT-ONE Project are acknowledged for their partial financial support. The work of H. Ottevaere was supported by the Flemish Fund for Scientific Research (FWO) which provided her with a 'Postdoctoraal Onderzoeker' fellowship. Mecoil Diagnosi Meccaniche srl is acknowledged for providing the lubricant oil samples.

References

1. I. Fecht, M. Johnson, 'Non-contact scattering independent water absorption measurement using a falling stream and integrating sphere', *Meas. Sci. Tech.*, vol. 10, 1999, pp. 612–618.
2. T. Jàvorfi, *et al.*, 'Quantitative spectrophotometry using integrating cavities', *J Photochem. Photobiol. B*, vol. 82, 2006, pp. 127–131.
3. E.S. Fry, *et al.*, 'Integrating cavity absorption meter', *Appl. Opt.*, vol. 31, 1992, pp. 2055–2065.
4. A.R. Alfano Ed., *The Supercontinuum Laser Source*, 2nd Ed, Springer, Berlin, 2006.
5. C.F. Kaminski, *et al.*, '*Supercontinuum* radiation for applications in chemical sensing and microscopy', *Appl. Phys. B*, vol. 92, 2008, pp. 367–378.
6. http://www.fianium.com/ - Detector: http://www.instrumentsystems.com/ – Integrating sphere: http://www.labsphere.com/
7. A.G. Mignani, et al., 'Optical fiber spectroscopy for measuring quality indicators of lubricant oils', *Meas. Sci. Tech.*, vol. 20, 2009, doi:10.1088/0957-0233/20/3/034011

SYSTEMS

A DIFFERENTIAL DIFFERENCE CURRENT-CONVEYOR (DDCCII) BASED FRONT-END FOR INTEGRABLE AND PORTABLE SENSOR APPLICATIONS

A. DE MARCELLIS[1], C. DI CARLO[1], G. FERRI[1], V. STORNELLI[1], A. D'AMICO[2], C. DI NATALE[2] AND E. MARTINELLI[2]

[1]*Department of Electrical and Information Engineering, University of L'Aquila, Località Monteluco di Roio, 67040 L'Aquila, Italy*
[2]*Department of Electronic Engineering, Roma Tor Vergata, Via del Politecnico, Roma, 00130, Italy*

Abstract. In this work we propose a very low voltage low power Differential Difference Amplifier based current conveyor (DDCCII) for the readout of the voltage value coming from potentiometric sensors and to be used in a self-powered wireless sensor network for monitoring the presence of contaminants in hydric systems. Since the DDCCII is an extremely versatile block, it can be configured to give either an output voltage or a current directly proportional to the voltage sensor value. The whole circuit has been implemented in a 0.18 μm standard technology with a total supply voltage of 1 V and a standby current and power of about 10 μA and 10 μW, respectively.

1. Introduction

Recently there was a growing proliferation of wireless devices for many Industrial and commercial applications for security, safety, surveillance and monitoring. Although the wiring costs for communications and power cables have been eliminated with wireless devices, these outgoing are replaced with the hand out of battery maintenance. This includes battery purchasing, protection from potential leakage and potential equipment downtime due to unexpected battery failure [1]. In this sense a critical issue is the ability to scavenge energy from the environment and to use this energy for power remote distributed sensors, actuators, processors, memories and display elements [2]. In order to realize a self-powered operation, a generic circuit must operate without batteries or any separate voltage source with low power consumption conditions and should be robust to power supply drift [3]. In this sense, analog circuits based on the current mode (CM) approach have been demonstrated (with respect to those based on op-amps) to be particularly suitable for low voltage and low power applications and also benefit of the following advantages: wide bandwidth, high slew rate and simpler circuitry [4, 5].

In this work we propose a very low voltage low power Differential Difference Amplifier (DDA) based CCII (DDCCII) for the readout of the voltage value coming from a potentiometric sensors. This is the first step in the direction of

P. Malcovati et al. (eds.), *Sensors and Microsystems: AISEM 2009 Proceedings*, Lecture Notes in Electrical Engineering 54, DOI 10.1007/978-90-481-3606-3_52,

developing a self-powered wireless sensor network (showing a very low power dissipation) for monitoring the presence of ions contaminants in a sample liquid.

2. DDCCII basic concepts

Concerning the Differential Difference Current Conveyor (DDCC) [6], it enjoys the advantages of the CCII and DDA such as larger signal bandwidth, greater linearity, wider dynamic range, simple circuitry, low power consumption, high input impedance and arithmetic operation capability.

The DDA is a five-terminal device, shown in Fig. 1. It has two differential input ports, (V_{pp}–V_{pn}) and (V_{np}–V_{nn}), and a single-ended or differential output. The output of the DDA can be expressed as:

$$V_0 = V_{0p} - V_{0n} = A_0[(V_{pp} - V_{pn}) - (V_{np} - V_{nn})] \tag{1}$$

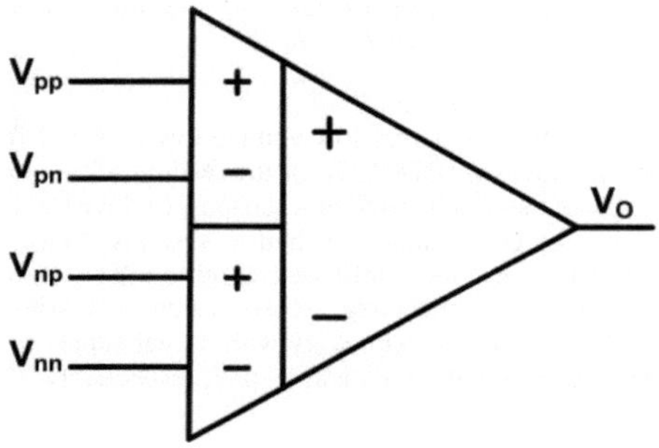

Figure 1. DDA block scheme.

Open-loop gain A_0 is required to be as large as possible in order to improve the DDA performance. At circuit level, like the op-amp, the DDA consists mainly of three stages: a differential-input stage (differential pair with active load) and two stages that typically are gain stages necessary to achieve a high open loop gain. Considering the theory developed in [7], where it has been demonstrated that for lower open loop gain it is better to use a current mode approach instead of classical op-amp based circuits for implementing high accuracy amplifiers, in this work the third gain stage of the DDA has been substituted directly by a buffer so to develop a fully differential DDCCII characterized by the following matrix representation:

$$\begin{bmatrix} V_X \\ I_{Y_1} \\ I_{Y_2} \\ I_{Y_3} \\ I_Z \end{bmatrix} = \begin{bmatrix} 1 & -1 & 1 & 0 \\ 0 & 0 & 0 & 0 \\ 0 & 0 & 0 & 0 \\ 0 & 0 & 0 & 0 \\ 0 & 0 & 0 & \pm 1 \end{bmatrix} \cdot \begin{bmatrix} V_{Y_1} \\ V_{Y_2} \\ V_{Y_3} \\ I_X \end{bmatrix} \tag{2}$$

Y_1 and Y_2 terminals are high impedance terminals while X_1 terminal is a low impedance one due to the buffer output stages and the negative feedback. The differential input voltage V_{Y12} applied between Y_1 and Y_2 terminals is conveyed to a single-ended differential voltage V_X ($V_X = V_{Y12}$). The input currents flowing into the X terminal are conveyed to the Z terminal. The Z terminal is a high impedance node suitable for current outputs (Fig. 2).

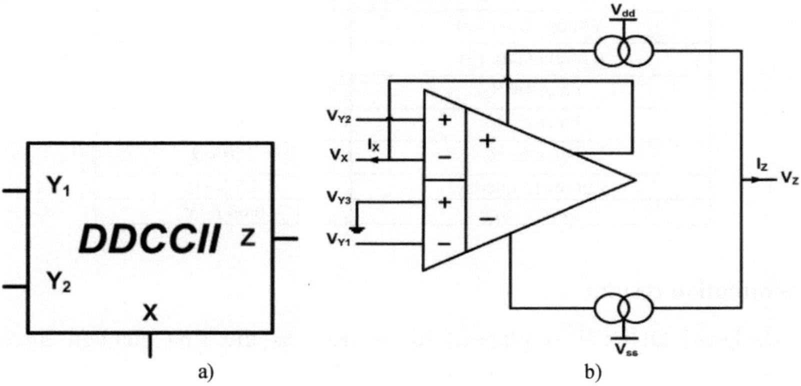

Figure 2. (**a**) DDCCII block scheme; (**b**) DDCCII basic internal implementation.

An internal DDCCII implementation at transistor level, using a CMOS 0.18 µm SMIC technology, is shown in Fig. 3 while Table 1 shows its main characteristics.

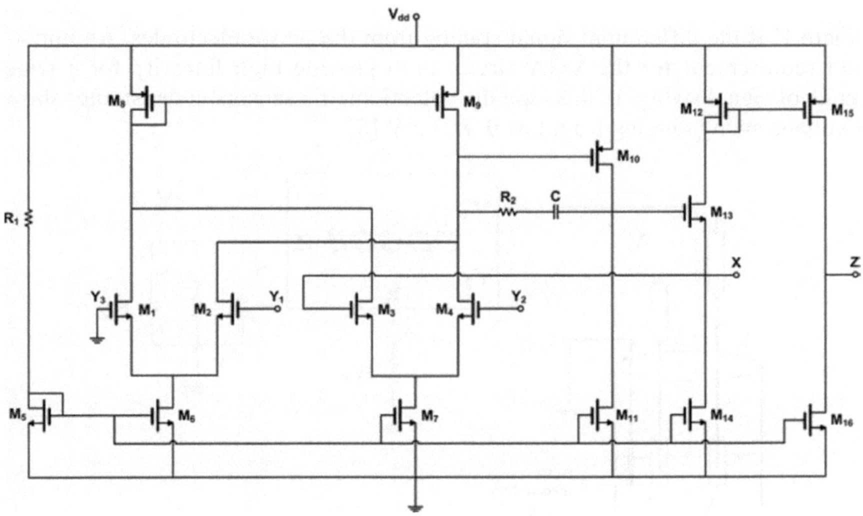

Figure 3. *DDCCII* circuit schematic at transistor level.

Table 1. *DDCCII* main characteristics.

DDCCII main characteristics	
Data	Value
Voltage Supply	1 V
Power Consumption	10 μW
Voltage Gain (α)	0.95
Current Gain (β)	0.99
Parasitic R_X	0.1 Ω
Parasitic L_X	0.01 μH
Parasitic R_Z	210 kΩ
Cut off frequency	3.5 MHz
Layout area	0.09 mm^2

3. Simulation results

The designed DDCCII is planned to be used as the first element directly connected to the sensor electrodes as shown in Fig. 4. From this figure it results that the DDCCII can be used as a Variable Gain Amplifier (VGA) in order to set the system sensitivity. In fact the Z node voltage output is given by:

$$V_0 = V_i \frac{R_z}{R_x} \tag{3}$$

where V_i is the differential signal coming from the sensor electrodes. An important requirement for the VGA circuit is to provide high linearity for a wide range of signal swing. In this case the potentiometric sensors under studies show a voltage swing ranging from few 0–800 mV [8].

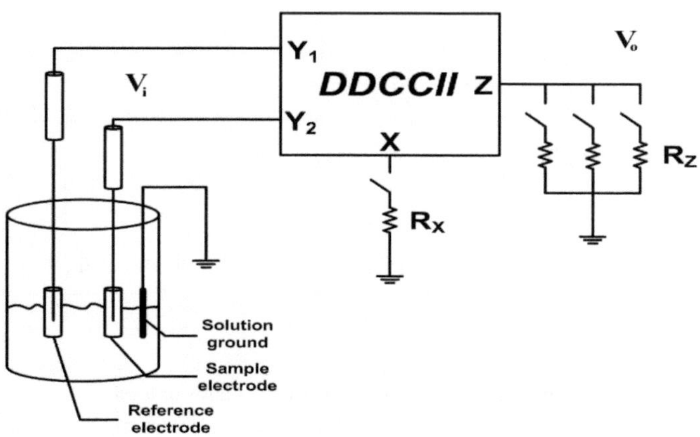

Figure 4. Ion-selective electrodes (ISE) interfacing circuit using the designed DDCCII in VGA configuration.

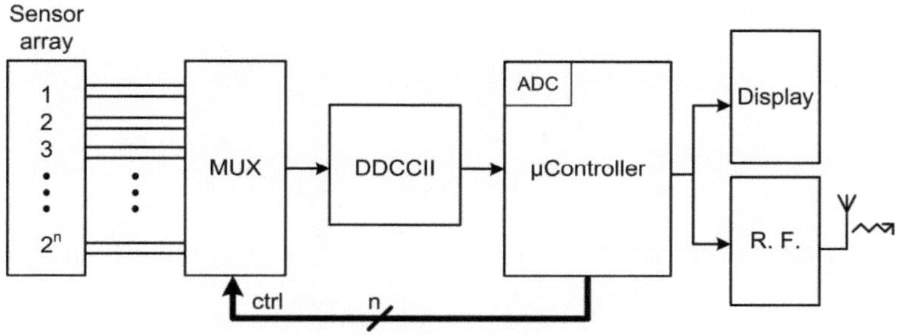

Figure 5. Sketch of the sensorial system where the DDCCII is planned to use as electronic interface.

4. Conclusions

We have presented a Differential Difference Amplifier based current conveyor (DDCCII) for the readout of the voltage value resulting from potentiometric sensors. The use of the current conveyor allows to implement a first analog front end with very low voltage low power characteristics. This interface represents the main part of a low cost artificial sensorial system that allows to realize a low price sensor network where the information of each single node is characterized by a response patterns of the sensor array.

References

1. M. A. Spohn, S. Sausen, F. Salvadori, M. Campos, "Simulation of Blind Flooding over Wireless Sensor Networks Based on a Realistic Battery Mode", Seventh International Conference on Networking (ICN), 13–18 April 2008, pp. 545–550.
2. H. Kulah, K. Najafi, "Energy Scavenging From Low-Frequency Vibrations by Using Frequency Up-Conversion for Wireless Sensor Applications, IEEE Sensors Journal, Volume 8, issue 3, March 2008, pp. 261–268.
3. C. Falconi, E. Martinelli, C. Di Natale, A. D'Amico, P. Malcovati, A. Baschirotto, V. Stornelli, G. Ferri, "Electronic interfaces", Sensors and Actuators B 121, January 2007, pp. 295–329.
4. C. Falconi, C. Di Natale, A. D'Amico, "Dynamic Op Amp Matching: a new approach to the design of accurate electronic interfaces for low voltage/low power integrated sensors systems", Proceedings of Eurosensors 2002, Prague, Czech Republic.
5. G. Ferri, N. Guerrini, "Low voltage low power CMOS current conveyors", Kluwer Academic Publisher, Boston, 2003, p. 226, ISBN 1-4020-7486-7.
6. S.-C. Huang, M. Ismail, S. R. Zarabadi, "A wide range differential difference amplifier: a basic block for analog signal processing in MOS technology", IEEE Transactions on Circuits and Systems II, Vol. 40, no. 5, May 1993, pp. 289–301.
7. C. Falconi, G. Ferri, V. Stornelli, A. De Marcellis, A. D'Amico, D. Mazzieri, "Current Mode, High Accuracy, High Precision CMOS Amplifiers", IEEE Transactions on Circuits and Systems II, vol.55, no. 5, May 2008, pp. 394–398.
8. M. Habara, H. Ikezaki, K. Toko, Biosens, "Study of sweet taste evaluation using taste sensor with lipid/polymer membranes" Bioelectron, 2004, 19, 1559.

A NEW FAST-READOUT FRONT-END FOR HIGH RESISTIVE CHEMICAL SENSOR APPLICATIONS

A. DEPARI[1], A. FLAMMINI[1], D. MARIOLI[1], E. SISINNI[1], A. DE MARCELLIS[2], G. FERRI[2] AND V. STORNELLI[2]

[1]*Department of Electronics for Automation, University of Brescia, Via Branze 38, 25123 Brescia, Italy*
[2]*Department of Electrical and Information Engineering, University of L'Aquila, Località Monteluco di Roio, 67040 L'Aquila, Italy*

Abstract. Metal oxide (MOX) gas sensors show a resistive value varying in a wide range, from hundreds of kilohms up to tens of gigohms. For this reason, oscillating circuits are often used to interface these sensors. Simple oscillators need a measuring time which is directly proportional to the resistive value; when large resistive value (gigohms) occur, the output updating time is on the order of seconds. In addition, the sensor is excited with a square wave, but the variable frequency can affect the measurement accuracy. In this paper, we propose a new oscillating circuit having a limited oscillating time regardless the resistive value. This behavior is obtained by means of self-moving thresholds, the slower limiting the maximum measuring time. The proposed circuit has been simulated and experimentally tested with commercial resistors (values between 1 MΩ and 100 GΩ).

1. Introduction

One of the main aims of actual sensor research is the design of full electronic systems, formed by the sensor, its first analog interface and the processing circuit, possibly in a miniaturized microelectronic environment. The sensor front-end has to be able to adapt itself to different kinds of gas responses, through suitable electronic circuits.

Some chemical sensors, as Metal Oxide (MOX) sensors, according to reagent substance concentration, strongly vary their resistance, from tens of kilohms to tens of gigaohms. In particular, they react under the presence of a target gas by varying their conductivity, so to be considered as resistive sensors, even if, due to the particular processes used to implement the sensor and to the presence of a heater element together with the sensitive film, a parasitic capacitance effect is created. This capacitance, C_{sens}, typically modeled in parallel with the sensor resistive element, R_{sens}, can affect the resistance estimation if it is not correctly taken into account.

The analog interfaces proposed in the literature for such sensors are especially based on a Resistance-to-Time Conversion (RTC), so to estimate also the

P. Malcovati et al. (eds.), *Sensors and Microsystems: AISEM 2009 Proceedings*, Lecture Notes in Electrical Engineering 54, DOI 10.1007/978-90-481-3606-3_53,

sensor parasitic capacitance C_{sens}, typically of few picofarad [1–6]. In these circuits, measuring time is proportional to the resistive sensor value.

The aim of this work is to propose a new circuit exploiting the RTC technique to estimate both sensor resistive and parasitic capacitance components, but keeping the measuring time in the order of tens of milliseconds also when very high values of the sensor resistance occur.

2. Proposed novel interface

The scheme in Fig. 1 has been recently proposed as a suitable interface able to estimate the sensor resistance value R_{sens} over a wide range together with the sensor parasitic capacitance C_{sens} [7]. The working principle is based on the integration of the current flowing into the sensor by means of an operational amplifier (*Int*) with a very low input bias current. The triangular waveform at the integrator output is due to the use of a comparator ($Comp_{th}$), tuned on a suitable threshold V_{th}, which determines the correct switching of the sensor excitation voltage V_{exc} to realize the oscillator. The oscillation period T depends on the current flowing to the integrator and, as a consequence, on the sensor resistive value. Particularly, if the sensor resistance R_{sens} is on the order of some gigohms, the measuring time is more than 1 s. The effect of the parasitic capacitance can be taken into account by adding a second comparator ($Comp_0$) tuned on a different threshold. It should be noticed that, differently from $Comp_{th}$, $Comp_0$ is a simple threshold without hysteresis, therefore its noise can have a strong effect on this comparator. With the circuit depicted in Fig. 1, both sensor resistive and parasitic capacitance values can be estimated by measuring the period T and the duty-cycle T_{on}/T. This interface has been successfully implemented in a standard CMOS technology as an integrated circuit. It has shown good performances in terms of linearity, with a weighted least mean square (WLMS) error less than 5% in the range from 100 kΩ to 100 GΩ.

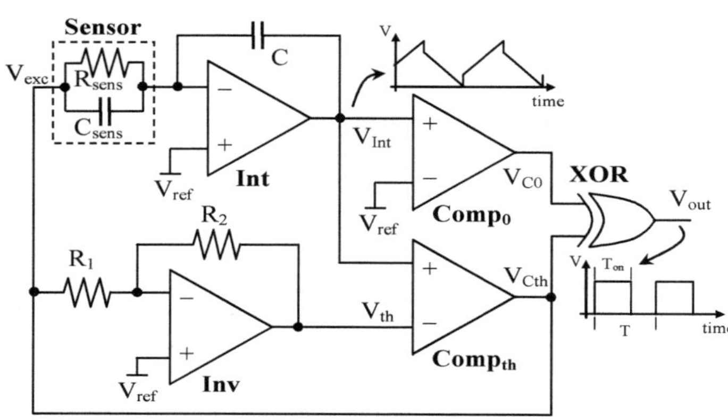

Figure 1. Previously proposed oscillating circuit for the estimation of sensor resistive and capacitive components.

Since the main limits related to this front-end include unstable behavior of $Comp_0$ comparator due to noise and, especially, the variable and long measuring time, this circuit seems unsuitable for applications where the sensor response must be well characterized during fast transients, due to gas sensing or thermal profiles or the modification of the sensor excitation.

For all these reasons we have developed a new circuit interface whose block scheme is shown in Fig. 2. The main difference with respect of the circuit previously proposed is in the comparator thresholds used to generate the output waveforms. In fact, in this new solution, such thresholds are not fixed voltages, but are implemented through two ramp voltages which move in the opposite direction with respect to the integrator ramp. In particular, *Op1* is the integrator, *Op2* and *Op3* are the threshold ramp generators, *Op4* is an inverting amplifier to switch the threshold ramp slope with respect to the integrator ramp and *Op5* is a buffer to decouple the circuit from the capacitive effect of the cables connecting the sensor. In this manner, timings related to the oscillating circuit do not depend on the sensor resistive value only, but on the moving threshold slopes as well as shown in the timing diagram in Fig. 3. By this solution, the two main limits of the previous circuit, that is the long measuring time for high resistive values and the noise at the comparator level, are overcome.

The following equations (1, 2) show formulas to determine R_{sens} and C_{sens}, being k the current commutation number and $\alpha = R_{tx}/R_{ty}$ (see Fig. 3).

$$R_{sens,k} = \alpha \frac{C_t R_t}{C_r} \frac{T_{y,k} - T_{x,k}}{T_{x,k} - \alpha T_{y,k}} \tag{1}$$

$$C_{sens,k} = \frac{C_r}{2C_t R_t} \left(2C_t R_t \frac{V_t}{V_{exc}} - \frac{1-\alpha}{\alpha} \frac{T_{x,k} T_{y,k}}{T_{y,k} - T_{x,k}} - T_{y,k-1} \right) \tag{2}$$

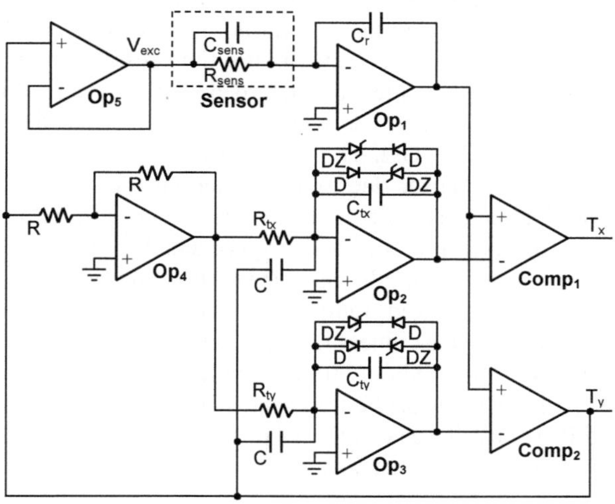

Figure 2. The scheme of the novel proposed circuit.

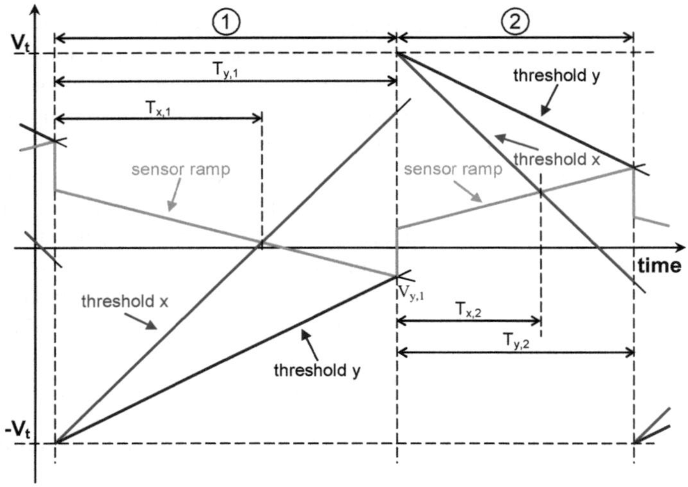

Figure 3. Time diagram of the proposed solution to estimate the sensor resistive value and parasitic capacitance by means of the commutation time measurement; Vt is equal to the sum of the direct voltage of the diode D and the reverse voltage of the Zener diode DZ.

3. Simulation results

PSpice simulations have been conducted assuming the following values: $\alpha = 0.17$, timing resolution of 100 ns, measuring time of 40 ms. We used a low input bias current operational amplifiers OPA350 and fast response comparators TLC3702, all from Texas Instruments, with a ±3.3 V power supply. Table 1 shows the obtained results. The estimation of R_{sens} has been performed using the weighted least mean square (WLMS) line, which is better than the usual least mean square line when a wide range of variation is considered. Except for the lowest resistance value, where the limit is the time resolution in the measurement (100 ns), as the high value of standard deviation clearly highlights, the relative linearity error is below 4% over more than five decades of resistance variation, from 1 MΩ to 100 GΩ. The circuit has been furthermore used to test the behaviour of a TiO2 sensor due to a fast thermal transient. Figure 4 shows that the proposed circuit is able to track fast responses, allowing a better characterization, for example, of new experimental sensors (the maximum measuring time is now about 60 ms instead than more than 1 s with the traditional method using fixed ramps). In this case the sensor resistance variation is of about 2 decades in less than 1 s, but the proposed circuit, having a maximum measuring time of about 60 ms, is able to track the sensor behaviour, allowing a better description of the transient. It is also important to note that a fixed-threshold RTC circuit, see e.g. [7], takes more than one second to measure a 10 GΩ resistor so this thermal transient could not be correctly characterized.

Table 1. Preliminary experimental results.

R_{sens} (MΩ)	<R_{sens}> WLMS (MΩ)	err. WLMS (%)
1.00E−01	1.287E−01	28.7
1.00E+00	1.003E+00	0.3
1.00E+01	9.619E+00	−3.8
1.00E+02	1.021E+02	2.1
1.00E+03	1.019E+03	1.9
1.00E+04	1.038E+04	3.8
1.00E+05	9.671E+04	−3.3

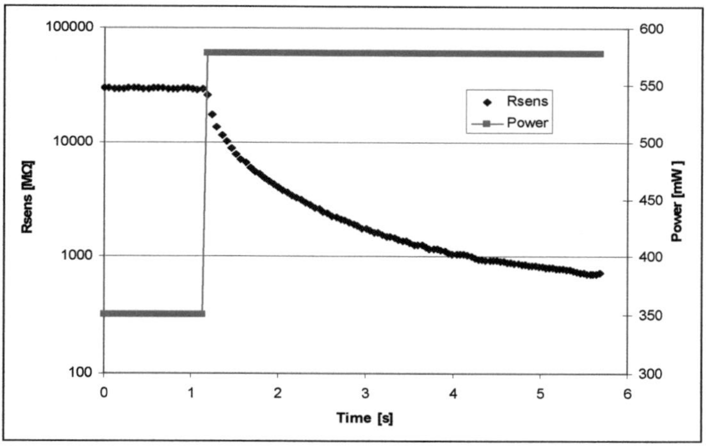

Figure 4. Response of a TiO2 MOX sensor to a fast thermal transient.

4. Conclusions

In this paper we have designed and investigated a new resistive sensor interface exploiting the RTC technique to estimate both resistive and capacitive components, but keeping the measuring time in the order of tens of milliseconds.

In order to validate it, the proposed circuit has been simulated and experimentally tested both with commercial resistors (values between 1 MΩ and 100 GΩ) and fast thermal transient TiO2 sensors with good results.

References

1. G. Sberveglieri, C. Baratto, E. Comini, G. Faglia, M. Ferroni, A. Ponzoni, A. Vomiero, Sens. & Act. B, vol. 121, Jan. 2007.

2. A. Flammini, D. Marioli, A. Taroni, IEEE Trans. Instrum. and Meas., vol. 53 no. 4, Aug. 2004.
3. A. Depari, A. Flammini, D. Marioli, S. Rosa, A. Taroni, IOP Meas. Sci. and Tech., vol. 17, no. 2, Feb. 2006.
4. A. De Marcellis, A. Depari, G. Ferri, A. Flammini, D. Marioli, V. Stornelli, A. Taroni, Sens. & Act. B, vol. 132, June 2008.
5. C. Falconi, E. Martinelli, C. Di Natale, A. D'Amico, P. Malcovati, A. Baschirotto, V. Stornelli, G. Ferri, Sens. & Act. B, vol. 121, no.1, Jan. 2007.
6. U. Frey, M. Graf, S. Taschini, K.U. Kirstein, A. Hierlemann, Proc. of IEEE Sensors 2008, Lecce, Italy.
7. C. Di Carlo, A. De Marcellis, V. Stornelli, G. Ferri, A. Flammini, A. Depari, Proc. of ISCAS 2009, May 24–27, 2009, Taipei, Taiwan, in press.

A NOVEL CALIBRATION-LESS CCII-BASED RESISTANCE-TO-TIME FRONT-END FOR GAS SENSOR INTERFACING

A. DE MARCELLIS[1], C. DI CARLO[1], G. FERRI[1], V. STORNELLI[1], A. DEPARI[2], A. FLAMMINI[2] AND D. MARIOLI[2]

[1]*Department of Electrical and Information Engineering, University of L'Aquila, Località Monteluco di Roio, 67040 L'Aquila, Italy*
[2]*Department of Electronics for Automation, University of Brescia, Via Branze 38, 25123 Brescia, Italy*

Abstract. We present a novel current-mode fully-integrable oscillating circuit for the interfacing of resistive gas sensors, using a reduced number of active blocks, in particular low voltage low power Second Generation Current Conveyors (CCIIs), to be used in portable applications. The proposed front-end, operating a Resistance-to-Time (R–T) conversion, can be utilized for AC-excited resistive sensors and does not need any initial calibration. Simulation and preliminary measurement results on a prototype board performed on the whole system have shown good linearity and agreement with theoretical expectations.

1. Introduction

The development of miniaturized integrated electronic circuits and the micro-electronic technologies progress have brought, in sensors environment, to the design of complete analog circuits implemented by both the sensor interface and the elaboration system. In this sense, the implementation of an appropriate front-end, which can be adapted to different kinds of gas sensors, through a suitable electronic topology, is one of the recent targets [1, 2].

In resistive sensor analog interfaces, the use of oscillating circuits performing an R–T conversion is advisable when sensor baseline or its response to reagent substances changes of some order of magnitudes [3]. This avoids the use of front-ends with a wide output range only through the use of either scaling factors or high-resolution pico-ammeters. In order to design suitable oscillators with low-voltage low-power performance for portable and compact sensor interfacing, the current-mode approach (and the use of CCII, the main analog current-mode device) is an open challenge. In fact, electronic circuit trend is toward low-voltage low-power SoC and future compact electronic noses should follow the same tendency.

In this paper, we propose a novel, compact, low-voltage, low-power, low cost interface based on a suitable CCII, designed at transistor level in a standard

P. Malcovati et al. (eds.), *Sensors and Microsystems: AISEM 2009 Proceedings*, Lecture Notes in Electrical Engineering 54, DOI 10.1007/978-90-481-3606-3_54,

CMOS technology, employing an R–T conversion utilizing a reduced number of blocks. Simulation results as well as preliminary measurements, performed through a prototype laboratory board (using commercial component AD844 as CCII and sample resistances) have demonstrated the validity of the proposed interface and a good agreement with theoretical expectations.

2. CCII basic theory and CMOS implementation

The basic current-mode analog device is the CCII, whose block scheme is in Fig. 1a, which behaves as both current and voltage buffer. CCII operation is the following (Fig. 1b): if a voltage is applied to *Y* node, an equal voltage is obtained at *X* terminal and the current flowing into *X* node is equal (or opposite) to the current flowing into *Z* terminal. Figure 2 shows the internal equivalent circuit of a CCII where node parasitic elements have been highlighted.

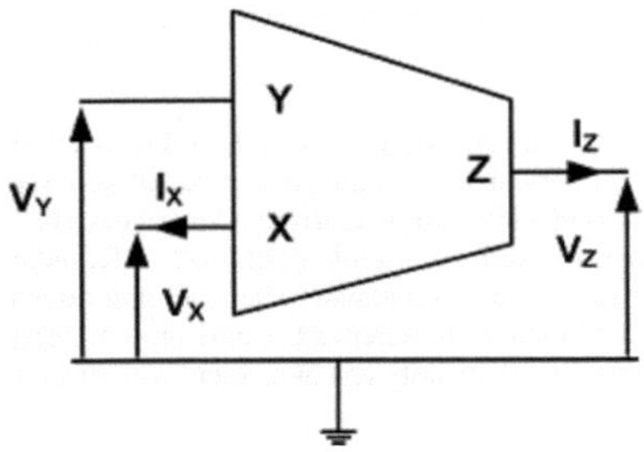

a)

$$\begin{bmatrix} I_Y \\ V_X \\ I_Z \end{bmatrix} = \begin{bmatrix} 0 & 0 & 0 \\ 1 & 0 & 0 \\ 0 & \pm 1 & 0 \end{bmatrix} \cdot \begin{bmatrix} V_Y \\ I_X \\ V_Z \end{bmatrix}$$

b)

Figure 1. CCII block scheme: (**a**) ideal, (**b**) ideal nodes relationships.

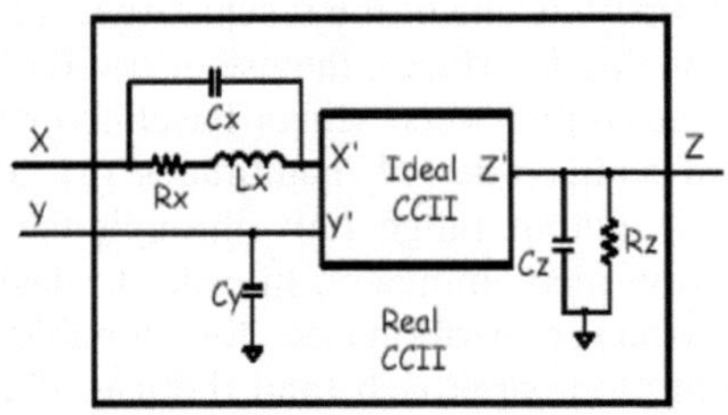

Figure 2. Internal equivalent circuit considering also its parasitic elements.

In Fig. 3 the schematic of the designed CCII is shown. The circuit has been implemented using a standard AMS CMOS 0.35 μm process. It is formed by a differential input stage (M_1–M_7) with a class AB output stage (M_8–M_{11}; R_1, R_2;

$M_{16}-M_{17}$) and a low voltage cascode current mirror ($M_{12}-M_{15}$; $M_{18}-M_{21}$). The class AB output stage allows to decrease X parasitic impedance, while cascode current mirror increases the Z one. The class AB output stage provides the current required by the X node load. This current is mirrored by using cascode current mirror that drives Z node load. The proposed CCII has a low voltage supply of ±0.75 V and 118 μW static power consumption. Bandwidth is about 10 MHz. Parasitic components have been minimized. Voltage and current gains are very close to 1.

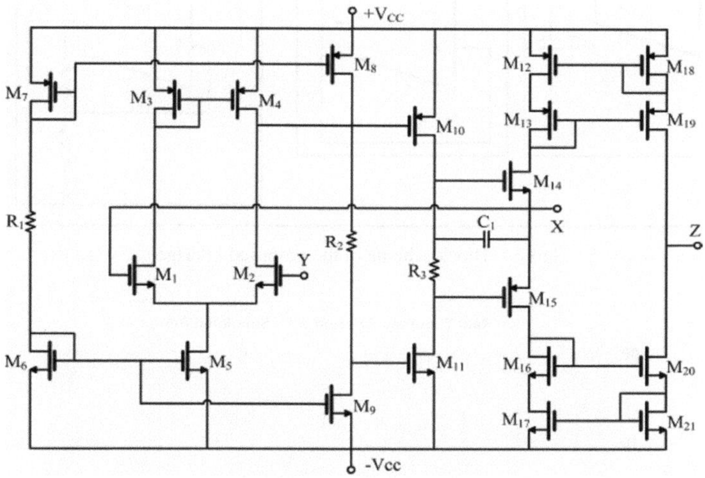

Figure 3. CCII schematic.

3. Proposed interface and simulation results

Figure 4 shows the basic block scheme of the proposed front-end. The expressions for the ideal and real periods T (corresponding to ideal and real CCIIs) of the generated output square wave signal, revealed at *Vout* node, as a function of the sensor resistance R_{SENS}, are reported in (1) and (2), respectively. Simulation results, considering sensor resistances ranging from 700 kΩ to 70 MΩ (these values are typical of some gas sensor resistances as Metal Oxide sensors [4–8]) and the following values for external passive components: R_1 = 5 kΩ, R_2 = 50 kΩ, R_3 = 10 kΩ, C = 100 pF, have been found to be in a good agreement with theoretical expectations (see Fig. 5), considering that CCII parasitic elements (that modify output period expression) show very different values, according to CCII node operating points. The interface sensitivity has been set, including CCII parasitic elements, to about 0.17 ms/MΩ.

$$T = 4R_{SENS}C\left(\frac{R_2 - R_1}{R_2 + R_3}\right) \qquad (1)$$

$$T = 4\frac{(R_{SENS}+R_X)(C+C_Z)}{\alpha\beta}\cdot\frac{\alpha R_2 R_Z-(R_2+R_3+R_Z)(R_1+R_X)}{R_Z(R_2+R_3)} \tag{2}$$

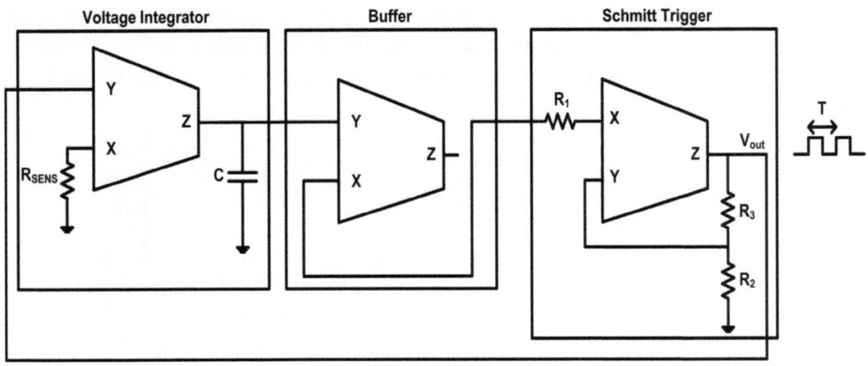

Figure 4. Block scheme of the proposed interface.

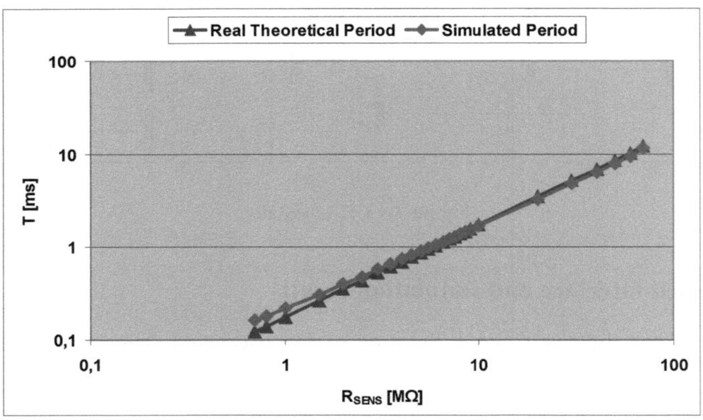

Figure 5. Real theoretical and simulated periods versus sensor resistance, showing a reduced relative error (lower than 10% for Rsens >2.5 MΩ) between the two curves.

Waiting for the layout design and the final chip fabrication, preliminary measurements on a prototype board (see Fig. 6), using the commercial component AD844 as CCII, have demonstrated the practical validity of the proposed interface and, once again, a good agreement with theoretical expectations. Data have been achieved using commercial resistors as resistive sensor, ranging from 18 kΩ to 1.8 MΩ. In this case, the interface sensitivity has been set to about 50 µs/MΩ. Figure 7 shows the measured periods compared with the theoretical ones, as a function of the sensor resistance, obtained with the following experimental values: $R_1 = 470$ Ω, $R_2 = 2.2$ kΩ, $R_3 = 4.7$ kΩ, $C = 47$ pF, $V_{ALIM} = \pm 15$ V.

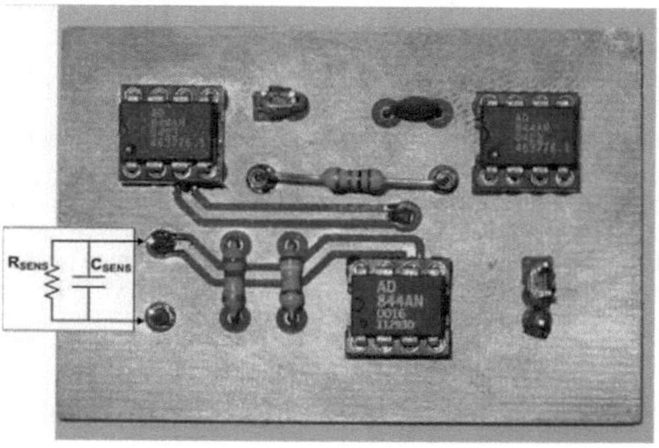

Figure 6. Prototype discrete element board.

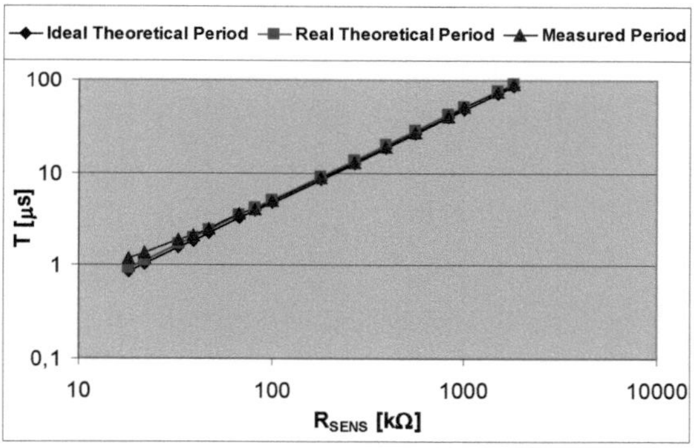

Figure 7. Measured and theoretical output period T vs. R_{SENS}.

4. Conclusions

A new current-mode oscillating circuit is proposed to estimate the resistance (and parallel parasitic capacitance) of chemical sensors. Simulation results and preliminary prototype measurements confirm the correct interface operation. The design of compact low-voltage low-power integrated CCIIs, at transistor level, in a standard CMOS technology, makes the proposed circuit particularly suitable for portable applications (e.g., consumer).

References

1. G. Ferri, V. Stornelli, W. Cappucci, C. Cantalini, Sensors & Actuators B, Vol. B 118, pp. 269-275, October 2006.
2. G. Ferri, V. Stornelli, A. De Marcellis, A. Flammini, A. Depari, Sensors & Actuators B, Vol. 130, pp. 207–215, March 2008.
3. C. Falconi, E. Martinelli, C. Di Natale, A. D'Amico, P. Malcovati, A. Baschirotto, V. Stornelli, G. Ferri, Sensors & Actuators Vol. B 121, pp. 295–329, 30 January 2007.

4. G. Sberveglieri, C. Baratto, E. Comini, G. Faglia, M. Ferroni, A. Ponzoni, A. Vomiero, Sensors &Actuators B, Vol. 121, January 2007.
5. T. Sahm, L. Mädler, A. Gurlo, N. Barsan, U. Weimar, A. Roessler, S. E. Pratsinis, Proc. of Eurosensors XIX, Barcelona, Spain, Sept. 11-14, 2005.
6. G. Neri, A. Bonavita, G. Rizzo, C. Baratto, G. Faglia, G. Sberveglieri, Proc. of Eurosensors XIX, Barcellona, Spain, September 11–14, 2005.
7. A. Fort, S. Rocchi, M. Serrano-Santos, R. Spinicci, N. Ulivieri, V. Vignoli, Proc. of IMTC 2004, pp. 599–604, Como, Italy, 18–20 May 2004.
8. I. Sayago, Md. C. Horrillo, S. Baluk, M. Aleixandre, M. J. Fernandez, L. Ares, M. Garcia, J. P. Santos, J. Gutierrez, IEEE Sensors Journal, Vol. 2, no. 5, pp. 387–393, October 2002.

HIGH-EFFICIENCY FRONT-END INTERFACE FOR THE VIBRATING-STRING STRAIN GAUGE SENSORS

A. SIMONETTI AND A. TRIFILETTI

Department of Electronic Engineering, University of Rome "La Sapienza" I-00184 Rome, ITALY, e-mail: [simonetti, trifiletti]@die.uniroma1.it

Abstract. Sensor networks are becoming a reality in different applications. As matter to fact, in mobile equipments and battery-operated instrumentation, the design of low-power systems becomes critical, and the vibrating-string strain gauges generally doesn't work with low tensions of the commercial wi-fi or zegbee modules. In addition, the reduction of the power-supply voltage also degrades the power of the desired signal, while the noise power remains noticeable. So a special interface is necessary. In this work, we propose a high-efficiency method based on the gating of the electromagnetic coil of the vibrating-string at the minimum energy to overcome the mechanical damping. Experimental waveforms for a low-cost interface implementation are also given to demonstrate its validity.

1. Introduction

Different kinds of the vibrating-string strain gauges, are used in the damage detection of the buildings such as bridges and tunnels. These sensors operate on the principle that a tensioned wire between two mobile flanges of their body, when excited by an electromagnetic circuit, vibrate to resonant frequency that is proportional to the tension in the wire, reflecting the strain of the reinforced concrete or the mass concrete [1]. Several advantages as the low cost and the accuracy and also the stability and the long duration make them competitive, although they are hard to apply in the pre-existing structures. However, a wired network increases the expensiveness due to the complexity of the whole system, and decreases the sensors quantity. Manufacturers themselves, don't provide appreciable solutions, because of the maintenance requirements and the poor autonomy of the alternative radio-frequency connections. Among these problems, reduction of the power-supply voltage degrades the power of the desired signal, while the noise power remains noticeable. Therefore, the fast growing interest for a cordless mesh made possible by the commercial wi-fi or zegbee modules, has led to an increasing demand for the low power interfaces that avoids both the performances penalty and the autonomy reduction [2, 3]. This work demonstrate that a simple AR model of the sensor response it is sufficient to reduce the activity and the power consumption of the interface by exciting the strain gauge sensor with a very low-voltage circuit.

P. Malcovati et al. (eds.), *Sensors and Microsystems: AISEM 2009 Proceedings*, Lecture Notes in Electrical Engineering 54, DOI 10.1007/978-90-481-3606-3_55,

2. Basic principle

Considering the energy-saving perspective, the worst case is reached in the undamaged condition, where the measurement is repeated a lot of times, with a small set of persistent values. In this signal processing chain, the accuracy of the front-end interface determine the minimum damage perception. Nevertheless, since the operating frequency is almost unchanged, it is possible to reduce the power consumption by exciting the sensor at the minimum energy to overcome the mechanical damping and the noise environment. Indeed, is very important the mean deviation of the measured values between several days or months, rather than the absolute precision of the single acquisition. The system behavior can be modeled by considering the frequency measurements of the unstressed sensor (with little deformations) with the noise environment in a representative condition of most practical applications. By superimposing various noise and stress constraints, each data-set solves an auto-correlation matrix of a Yule-Walker problem, and provide a fourth-order autoregressive (AR) series. The order of the model, is experimentally determined for a partial recovery in the series variance imposed by a truncation problem in the coefficients representation of the microcontroller-oriented software. So, by inverting the previous AR model, it is possible to determine a same order finite impulse response (FIR) filter of the error prediction [4], which is valid, with a good approximation, for a wide range around the resonant frequency of the unplugged strain-gauge. This bank of filters of the error prediction, is used both in the excitation and sensing phases. The square wave generator, emulated by a microcontroller, initially gates the coil of the sensor to obtain a stable oscillation condition, which is maintained by the inertia of the mechanical system. The damping factor of the latter, determines the successive variations in the duty-cycle of the sensor activation (not in its excitation frequency). By sensing the frequency decay, the filter enables a new least excitation cycle. In case of a significant variation in the fundamental frequency (i.e. 15%), the algorithm tests a different coefficients set on the look-up table. The new coefficients are used as a default in the next startup. Obviously, this modeling approach, is optimized only for a specific strain-gauge vibrating string, but by embedding this sensor in the concrete, it becomes irreplaceable, thus a dedicated interface, doesn't reflect a real problem.

3. Interface description

The low-cost interface has been built with one microcontroller (PIC-16F84 or equivalent), two low-noise double op-amp (LMV722), and one switching power-supply IC (MAX-856), for stable 3.3 V operations. The analog signal has been handled with a cascade of two differential amplifiers to reach a very high gain (5,000 V/V), and a good noise rejection. This preamplifier is finalized with a Schmitt trigger, which provides a stable square wave for the capture function. By processing the external interrupt connected to the squarer, the microcontroller enables a simple but precise frequency-counter system. Similarly, in the excitation phase, two pins of a general purpose I/O port directly gate the sensor by alternating their input and output configuration. This approach enables a wave generator with a minimum circuitry and doesn't subtract energy from the sensor during the I/O swap.

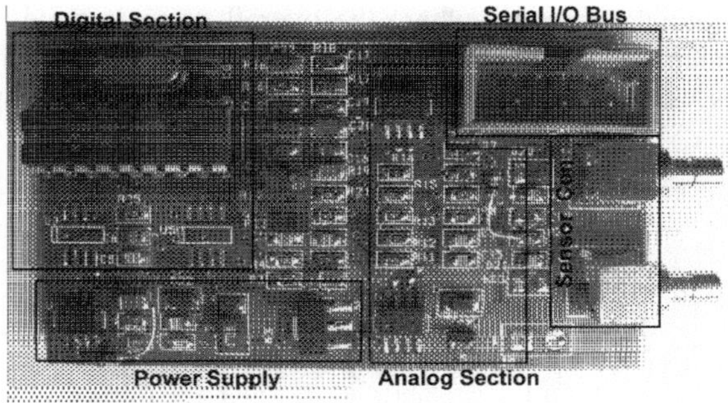

Figure 1. The proposed interface.

Figure 1 shows the circuit assembled in a small size board with the plug-in compatibility with the bus expansions of the TMote-Sky zegbee module. The bandwidth of this interface, is tuned to frequency response of various sensors, and is strictly contained from 300 Hz to 4 kHz.

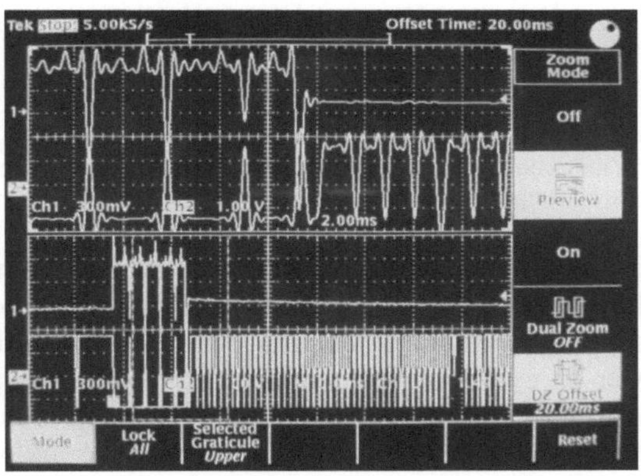

Figure 2. Device waveforms. Lower graph: waveform at the output of the signal processing chain (ch. 2) and the duty-cycle excitation of the sensor (ch. 1). Upper graph: horizontal zoom ×5.

Figure 2 shows the main (lower grid) and the enlarged (upper grid) horizontal resolution of the device waveforms in the unstressed condition for a commercial sensor with a 47 μH coil and 150 Ω loss. The excitation is performed with 13 ms sequence of a 500 Hz square-wave tone with 95% duty-cycle and 900 mVp-p amplitude. The sensing phase, produces a constant amplitude signal of 800 ms length (efficiency: 98%), without further energy requirements (the waveform period has been truncated to increase the horizontal resolution).

4. Experimental results

In this work we have tested the proposed system, in a real environment "A1 highway" (Nazzano, Italy), with six strain gauge sensors (marked S1...S6) already placed in a tunnel basement, where the overall noise is quite high. A professional instrumentation is also used to compare the results. In each prototype, the acquired data, is transferred to a wireless module (not shown in this work) by a serial bus.

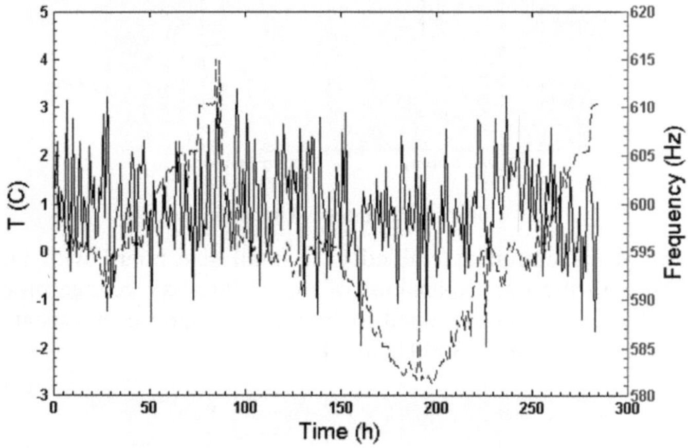

Figure 3. Example of data set of sensor S6. Observation period: 3 day (285 h). Effective measures: 2,850 (10/h). Power efficiency (estimated): 95%. Strain error (estimated): ±4%. Sensor frequency: (585 ± 5) Hz. Temperature correction can help to understand real changes in stress due to expansion and contraction caused by temperature changes.

Figure 3 demonstrate that the proposed system allow a little error in the mean value of a lot of measures (±4%) but the standard deviation is almost stable. Since this front-end support the standby function, the remote module also controls its own the power-on cycles of the interface. Moreover, by applying a more sophisticated algorithms [5], these wireless stations enable a dedicated power-saving functions that extend the autonomy and maximize the performances by reducing the activities of the broadcast and the listening periods of the RF-IC in the modules themselves.

References

1. R. H. Bishop, "The Mechatronics Handbook" ISA,CRC-press, 2002. ISBN 0849300665.
2. G. Werner-Allen, J. Johnson, M. Ruiz, J. Lees, M. Welsh, "Monitoring volcanic eruptions with a wireless sensor network," *Proc. of the Second European Work. on Wireless Sensor Networks*, pp. 108–120, 31 Jan.–2 Feb. 2005.
3. K. Chintalapudi, E. A. Johnson, R. Govindan, "Structural damage detection using wireless sensor-actuator networks," *Proc. of the IEEE International Symposium on Intelligent Control, Mediterranean Conference on Control and Automation* , pp. 322–327, 27–29 June 2005.

4. C. Byoung-Seon, "An algorithm for solving the extended Yule- Walker equations of an autoregressive moving-average time series (Corresp.)," *IEEE Transactions on Information Theory*, vol. 32, no. 3, pp. 417–419, May 1986.
5. D. Fagen, P. A. Vicharelli, J. Weitzen, "Automated wireless coverage optimization With controlled overlap," *IEEE Transactions on Vehicular Technology*, vol. 57, no. 4, pp. 2395–2403, July 2008.

SIGNAL CONDITIONING SYSTEM ANALYSIS FOR ADAPTIVE SIGNAL PROCESSING IN WIRELESS SENSORS

LEONARDO BARBONI AND MAUNZIO VALLE
Department of Biophysical and Electronic Engineering, University of Genoa, Via Opera Pia 11A, 16146-Genoa, Italy
{l.barboni, maurizio.valle}@unige.it

Abstract. In this work it is deduced an extended signal-to-noise (SNR) expression for embedded wireless sensor nodes. The expression can be included onto system level design methodologies for adaptive power-aware sensor acquisition design. In contrast to the conventional approach, it enables designers to estimate the achievable SNR regardless of sensor signal variations and battery voltage droop due to the depletion process. It is offered the theoretical discussion and expression usage examples.

1. Introduction

Wireless Sensor Networks are built with embedded wireless sensor nodes. Each sensor node includes a microprocessor, sensors and circuitry for sensor signal conditioning, AD converters, radio transceiver and batteries for power supply. In embedded wireless sensor nodes, sensor data acquisition is an important energy dissipation source. Hence, the node must be aware and adaptive for using sensors and data acquisition circuitry, without wasting energy. For instance, the node should work with the ADC resolution as low as possible as well as with adaptive frequency sampling.

On the other hand, due to recent advances in CMOS technology scaling, the tendency is to implement the entire node on a single chip. It results in embedded wireless sensor nodes powered by reduced voltages (e.g. below IV). However, the penalty for using reduced voltage power supply is the SNR degradation of sensor signals during the acquisition and conditioning stage, before the digital processing [1].

Clearly, sensor nodes offer an adverse scenario for sensing and low-energy signal processing. System level design methodologies and strategies for adaptive energy-aware circuits have started to be studied and benefits arise.

In [2] authors present a low-power 8-bit current mode interleaved successive approximation ADC. It can work with different number of resolution bits and sampling rates, allowing to reach different power consumption levels. The circuit has been implemented in CMOS 0.18 μm technology. In [3] is discussed the performance of an ADC designed with scaled CMOS technology. Authors

P. Malcovati et al. (eds.), *Sensors and Microsystems: AISEM 2009 Proceedings*,
Lecture Notes in Electrical Engineering 54, DOI 10.1007/978-90-481-3606-3_56,

argue that the problem lies in the reduced signal amplitude. Alternatives to maintain suitable SNR values and tradeoffs are discussed. A noteworthy paper that highlights trade-offs exploration benefits is [4], where authors have studied the acquisition system parameters trade-offs to save power in a sigma-delta modulator ratiometric with respect to the power supply. It is discussed the effects of to use the system power supply as reference voltage instead of the usual reference voltage produced by a band-gap circuit. The energy dissipated by reference buffers and band-gap circuits is then saved.

In the previous works each case has been separately studied, therefore the mam drawback lies in absence of a methodological approach. The purpose of this work is to develop a novel extended expression to describe the acquisition circuitry SNR behavior at system level. Such expression provides insight for implementing adaptive countermeasures to avoid SNR degradation. This is because the addressed statistical approach visualize the dependence of the SNR on the acquisition and signal features for general cases.

2. Derivation of the extended SNR expression

In literature it is possible to found expressions for SNR estimation as (1) [5]:

$$\text{SVR(dB)}=6.02n-1.25+10\log_{10}\left(\frac{fs}{fw}\right) \quad (1)$$

where: f_w is the signal bandwidth, $f_s=1/T_s$ is the sampling frequency and n is the number of bit for data representation. Nevertheless, such expression only models simplified cases, under hypothesis that do not met actual acquisition systems and signal behaviour: (i) the input signal is assumed to be a deterministic sine wave with amplitude equals to the full-scale voltage range; (ii) the reference voltage is assumed to be fixed; (iii) the number of bits for data representation must be almost 4; (iv) it is assumed the existence of a low pass filter with bandwidth f_w for signal reconstruction.

In fact, Eq. (1) is used as figure of merit to characterize or to compare different ADC architectures under controlled testing and there are effects on the SNR that are not clearly captured by Eq. (1).

Due to energy consumption constraints, integrated voltage regulators are only used for RF transceivers voltage regulation (e.g. the MicaZ node [6]). If the battery voltage is used as full-scale voltage reference (V_{ref}), sensors and signal conditioning circuits experiment the progressive battery voltage decrease.

In order to explore this problem, an extended SNR expression has to be calculated. We describe the signal as a random ergodic processes with uniform PDF between two boundaries. The signal is composed by two components: the DC component named s_D that represents the sensor bias and the time-varying component defined as $s_v(t)$. It holds: $s(t)=s_v(t)+s_D$. The signal of interest $s_v(t)\in[-s_{min}, s_{max}]$or $s(t)\in[s_D-s_{min}, s_D+s_{max}]$ as is depicted in Fig. 1.

The probability density for $s_v(t)$ is $p_s=1/(s_{max}+s_{min})$ and constant for all time t.

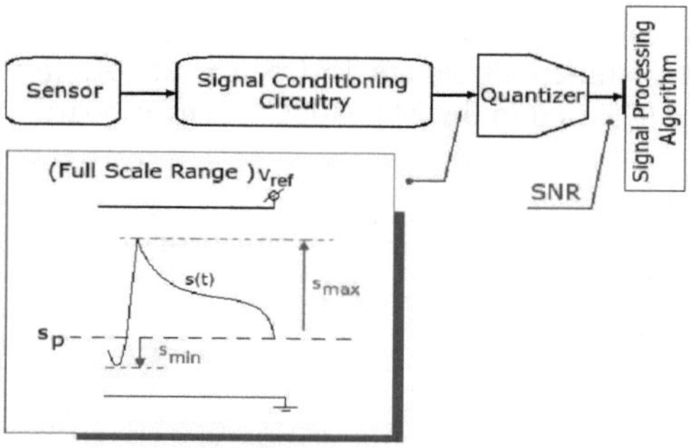

Figure 1. Considered signal acquisition architecture and signal model description.

In this situation we have deduced the following SNR expression for a generic non-oversampled AD converter:

$$SNR = \frac{S_{max}^3 + S_{min}^3}{2(N_L+1)\left(\frac{q_s}{2}\right)^3} - 1 \qquad (2)$$

where f_w is the signal bandwidth, f_s is the sampling frequency, $q_s = V_{ref}/2^n = LSB$ is the quantization step and N_L is the number of quantization levels required to represent the signal, it holds: $N_L q_s \leq S_{maxi} + S_{min} \leq (N_L+2)q_s$.

3. Analysis: SNR surfaces levels

Equation (2) shows the existence of SNR level surfaces for different system parameters, i.e the SNR constant value contour. To have better insight, a numerical example is depicted in Fig. 2 and the associated numerical values are offered in Table 1. For instance, the curve C_4 in Fig. 2 is the SNR = 45 dB level surface of Eq. (2) for $V_{ref} = 1.8$ V and $n = 8$. The marked point D on such contour has signal and ADC converters features shown in Table 1. Similarly, the curve C_5 *is* the SNR = 45 dB level surface of (9) for $V_{ref} = 2.7$ V and $n = 8$.

Table 1 Features and points marked in the SNR contours of Figure 2.

	n	SNR [dB]	V_{ref}[v]	S_{max} [v]	S_{min} [v]
Point A on C_1	10	45	1.8	0.09	0.18
Point B on C_2	10	45	2.1	0.15	0.2
Point C on C_3	10	45	2.7	0.26	0.2
Point D on C_4	8	45	1.8	0.5	0.7
Point E on C_5	8	45	2.7	0.85	1.0

Hence, it makes it possible to modify on-the-fly the signal features by using adaptive companding techniques, in order to maintain (or to enhance) the SNR value after the acquisition process. An actual example of companding usage, is appreciated on the commercial device TAOS TSL2550D (ambient light sensor with companding A/D) used in the MTS400CA sensor board for the Mica2 node [6].

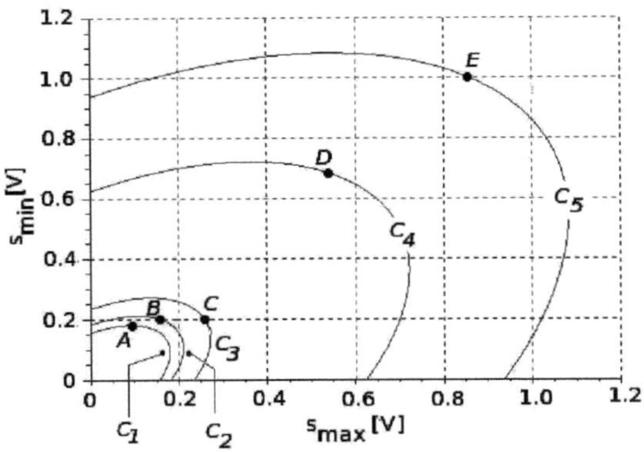

Figure 2. SNR level surface plots of Equation (2): loci of different signal features (axes) and different acquisition parameters (ADC voltage reference and number of bits) that achieve the same SNR value.

4. Conclusions

We derived an extended SNR expression that can be used to implement system level design methodologies for adaptive power-aware acquisition circuits design. It allows designers to estimate achievable SNR values with different signal characteristics and acquisition parameters, opening new possibilities for adaptive power-aware techniques. A further empirical validation has to be conducted in order to validate the extended SNR.

References

1. Y. Chiu, B. Nikolic, P.R. Gray "Scaling of analog-to-digital converters into ultra-deep-submicron CMOS" Proceedings of the IEEE Custom Integrated Circuits Conference, 2005. DOI: 10.1109/CICC.2005.1568684
2. R. Dlugosz, K. Iniewski "Flexible architecture of ultra-low-power current-mode interleaved successive approximation analog-to-digital converter for wireless sensor network" Hindawi Publishing Corporation VLSI Design Volume 2007, Article ID 45269, p. 13 DOI: 10.1155/2007/45269
3. Akira Matsuzawa "Design Challenges of Analog-to-Digital Converters in Nanoscale CMOS" EICE TRANS. ELECTRON., Vol.E90-C, No.4 April 2007 DOI: 0.1093/ietele/e90-c.4.7
4. E. Dallago, P. Malcovati, D. Miatton, T. Ungaretti, G. Venchi "Analysis of sigma-delta converters for MEMS sensors using power supply voltage as reference" Proceedings of Circuits, Devices and Systems, Vol. 153, October 2006 pp. 473–479 DOI: 10.1049/ip-cds:20060079
5. Rudy van de Plassche "CMOS Integrated Analog-to-Digital and Digital-to-Analog Converters" Kluwer Academic Publisher 2nd Edition Printed 2003 ISBN 1-4020-7500-6
6. Vendor: Crossbow Technology [Online] www.xbow.com

A 0.13μm CMOS FRONT-END FOR DRIFT CHAMBERS

S. D'AMICO[1,2], A. BASCHIROTTO[3], M. DE MATTEIS[1], F. GRANCAGNOLO[2], M. PANAREO[1,2], R. PERRINO[2], G. CHIODINI[2] AND A.CORVAGLIA[2]

[1]*Department of Innovation Engineering, University of Salento, Italy*
[2]*INFN Sezione di Lecce, Italy, e-mails: [stefano.damico, marcello.dematteis]@unile.it, andrea.baschirotto@unimib.it, [franco.grancagnolo, marco.panareo, roberto.perrino, gabriele.chiodini, alessandro.corvaglia]@infn.le.it*
[3]*Department of Phisics, University of Milan-Bicocca, Italy*

Abstract. In this paper a 0.13 μm CMOS front-end for drift chambers is presented. The front-end presents characteristics of low power consumption, which are well-suited in order to implement the cluster counting technique for particles identification. The front-end consists of a Variable Gain Amplifier (VGA) with 0, 10 and 20 dB gain steps and 1 GSa/s 6 bits Analog to Digital Converter (ADC). The VGA power consumption has been optimized according to the selected gain setting. The VGA power consumption is 8.4 mA, 9.4 Ma, 10.6 mA for 0, 10, 20 dB gain, respectively.

1. Introduction

The counting of the consecutive ionization clusters in a drift chamber is a very promising technique for particle identification purposes. Although this technique features a number of advantages, the bottleneck for its implementation is represented by the difficulties in realizing a low cost, small area occupancy, low-power consumption, high-speed electronic interface. Typical time separation between each ionization act in a helium-based gas mixture is from a few *ns* to a few tens of *ns*. Thus the read-out interface has to be able to process such a high-speed signals. Starting from these considerations, an integrated circuit solution seems to better fit these requirements. Moreover, the integrated circuit solution guarantee less parasitic capacitances which led to an improvement of the frequency and noise performance. In this paper, a CMOS 0.13 μm integrated readout circuit is presented [1, 2]. It includes a fast Variable Gain amplifier (VGA) with a −3 dB bandwidth of 160 MHz and three gain steps (0, 10, 20 dB), and a 6 bits 1 GSa/s analog to digital converter which is under testing.

2. Circuit description

The proposed VGA presents characteristics of very low current consumption (10.6 mA @ 20 dB gain setting). Moreover the current consumption is adjusted according to the selected gain setting. The VGA architecture is reported in Fig.

P. Malcovati et al. (eds.), *Sensors and Microsystems: AISEM 2009 Proceedings*, Lecture Notes in Electrical Engineering 54, DOI 10.1007/978-90-481-3606-3_57,

1. The external RC network guarantees an AC coupling and input matching. The VGA presents a fully differential architecture in order to be more robust against disturbs from supply. The signal coming from the drift tube is single ended. Therefore a single-ended to differential stage is mandatory. This stage is followed by a differential variable gain amplifier.

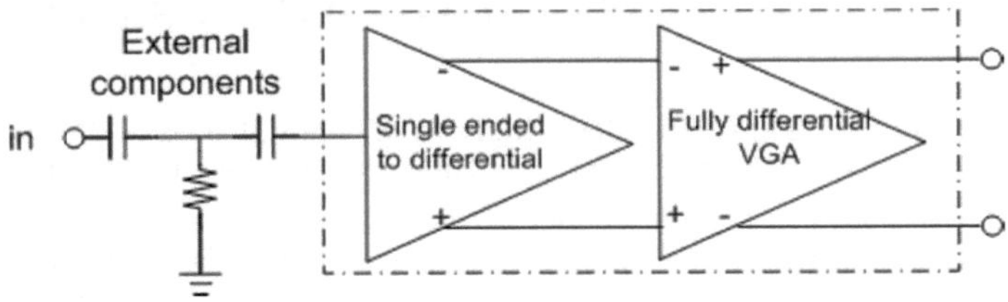

Figure 1. The VGA architecture.

Figure 2 shows the single ended to differential amplifier. The architecture of this stage is simple because it should guarantee high frequency of working and low power consumption.

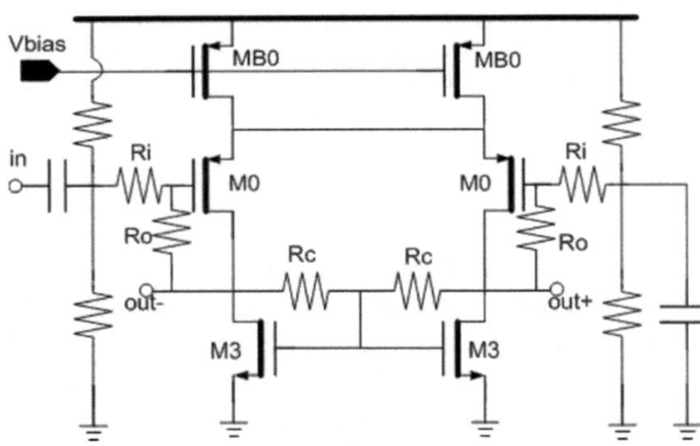

Figure 2. The single ended to differential stage.

Figure 3 shows the fully differential variable gain amplifier. The gain is set by reconfiguring the amplifier input stage. This allows to obtain a nearly constant bandwidth and a power consumption optimization.

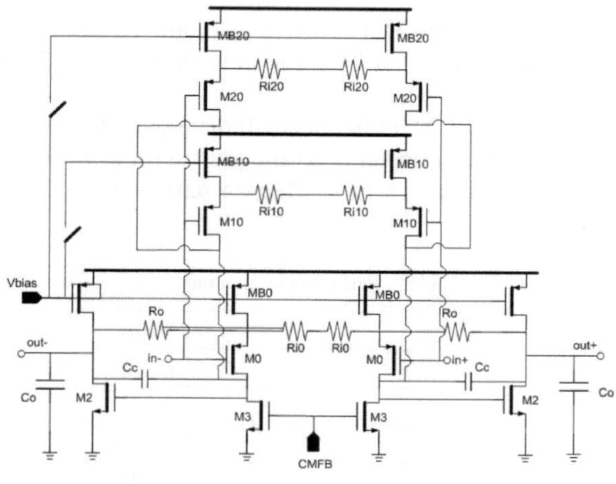

Figure 3. The fully differential VGA schematic.

3. Measurements results

The baseband receiver chain has been realized in a 0.13 μm CMOS technology with 1.2 V supply voltage. The active die size occupancy is 260 × 260 μm. Figure 4 shows the chip microphotograph.

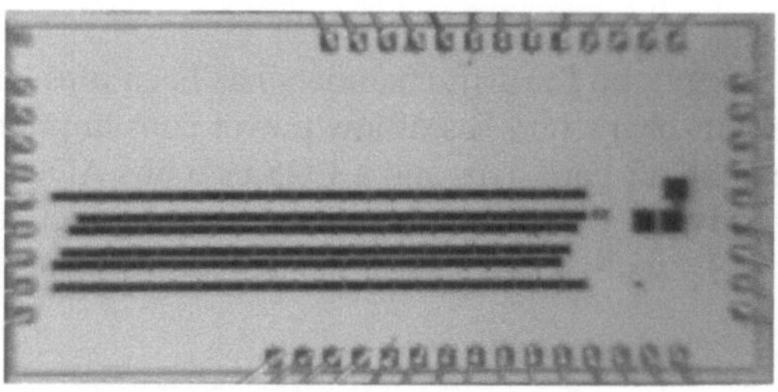

Figure 4. Frequency response for different gain settings.

Figure 5 shows the VGA frequency behavior. Three gain settings are possible: 0, 10, 20 dB.

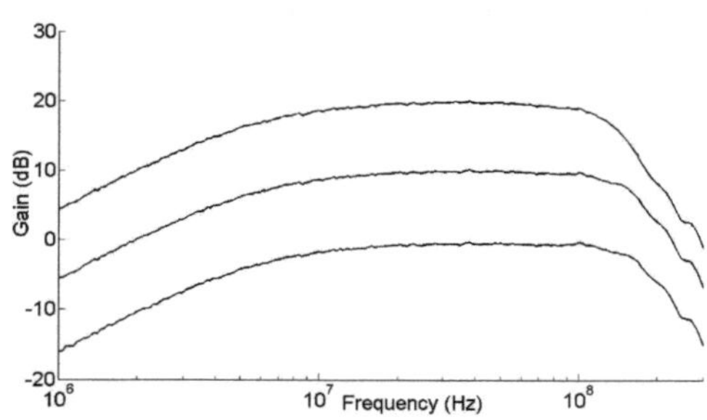

Figure 5. Frequency response for different gain settings.

The VGA cut-off frequency is 160 MHz at 0 and 10 dB gain settings, and 140 MHz at 20 dB gain setting. At 20 dB gain setting and 100 MHz the Output noise level is −132 dBm.

At 0 dB gain setting and 1 MHz frequency 0 dBm output sine, the THD is −36 dB. The current consumption depends on the gain setting. It goes from 8.4 mA at 0 dB gain setting up to 10.6 mA at 20 dB gain setting. Table 1 summarizes the VGA performance.

Table 1. Performance summary.

Parameter	**Value**		
Technology	200°C		
Power supply	1.2 V		
Die size	1.66 mm^2		
Output noise level (@20 dB)	−132 dBm		
THD (@0 dB)	−36 dB @ 0 dBm output signal level		
Gain	0 dB	10 dB	20 dB
Bandwidth	160 MHz	160 MHz	140 MHz
Current consumption	8.4 mA	9.4 mA	10.6 mA

4. Conclusions

A 0.13 μm CMOS front-end for drift chambers has been presented in this paper. The front-end presents characteristics of low power consumption. It consists of a VGA with 0, 10 and 20 dB gain steps and a 1 GSa/s 6 bits ADC. The VGA power consumption depends on the selected gain setting. In fact, the VGA power consumption is 8.4, 9.4 and 10.6 mA for 0, 10 and 20 dB gain, respectively.

References

1. A. Baschirotto, S. D'Amico, M. De Matteis, F. Grancagnolo, M. Panareo, R. Perrino, G. Chiodini, G. Tassielli "A CMOS high-speed front-end for cluster counting techniques in ionization detectors" Proc. of IWASI 2007.
2. T. Hui Teo, M. Annamalai Arasu, W. Gan Yeoh, M. Itoh "A 90nm CMOS variable-gain amplifier and RSSI design for wide-band wireless network application", Proc. of ESSCIRC 2006, pp. 86–89.

A NEW LASER TECHNOLOGY FOR AIR TRAFFIC MANAGEMENT

M. SALERNO[1], G. COSTANTINI[1], M. CAROTA[1], D. CASALI[1], D. RONDINELLA[2] AND M. V. CRISPINO[3]

[1]*Dipartimento di Ingegneria Elettronica, Università di Roma "Tor Vergata", Via del Politecnico, 1, 00133 Roma, Italy*
[2]*Olympus Engineering S.r.l., Via Appia Nuova, 666 – 00179 Roma, Italy*
[3]*Viale Bardanzellu, 8 – 00155 Roma, Italy*

Abstract. Laser detection and tracking of aircrafts based systems (LIDARs, LIgth Detection And Ranging systems) are emerging as a critical design trend in development of new generation ATM (Air Traffic Management) paradigms, of which they are the main innovations. The main goal of the presented project is therefore to develop a novel laser tracking technology (SKY-Scanner System) capable to detect and track of aircrafts up to at least 6 nautical miles from the Aerodrome Traffic Zone (ATZ) barycenter, namely a facility of enabling techniques, protocols, numerical prediction tools and devices specifically designed for the analysis of the laser systems performances in ATC applications, with the final target of defining a new generation ATM paradigm based on radar and laser tracking data fusion, and ground to air laser communications. The proposed methodology is considered at the frontier of technological research but it represents the only realistic way to put solid basis for the fabrication of effective radar and lidar integrated systems for incorporation in new generation ATM paradigms.

1. Introduction

Conventional methods for Air Traffic Management (ATM) that have worked until now cannot continue to cope indefinitely. As shown in [1], a new generation ATM is needed [2, 3]. Several methods have recently been proposed in order to improve both security and performance [4, 5].

Laser rangefinders have been be effectively used in conjunction to robot and autonomous systems [6, 7]; the same concept has be applied in our project for a system that will work in conjunction with radars.

The SKY-Scanner project work plan has been designed to meet this objective through the integration of different tasks addressing specific hardware (HW) and software (SW) items. The proposed technology, which is composed of four main sub-systems to be integrated:

- Montecarlo System Simulation
- Laser Sensor Array (LSA)

P. Malcovati et al. (eds.), *Sensors and Microsystems: AISEM 2009 Proceedings*,
Lecture Notes in Electrical Engineering 54, DOI 10.1007/978-90-481-3606-3_58,

- Sensor Management Computer (SMC)
- Command and Control Computer (C2C)

is completely novel, in the sense it has never been conceived to fulfill the proposed target.

Such approach has not been applied in lidar engineering.
Scientific objectives of the proposed research shall include:

- Control of the tracking of aircrafts by means of a rotating cylindrical laser range-finder array
- Development of mathematical models of aircraft collision probability and optimal decision on corrective actions (DSS, Decision Support System) based on data fusion between radar data and laser tracking data
- Definition of a new generation ATM paradigm based on data fusion between radar data and laser tracking data and ground to aircraft laser communications

The project will promote breakthrough knowledge on laser tracking of aircraft, new DSS models and ATM paradigms based on data fusion between radar data and laser tracking data and ground to aircraft laser communications, such to sustain the reliable development of new perspectives in the ATM world.

2. Description of the system

The SKY-Scanner laser sensor motion system will be composed by the following sub-systems:

- *Bedplate*: Electro-mechanical subsystem including clamp systems, mechanical joint rolling, driver azimuth motor and the tracking module, with control systems enclosed
- *Tracking module*: Subsystem that moves the platform, the laser itself and the relative drivers and position control systems (encoders)

The advantages for this proposed electromechanical solution are:

- *Zenith angles range*: The laser beam can track angular quantas from 0° (referring to the horizontal plane) to 90°, with a pointing resolution depending on encoder precision (5 axes).
- *Mechanical errors on zenithal pointing*: The mechanical errors are limited by the warm gear pair couple (in opposition) (item 4 and 5). In this case the warm gear pair is coupled on the same side (for every movement) and its back-lash is low. The warm-gear pair has a low back-lash for every constant-rate wear.
- *Mechanical errors on laser planarity*: The laser movement are in a plane and, as a consequence, the planarity plane is a fundamental factor for the precision scan system. With an appropriate chip-forming

machining, the planarity will be guaranteed and a thrust block will be mounted on the rear of the laser plate.

- *Vibrations and frictions*: Reduction of the vibration transmissions and friction effects will be achieved with the use of a belt for the coupling between the pulleys. The vibrations are generated by eccentric mass (mass movements).
- *Encoders*: Closed-loop control of the rotation of the electric motors.
- *Gear box subsystem*: Internal protection for the etching by saline atmosphere, dust, water and other.

Aircraft surveillance falls into three categories: primary radar, secondary radar, and satellite based systems:

- Monopulse Secondary Radar
- Primary Surveillance Radar (PSR) for approach/terminal
- ADS systems

Primary radar is based on the fact that objects reflect radio waves. Primary radars emit high power RF energy and detect the presence of an aircraft by detecting the energy reflected back by the target. Secondary radar is a combination of radar and a communication system. In contrast to primary radar, secondary radar does not use the passive echo reflected from a target, but uses an active transponder, which is located in the target aircraft.

SKY-Scanner innovation is based on the introduction of a new family of multi-function sensors with reference to the current ATM Paradigm: laser and radar data fusion at the RDPS and laser data displaying.

Regarding the antenna bedplate, the design will include:

- Electric motor with encoder
- Gear box (with belt system or gear)
- Encoder on the couplet axle
- Reset for the zero position angle
- Rotating joint for the connection between the mobile plate and fixed support of the bedplate antenna.

The principal focus of the design will be the control of the planarity of the azimuthal plate and the low axial oscillations referring the rotation axle. The Safety Critical Operational Features of the FDPS are:

- Short-term Conflict Alert (STCA):
  – Real-time radar-based alert
- Medium-term Conflict Detection:
  – Mathematically models flight path
  – Strategic tool for controllers to project flight paths and view potential conflicts before they occur

- Minimum Safe Altitude & Hazardous Airspace Warnings:
  - Provides alerts when aircraft is projected to descend below minimum safe altitudes
  - Reduced Vertical Separation Minima (RVSM)
  - ATM support for RVSM operations

A laser detection and ranging system based on a rotating cylindrical laser rangefinder array can be a complementary solution to the current aircraft detection and tracking techniques. The system is able to perform a high precision tracking of aircraft with a maximum range of 6 NMs (Nautical Miles) from the ATZ barycentre in order to contrast the classical limitations of radar systems and it represents a promising commitment in the field of the ATZ volume surveillance applications.

3. Conclusion

In this paper, the project for an innovative LIDAR technology for air traffic management is presented. The research covers many areas in the field of LIDARS, Air Traffic Managements, Decision Support Systems, data fusion between radar data and laser tracking data and ground to aircraft laser communications.

Part of the work has already been accomplished: the flat panel for the demonstrator is ready, and the first measurement session is started.

Acknowledgments This work is supported by European Union's Sixth Framework Programme for Research and Technological Development.

References

1. S. Zerkowitz, "Is the need for a new ATM operational concept a strategic necessity?," Aerospace and Electronic Systems Magazine, IEEE Volume 18, Issue 1, Jan. 2003 Pages: 10–12.
2. J. Kosecka, C. Tomlin, G. Pappas, S. Sastry, "Generation of conflict resolution manoeuvres for air traffic management", Intelligent Robots and Systems, 1997. IROS '97, Proceedings of the 1997 IEEE/RSJ International Conference on, Volume 3, 7–11 Sept. 1997 Pages: 1598–1603.
3. J.W. Jackson, S.M. Green, "Control applications and challenges in air traffic management", American Control Conference, 1998. Proceedings of the, 1998 Volume 3, 24–26 June 1998 Pages: 1772–1788.
4. T.J. Callantine, "Air traffic management system domain and control strategy analysis", IEEE International Conference on Systems, Man and Cybernetics, Volume 7, 10–13 Oct. 2004 Pages: 6268–6273.
5. I.R. de Oliveira, F.S. Carvalho, J. Batista Camargo, L.M. Sato, "Multi-agent tools for air traffic management", 11th IEEE International Conference on Computational Science and Engineering Workshops, 2008. CSEWORKSHOPS '08. 16–18 July 2008 Pages: 355–360.
6. L. Wang, J. Shu, T. Emura, M. Kumagai, "A 3D scanning laser rangefinder and its application to an autonomous guided vehicle" Vehicular Technology Conference Proceedings, 2000. VTC 2000-Spring Tokyo. 2000 IEEE 51st, Volume 1, 15–18 May 2000 Pages: 331–335.
7. H. Lamela, E. Garcia, "A low power laser rangefinder for autonomous robot applications", Industrial Electronics, Control, and Instrumentation, 1996, Proceedings of the 1996 IEEE IECON 22nd International Conference on, Volume 1, 5–10 Aug. 1996 Pages: 161–167.

A 100 MICROWATT ULTRA LOW-POWER CONTRAST-BASED ASYNCHRONOUS VISION SENSOR

N. MASSARI[1], M. GOTTARDI[1], S. A. JAWED[1] AND G. SONCINI[2]

[1]*Fondazione Bruno Kessler, via Sommarive, 18 38100 Trento, Ital*
[2]*Università degli Studi di Trento, Via Sommarive, 14 38100 Trento, Italy*

Abstract. An ultra-low power vision sensor is here presented. It consists of a 128 × 64 pixels array, embedding local spatial contrast extraction, binarization and data delivery, which is based on address-coding technique. Spatial contrast among a three-pixel kernel is extracted along the integration time and over 100 dB of dynamic range, thanks to the pixel auto-exposure capability. Contrast binarization is executed at pixel–level in order to convert the final image into a binary frame. This makes post-processing much easier and energy-efficient. The vision sensor consumes less than 100 μW at 50 fps with 25% of pixel activity. This value can be further reduced down to about 30 μW forcing the sensor to work in idle-mode, minimizing the activity at the output. The sensor has been fabricated in a 0.35 um technology and measures 11 mm^2.

1. Introduction

Recent trends toward wireless sensor networks necessitate an efficient way to extract visual data from a camera meeting the limited energy budget of the sensor node. Conventional scanned imagers are not able to fulfill these requirements due to their poor efficiency in the use of the signal bandwidth and the requirement for expensive video processing on the raw pixel data. Raster-scan architectures are mostly oriented toward image quality reproduction [1], but they do not meet some basic requirements, which are crucial for battery-operated sensor networks, such as low-energy features extraction and control strategies for output bandwidth reduction [2, 3]. In literature, several examples of vision sensors implementations are reported, targeted to low-power applications [4]. Even though most of them represent interesting implementations, the majority approaches the problem from the sensor perspective, without to take much into account the system-level data communication and energy issues. Here, the problem seems to be addressed limited to the chip design point of view rather from the system point of view.

The proposed sensor is the first example of a vision sensor capable of pre-processing the visual information at early-stage with tens of μW of average energy budget and equipped with additional operating functionalities, which make it suitable for sensor network applications. It is intended to become the

P. Malcovati et al. (eds.), *Sensors and Microsystems: AISEM 2009 Proceedings*,
Lecture Notes in Electrical Engineering 54, DOI 10.1007/978-90-481-3606-3_59,

main building block of a battery-operated wireless sensor network targeted to surveillance application such as people localization and tracking.

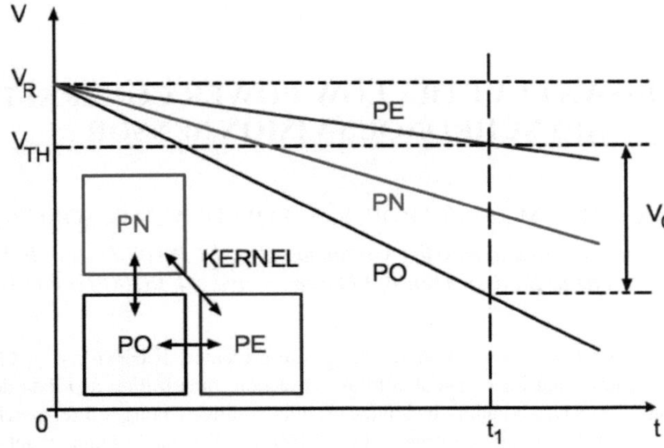

Figure 1. Contrast estimation. Principle of operation.

2. Principle of operation

Before describing the sensor architecture [5], it is worth explaining how visual processing has been embedded on-chip. Spatial contrast is one of the most useful visual features, allowing a target to be detected from the surrounding environment. Visual contrast extraction over high dynamic range images is a typical task requiring on-chip processing in order to be efficiently implemented. Several implementations have been proposed in the last years, aimed at demonstrating the benefits of a pixel-level approach, exploiting the pixel auto-exposure capabilities and/or the availability of local information from neighboring pixels. Let consider a generic pixel of the imager (P0) together with two neighbors PN and PE. Let PE be the less illuminated pixel and PO the most illuminated one. All photodiodes start discharging from reset voltage VR. As soon as the last one (PE) reaches the threshold voltage VTH, the maximum voltage variation is measured, which is linearly related with the Weber Contrast, as shown in Eq. (1):

$$V_C = V_{PE}(t_1) - V_{PO}(t_1) = (V_R - V_{TH})\left(\frac{I_{PO} - I_{PE}}{I_{PE}}\right); \qquad (1)$$

where I_{PE} and I_{PO} are the photo-generated currents of pixels PE and PO respectively.

3. Sensor architecture

The presented ultra low-power vision sensor integrates 128 × 64 pixels imager together with simple digital interface. The pixel is the core of the sensor, embedding most of the visual processing. As shown in Fig. 1, it consists of a photodiode feeding both a voltage comparator (Comp1) and a Contrast Block, implementing the contrast operation of Fig. 1. The photodiode signal is also available to the other two neighboring pixels for contrast estimation (Fig. 2).

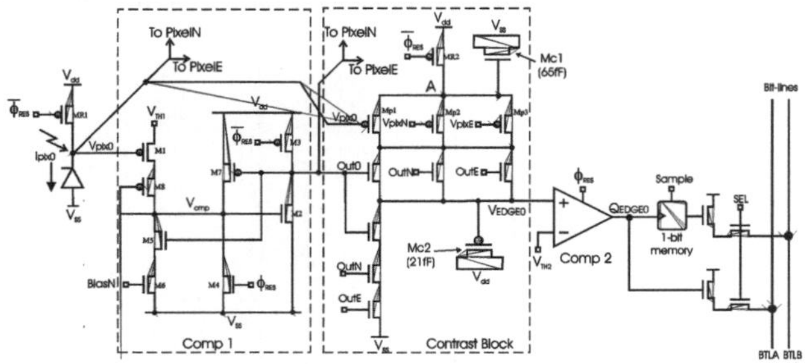

Figure 2. Block diagram of the pixel with embedded contrast extraction and binarization.

The analog contrast (VEDGE) is then binarized by comparator Comp2 and can be stored into the 1-bit memory of is directly available to the output bit-lines (BTLA, BTLB). The imager readout turns into a binary array readout, which is fast, simple and low power. Along the readout phase, some binary processing can be easily performed at column-level. In Fig. 3a, the complete sensor architecture is shown. After selecting a row, the Column Decoder checks for a disparity on each bit-line pair and dispatches the column address of the current pixel to the output of the chip. In this way, only the asserted pixels, those detecting a significant contrast, are delivered to the output with a drastic reduction of the data bandwidth and sensor power consumption. With this type of image compression, in normal situations the data bandwidth can be cut down by 70% with respect to a standard raster-scan image sensor.

Table 1. Key performance parameters.

Parameter	Typical value
Active pixels	128 (H) × 64 (V)
Pixel size and fill factor	26 × 26.5 μm; 20%
Shutter type	Pixel auto-exposure
Frame rate	4,000 frames/s (max)
Contrast sensitivity	10%
Power consumption	<100 μW (50 fps)

4. Experimental results

The sensor has been interfaced with a Field Programmable Gate Array (FPGA), driving the sensor in different operating modes, reading data and communicating with a PC through the USB link. The main sensor parameters (exposure time, frame rate and operating modes) are set through a LabView interface, which also reconstructs the final image, dispatched by the sensor. In Table 1, the main sensor characteristics are reported. Figure 3b, c shows some examples on how the sensor works in the two main operating modes: contrast mode and motion operation.

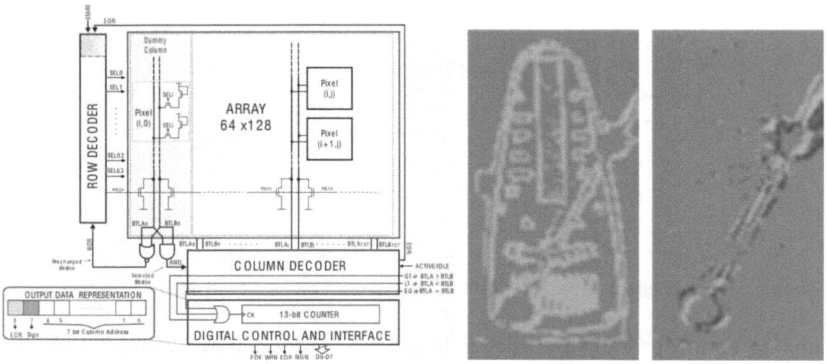

a) b) c)

Figure 3. Vision sensor architecture. Contrast-based image and motion-based image.

5. Conclusions

The presented vision sensor directly extracts and binarizes the spatial contrast over a high dynamic illuminating range. In order to minimize the amount of data, only the asserted pixels are dispatched, turning into a typical image compression of about 70%. Power consumption performance make this sensor suitable for battery-powered systems and sensor networks applications.

References

1. M. F. Snoeij, A. J. P. Theuwissen, K. A. A. Makinwa, and J. H. Huijsing, "A CMOS Imager With Column-Level ADC Using Dynamic Column Fixed-Pattern Noise Reduction," IEEE Journal of Solid-State Circuits, Vol. 41, No. 12, pp. 3007–3015, Dec. 2006.
2. E. Culurciello, R. Etienne-Cummings, and K. Boahen, "A Biomorphic Digital Image Sensor," IEEE Journal of Solid-State Circuits, Vol. 37, No. 2, pp. 281–294, Feb. 2003.
3. P.-F. Rüedi, P. Heim, F. Kaess, E. Grenet, F. Heitger, P.-Y. Burgi, S. Gyger, and P. Nussbaum, "A 128 × 128 Pixel 120-dB Dynamic-Range Vision-Sensor Chip for Image Contrast and Orientation Extraction," IEEE Journal of Solid-State Circuits, Vol. 38, No. 38, pp. 2325–2333, Dec. 2003.
4. R. Berner, P. Lichtsteiner, and T. Delbruck, "Self-timed vertacolor dichromatic vision sensor for low power face detection," Proceedings of ISCAS 2008, Seattle, WA.
5. M. Gottardi, N. Massari, and S. A. Jawed, "A 100uW 128 × 64 Pixels Contrast-Based Asynchronous Binary Vision Sensor for Sensor Network Applications," Journal of Solid-State Circuits, Vol. 44, NO. 5, pp. 1582–1592, May 2009.

A 32 × 32-CHANNELS CHIP FOR X-RAY PIXEL DETECTOR READ-OUT

M. GRASSI[1], V. FERRAGINA[1], P. MALCOVATI[1], S. CACCIA[2], G. BERTUCCIO[2], D. MARTIN[3], P. BASTIA[4], I. CAPPELLUTI[4] AND N. RATTI[4]

[1]*Department of Electrical Engineering, University of Pavia, Italy*
[2]*Department of Electronic Engineering and Information Science, Polytechnic of Milan, Italy*
[3]*ESA-ESTEC, Noordwijk, The Netherlands*
[4]*Thales Alenia Space, Vimodrone (MI), Italy*

Abstract. In this paper we present a spectrometer realized by bump-bonding a 300-μm-pitch, 32 × 32-pixel silicon X-ray detector chip to a 0.35-μm CMOS, 3-cm^2 read-out chip. The self-triggered, mixed analog-digital read-out chip, with 1,024 channels, digitizes the X-ray photon energy with 10 bits of resolution, provides the coordinates of the triggered pixels and achieves 34 e^-_{rms} of input referred noise, ±3.3 LSB of INL and ±0.2 LSB of DNL, while consuming 555 mW from a 3.3-V supply. Preliminary experimental results on the complete spectrometer are reported.

1. Introduction

Several applications in the industrial and scientific fields, especially on spacecrafts, nowadays require large format (thousands of pixels) X-ray spectroscopic detectors. One of the most challenging goals in this kind of applications is the implementation of spectroscopy-grade mixed analog-digital read-out circuits for large format pixel arrays, due to constraints in the area and power consumption available for each channel as well as interferences and crosstalk among channels and between analog and digital circuits, in view of the extremely weak signals (few tens of electrons) to be detected.

The proposed spectrometer consists of two chips: a detector chip and a read-out chip. The detector chip, consisting of a 32 × 32-pixel silicon detector with 300 × 300 μm pitch, is connected by bump-bonding on top of the read-out chip. The sandwich of the two chips is then glued on a hybrid substrate and connected by wire-bonding, as shown in Figure 1. Finally, the hybrid circuit, which includes the chips and some passive components (resistors and decoupling capacitors) is packaged in a ceramic case.

P. Malcovati et al. (eds.), *Sensors and Microsystems: AISEM 2009 Proceedings*,
Lecture Notes in Electrical Engineering 54, DOI 10.1007/978-90-481-3606-3_60,

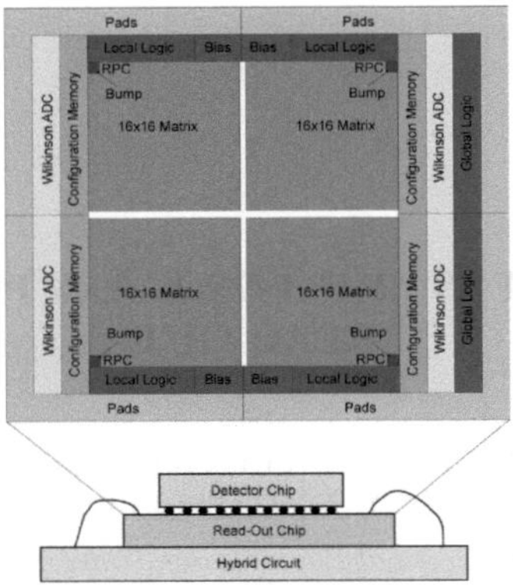

Figure 1. Structure of the proposed X-ray spectrometer.

2. Read-out chip

The read-out chip (StarX32) consists of 32 × 32 300 × 300 μm read-out pixel cells (RPCs), containing a pad for bump bonding, arranged in four identical 16 × 16 arrays (quadrants), as shown in Figure 1.

Each 16 × 16 array is connected to its configuration memory, which contains the programmable parameters of the RPCs, to the trigger logic, which manages the trigger signals produced by the RPCs (self triggered operation), and to 16 Wilkinson (single ramp) 10 bit ADCs, as shown in Figure 2.

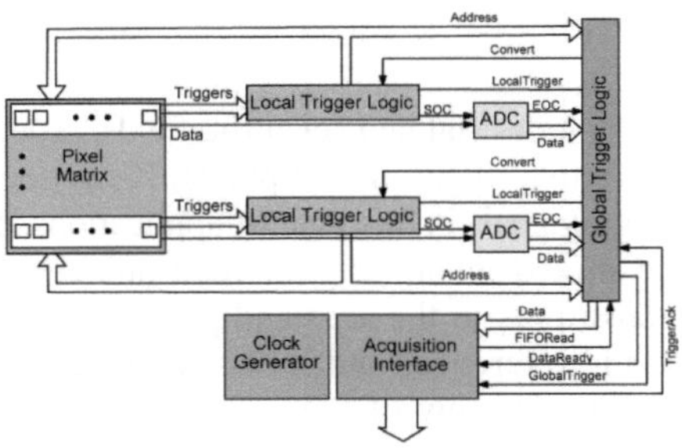

Figure 2. Block diagram of one quadrant of the read-out chip.

The outputs of the trigger logic and of the ADC coming from the four quadrants are collected by a global logic block, which manages the communication with the external acquisition system. The system is equipped with a probe network, which allows the internal nodes of each RPC to be monitored, and a calibration network, which allows us to stimulate each RPC independently for testing purposes.

The RPC 0, whose block diagram is shown in Figure 3, consists of a charge sensitive preamplifier, a second-order RC-CR shaper, a peak stretcher, an output buffer, a current mode amplitude discriminator, a voltage mode peak discriminator and a logic circuit for reset and pulse pile up rejection. The RPC, through the configuration memory, features programmable shaping time (1–10 μs with 1 μs steps) and amplitude discrimination threshold (8 bits), as well as selective enabling and disabling of both the preamplifier and the amplitude discriminator.

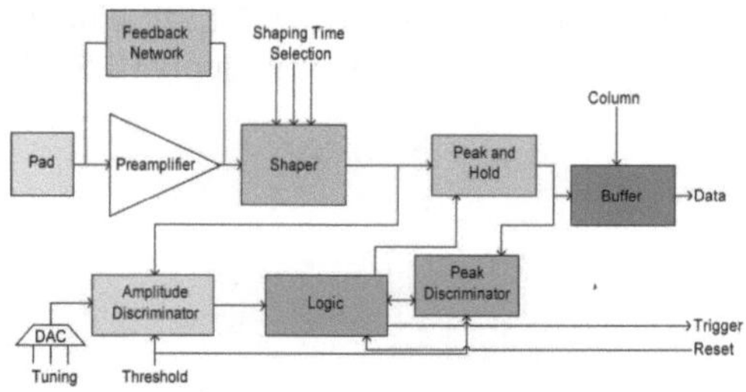

Figure 3. Block diagram of the read-out pixel cell (RPC).

The Wilkinson ADC 0, whose block diagram is shown in Figure 4, consists of an input buffer, a reference voltage source matched to the RPC steady-state output voltage, a comparator, an output register, a ramp generator, a counter and a 10 MHz clock generator. The ramp generator, the counter and the clock generator are common to a whole quadrant, while there is an input buffer, an output register and a comparator associated with each row of the 16 × 16 arrays. The comparator compares the output signal of the selected RPC within the row, with the signal produced by the ramp generator. When the output register is enabled by the SOC signal and the comparator changes state because the ramp crossed the input signal, the content of the counter is copied in the output register. When the counter reaches full scale, the output register produces the EOC signal.

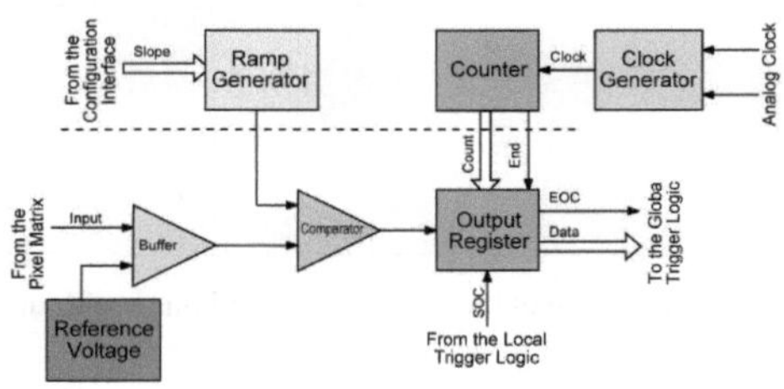

Figure 4. Block diagram of the Wilkinson ADC.

3. Experimental results

Figure 5 shows a photograph of the hybrid circuit with the detector chip bonded on top of the read-out chip as well as a detail of the 0.35 μm CMOS, 18.4 × 16.2 mm read-out chip. A typical acquisition obtained using the on-board calibration network is shown in Figure 6, which shows the overall acquired spectrum as well as the spectrum acquired by a specific pixel. The gray-scale map shows graphically the event-count of each pixel. The measurement has been carried-out with a dedicated test setup, which allows us, through a PC interface, to program all the parameters of the chip and to collect and analyze the data. The complete acquisition chain achieves a minimum input referred noise of 34 e^{-}_{rms} with a detector capacitance of 400 fF and a detector leakage current of 2.5 pA, an input range equal to 120e-rms-12000 e^{-}_{rm}, an integral non-linearity (INL) of ±3.3 LSB and a differential non-linearity (DNL) of ±0.2 LSB with 10 bits of resolution. The chip consumes 555 mW from a 3-V power supply (0.5 mW per pixel).

Figure 5. Photograph of the hybrid circuit with the detector chip bonded on top of the read-out chip.

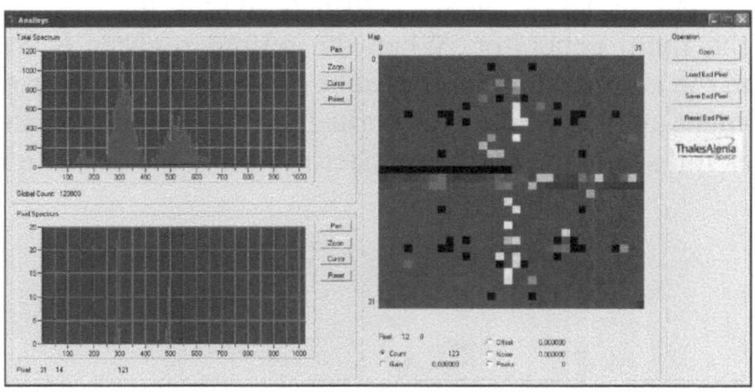

Figure 6. Typical acquisition obtained using the on-board calibration network.

References

1. S. Caccia, G. Bertuccio, D. Maiocchi, P. Malcovati, N. Ratti, and D. Martin, "A Mixed-Signal Spectroscopic-Grade and High-Functionality CMOS Readout Cell for Semiconductor X-Gamma Ray Pixel Detectors", *IEEE Trans. Nuclear Science*, vol. 55, pp. 2721–2726, Oct. 2008.
2. A. Rossini, S. Caccia, G. Bertuccio, F. Borghetti, V. Ferragina, P. Malcovati, D. Martin, P. Bastia, I. Cappellutti and N. Ratti, "A Complete Read-Out Channel with Embedded Wilkinson A/D Converter for X-Ray Spectrometry", *IEEE Trans. Nuclear Science*, vol. 54, pp. 1216–1221, Aug. 2007.

MENTAL TASKS RECOGNITION FOR A BRAIN/COMPUTER INTERFACE

G. COSTANTINI[1], D. CASALI[1], M. CAROTA[1], G. SAGGIO[1], L. BIANCHI[2], M. ABBAFATI[2] AND L. QUITADAMO[2]

[1]*Dipartimento di Ingegneria Elettronica, Università di Roma "Tor Vergata", Via del Politecnico, 1- 00133 Roma, Italy*

[2]*Dipartimento di Neuroscienze, Università di Roma "Tor Vergata", Via Montpellier 1 - 00133 Roma, Italy*

Abstract. In this paper, a brain/computer interface is proposed. The aim of this work is the recognition of the will of a human being, without the need of detecting the movement of any muscle. Disabled people could take, of course, most important advantages from this kind of sensor system, but it could also be useful in many other situations where arms and legs could not be used or a brain-computer interface is required to give commands. In order to achieve the above results, a prerequisite has been that of developing a system capable of recognizing and classifying four kind of tasks: thinking to move the right hand, thinking to move the left hand, performing a simple mathematical operation, and thinking to a carol.

1. Introduction

Brain electrical activity can be easily observed by simply placing a set of wet electrodes on the surface of the head. Every kind of task or thought the human being can perform causes electrical activities in different parts of his/her brain; therefore, the recognition of this activity could be considered as a desirable machine learning application. The task is not so trivial, for many reasons: first, we cannot know the state of all neurons in the brain, but just a mean value of it in some zones of the outer part of the brain. Second, the electrical activity is not limited to a single zone, depending on the task a person is performing: in most cases, it involves the whole brain. The difference among different tasks is mainly in the way electrical waves move from one zone to another. A third problem is that there is always a lot of electrical activity in the brain, also when we are thinking or doing "nothing". This kind of activity, including breathing and all involuntary movements, is always present and can eventually mask the task we intend to monitor. All of this activities represent for us a "noise" that is often bigger than the "signal" we need to detect.

Because of these reasons, the main challenge we are going to face is the classification of the dataset that we collected from the electrodes.

P. Malcovati et al. (eds.), *Sensors and Microsystems: AISEM 2009 Proceedings*, Lecture Notes in Electrical Engineering 54, DOI 10.1007/978-90-481-3606-3_61,

A reasonable classification method for this kind of data relies on artificial neural networks. In this work, we used Support Vector Machines: a tool that is very similar to a neural network, but with the advantage that it can support datasets with a huge number of components; therefore there's no need of a reduction of the feature space. Moreover, it has a training algorithm that is much better than the "back-propagation" rule usually used in neural networks.

2. Sensor system

The sensor system is composed of 61 electrodes that are placed at the surface of the head of the subject, according to a standard disposition used in these kind of applications [1, 2], as shown in Fig. 1.

Electrodes are connected to the computer by means of fiber optics, in order to provide necessary electrical insulation that guarantee the subject by any risk of electrical shock.

Signals are sampled at a sampling rate of 256 Hz.

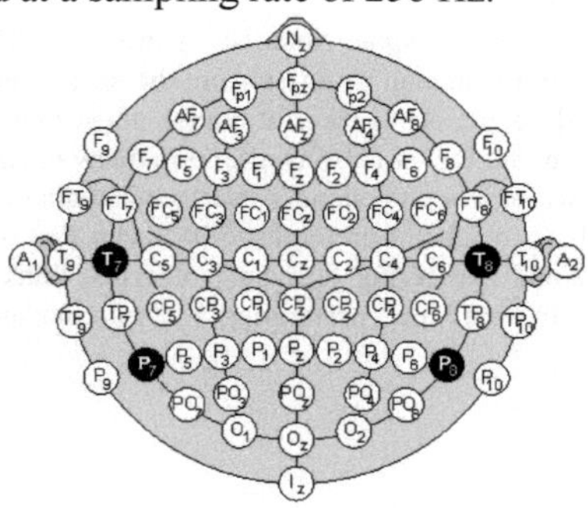

Figure 1. Position and names of the electrodes.

3. Preprocessing

Different types of brain activity are often related to the frequency of the waves that we can find in EEG signals, such as alpha waves (8–12 Hz), beta waves (12–19 Hz), gamma waves (around 40 Hz), delta waves (1–4 Hz) that are associated to weakness, sleep, REM, and other kind of brain states [3–5].

For that reason, we suppose that analyzing the dataset in the frequency domain could be useful for our purpose. For every task, we calculated the FFT in three windows of 256 samples. Finally we calculated the mean value of the three windows, for every frequency channel. Of course we considered only the first half of every FFT window, because the second half is symmetric and doesn't give any further information. Channels from 1 to 127 represent frequencies from 0–127 Hz. Channel 0 represents frequency 0, and it is omitted.

In addition to the 91 electrodes, we the mean value of all electrodes, and a "marker" signal used as time reference.

So we have a total number of 8,001 data points for every task that we want to classify.

4. Classifier

In the past few years, SVMs aroused the interest of many researchers being an attractive alternative to multi-layer feed-forward neural networks for data classification and regression or PCA [6, 7].

The basic formulation of SVM learning for classification consists in the minimum norm solution of a set of linear inequality constraints. So, it seems useful to exploit the relation between these two paradigms in order to take advantage of some peculiar properties of SVMs: the "optimal" margin of separation, the robustness of the solution, the availability of efficient computational tools. In fact, the SVM learning problem has no non-global solutions and can be solved by standard routines for quadratic programming (QP); in the case of a large amount of data, some fast solvers for SVMs are available, e.g. SVMlight [8]. In the following paragraph we will give a brief description of an SVM.

5. Experiments and results

The experiment involved a set of six subjects for 2 days. Every day a subject performed two sessions. During a session, the subject was asked to perform 400 tasks randomly selected among the following: thinking to move the right hand, thinking to move the left hand, performing a simple mathematical operation, and thinking to a carol. Every task lasted 3 s, hence the entire session was 20 min long.

Our objective was to operate discriminations between every couple of task: left hand vs. right hand, mathematical operation vs. carol, right hand vs. mathematical operation, left hand vs. carol, right hand vs. carol and left hand vs. mathematical operation. Hence we prepared six kinds of datasets, one for every possible combination of the six tasks. The whole datasets was divided in training sets (75% of the datasets) and test sets (25% of the datasets). Accuracy results for training set was always 100%, accuracy results for test set are shown in Table 1.

For every subject (denoted with a number, for privacy reason), we reported the mean values related to the four sessions. In the last line of Table 1, we reported mean value for all the subjects. As we can easily note, difference from left and hand, at least with our method is almost unnoticeable, while if we want to discriminate between left hand and a carol, the accuracy is relatively high. The cause of this big difference may be that hand movement and thinking to a carol are completely different tasks, which involves different areas of the brain, while left and right hand movements activate both sides of the brain, and not only right side and left side respectively.

6. Conclusions

A brain/computer interface is presented, which is able to discriminate among different kind of mental tasks that a subject is performing. It is based on a SVM classifier, which is trained by the power spectrum of the EEG signals coming from 61 electrodes set in the surface of the head.

An experimental test showed quite good results in case of discriminating between the thought of a carol and a mathematical operation, or mathematical operation or carol and hand movements, while results have been very poor in case of discriminating between movement of right hand and left hand. A greater difference could probably reside in the activation time of the involved areas. Future development will include an analysis in the time domain, in addition to the data considered in this paper. Moreover, we found an high difference in performance according to different subjects: for example, as you can see in the table, with subject 2 we obtained more than 80% in almost all tasks, while with subject 5 we was just a little over 50%.

Table 1. Accuracy results.

Subjects	Left/right (%)	Math/carol (%)	Right/math (%)	Left/carol (%)	Right/carol (%)	Left/math (%)
1	59.00	70.75	64.92	65.00	53.25	67.40
2	65.90	70.39	81.40	85.64	77.55	88.68
3	53.50	59.50	63.96	65.77	56.75	62.19
4	57.60	65.39	70.25	81.91	72.92	75.88
5	55.90	61.29	70.93	67.44	72.00	71.70
6	57.50	80.27	66.54	74.77	68.00	55.76
Average	58.24	68.35	69.66	73.42	66.77	70.27

References

1. Sharbrough F, Chatrian G-E, Lesser RP, Lüders H, Nuwer M, Picton TW, "American Electroencephalographic society guidelines for standard electrode position nomenclature." J. Clin. Neurophysiol 8 (1991): 200–202.
2. Sellers EW, Krusienski DJ, McFarland DJ, Vaughan a TM, Wolpaw JR, "A P300 event-related potential brain–computer interface (BCI): The effects of matrix size and inter stimulus interval on performance" Bio. Psychol. 73 (2006): 242–252.
3. Brazier MAB, The Electrical Activity of the Nervous System, Pitman, London, 1970.
4. Ward LM, Doesburg SM, Kitajo K, MacLean SE, Roggeveen AB, "Neural synchrony in stochastic resonance, attention, and consciousness." Can. J. Exp. Psychol. 60(4) (2006): 319–26.
5. Walker, Chambers Dictionary of Science and Technology, Chambers Harrap Publishers, 2nd ed., p. 312, P.M.B., 1999.
6. Jolliffe IT, Principal Component Analysis, 2nd ed., Springer, NY, 2002.
7. Burges CJC, "A tutorial on support vector machines for pattern recognition", in Data Mining and Knowledge Discovery 2, Kluwer, pp.121–167, 1998.
8. Joachims T, "Making large scale SVM learning practical", Advances in Kernel Methods-Support Vector Learning, B. Scholkopf, CJC Burges and AJ Smola Eds., MIT Press, Cambridge, MA, 1999, pp. 169–184 (http://svmlight.joachims.org/).

SILICON INTEGRATED MICRO-BALANCES ARRAY FOR DNA HYBRIDIZATION ELECTRONIC DETECTION

G. BARLOCCHI, U. MASTROMATTEO AND F.F. VILLA
STMicroelectronics, Via Tolomeo 1, 20010 Cornaredo, Italy

Abstract. A novel biosensor based on silicon integrated micro-balances array for DNA hybridization electronic detection is presented. The microbalances have been implemented on a silicon substrate putting together low cost silicon membranes and piezoelectric resonator made up of an AlN film sandwiched between two metal layers. The mass change, occurring during the DNA hybridization process, is converted into resonant frequency change that can be easily measured. Mass sensitivities around 75 Hz/pg for a $200 \times 200\ \mu m^2$ membrane area and 300 Hz/pg for $100 \times 100\ \mu m^2$ area are expected.

1. Introduction

The detection of specific DNA sequences is of significance in many areas including clinical, environmental and food microbiology diagnosis. The analysis of gene sequences plays a fundamental role in rapid detection of genetic mutations. This means it is possible to make reliable diagnosis of diseases even before any symptoms appear. Hybridization, the pairing of separated strands of DNA with complementary DNA strands that act as probe, is an important tool in the field of genetic analysis. In hybridization-based detection one of the strands is known (a DNA probe) and the other unknown. The hybridization bond is specific since it occurs only when there is a match of complementary strands. The presence of a double strand in the mixture is indicative of a match; hence hybridization serves as a sequence detection mechanism. In *standard microarray* the probes are attached to a solid surface (glass or silicon) using a linker molecule. Probe-target hybridization is usually detected and quantified by fluorescence-based detection. This method, however, has several significant drawbacks, since it requires preventive manipulation of the analyte to introduce the optical markers and need expensive and power consuming instrumentation for optical reading. Typical commercial equipments are systems with photomultiplier tubes (PMTs) or charge-coupled devices (CCD). Recently to solve these problems a quartz crystal microbalance (QCM) which does not require radioisotopes or fluorophores as labels has been proposed. Quartz crystal microbalance consists of a quartz disc with gold electrodes patterned on the opposite sides, one of which covered with a sensing layer capable of suitable interactions with the analyte of interest. In practice the microbalance is a miniaturized resonator and placing it in the proper

P. Malcovati et al. (eds.), *Sensors and Microsystems: AISEM 2009 Proceedings*, Lecture Notes in Electrical Engineering 54, DOI 10.1007/978-90-481-3606-3_62,

site in an electronic oscillating circuit, the electric field across the electrodes triggers mechanical vibrations, whose frequency is related to the acoustic load of the sensor. In this way piezoelectric quartz crystal allow the direct monitoring of nucleic acids interactions by measuring the quartz frequency variation as result of the superficial mass increase. The mass change (occurring during the DNA hybridization process) is converted into resonant frequency change that can be easily measured. In order to use in *standard microarray* this technique it is mandatory increase the sensibility like this to detect mass variation around one picogram (comparable with the mass variation normally present during DNA hybridization process). Now, as it known [1, 2], for a given piezoelectric crystal the sensitivity of a microbalance can be increased decreasing the dimensions of microbalance itself. Therefore, miniaturizations (silicon integration) of crystal balance represent the best way to reach this goal.

2. New approach description

The approach proposed by *ST* to improve the sensibility of crystal balance consists to generate, for each DNA target probes of standard microarray, a corresponding array of silicon integrated microbalances. The microbalances have been implemented on < 100 > silicon substrate putting together low cost silicon membranes and piezoelectric stack made up of an AlN film sandwiched between two metal layers that serve as electrical terminals (Fig. 1).

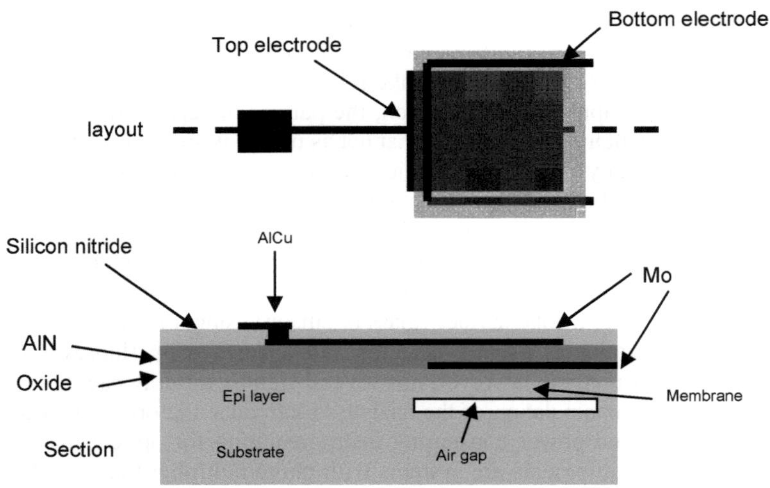

Figure 1. Silicon microbalance cross section: membrane area = 200 × 200 μm^2, membrane thickness = 20 μm, resonance frequency = 200 MHz

The structures are robust in construction and miniaturized in size and the technology involved in their fabrication is fully compatible with those of the silicon integrated circuits. Detail of process flow steps and simulation preliminary results are indicated in the next two subsections.

2.1. *Process flow steps*

The first technological step for the device fabrication consists in making thin silicon membranes on a silicon substrate The mono-silicon membrane is obtained with just one mask. The creation of the embedded open spaces (cavity in Fig. 1) in the silicon starts with a silicon substrate dry-etch to generate (Fig. 2a) a honeycomb lattice (silicon columns).

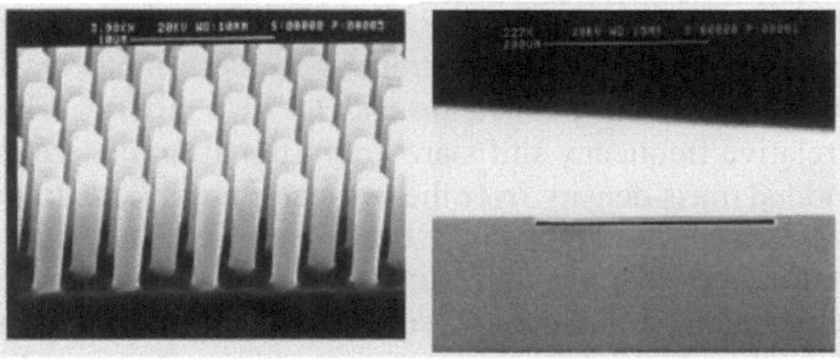

a) b)

Figure 2. (**a**) honeycomb lattice generated by silicon dry etch (substrate); (**b**) silicon membrane.

After the trenches are formed, the substrate is placed in an ASM single wafer epitaxial reactor to perform an epitaxial growth. Consequently, a silicon layer grows on top of the silicon columns and close, at the top, the trenches trapping hydrogen molecules. An annealing step is then carried out in N2 atmosphere (with silicon oxide cap) using conventional furnace or in hydrogen furnace. This annealing causes a migration of the silicon atoms, which tend to arrange themselves so as to minimize the surface energy. Consequently, because the columns are arranged close together in the suitable way, the silicon atoms migrate completely and form a sensor cavity closed at the top by a membrane (Fig. 2b). The end result is very similar to what it is possible to get with bulk micromachining wafer to wafer bonding. But with the big advantage to have, at low cost, thinner, smaller and mechanically more robust chip. Moreover, the sealing of the cavity doesn't require any wafer-to-wafer bonding and thus the reliability of the sealing joint is definitely higher. The next process steps to complete the silicon microbalances are:

1. Deposition of 100 nm of molybdenum layer (ground electrode)
2. Deposition of 1 μm of AlN
3. Deposition of second molybdenum layer (bias electrode)
4. Al/Cu deposition
5. Protection layer, assembly, and wire bonding

At this point the microbalances are ready for DNA spotting phase and analysis.

2.2. *Simulation results*

Some simulation are performed in order to predict the resonance frequency of microbalance, estimate the sensitivity of frequency to the added mass (occurring during the DNA hybridization process) and carry out an analysis of the resonator to process dispersions (mainly thickness dispersions). Simulations are based on Mason's model; material constants have been obtained by acoustic characterizations of the materials deposited at CEA-Leti. A first simulation was performed to estimate the resonant frequency of the microbalance (200×200 μm^2): the value obtained is about 200 MHz. Moreover, to the first order, the frequency shift induced by an added layer is proportional to the added mass. Therefore, it remains independent of the elasticity of the material involved. This ensures the operation principle of the microbalance. Determined mass sensitivities are: 3.0 kHz mm^2/ng without passivation and 2.9 kHz mm^2/ng with passivation. Note that the relative frequency shifts are proportional to relative mass changes; therefore the added mass density over the resonator area needs to be considered, not directly the mass itself. To estimate the sensitivity of resonant frequency to process dispersion, we have evaluated the influence of small thickness changes (in nm) on the resonance frequency. Results show that to the first order, the two Mo electrodes lead to the highest frequency sensitivity. Nevertheless, if relative thickness dispersions are considered, silicon membrane thicknesses dispersions (± 1 μm) are the main source for frequency dispersions, since dispersions for the other layers remain moderate (a few percents of small thicknesses). Shifts caused by thickness variations of the silicon membrane may leave resonance frequency outside of the 200 MHz $\pm$ 10%. To come closer to the targeted resonance frequency (200 MHz), the thickness of the silicon membrane can be decreased to 18 μm. Estimated mass sensitivities are in this case: 3.5 kHz mm^2/ng without passivation and 3.4 kHz mm^2/ng with passivation.

3. Conclusions

The Silicon Integrated MultiMicrobalances appear to be a suitable and convenient tool for monitoring hybridization of complementary strands of oligonucleotides. Label-free strategy not only guarantees the accuracy of the test, but also shortens the detection time and reduces the detection cost; expensive instrumentation for optical reading is not more necessary. Mass sensitivities around 3 kHz mm^2/ng are expected (i.e. 75 Hz/pg for a 200×200 μm^2 resonator and 300 Hz/pg for a 100×100 μm^2 resonator).

References

1. G. Z. Sauerbrey, Z Phys. 155, 206 (1959).
2. C. K. O' Sullivan, G. G.Guilbault, "Commercial quartz crystal microbalances-theory and applications", Biosensors & Bioelectronics 14, 663–670, (1999).

A FULLY INTEGRATED SYSTEM FOR SINGLE-SITE ELECTROPORATION AND ADDRESSED CELL DRUG DELIVERY

L. ODORIZZI[1], C. COLLINI[1], E. MORGANTI[1], R. CUNACCIA[1], C. RESS[1], L. LORENZELLI[1], A. GIANFELICE[2], E. JACCHETTI[2], C. LENARDI[2] AND P. MILANI

[1]*FBK, Materials and Microsystems Center – BIOMEMS Unit, Via Sommarive 18, 38123 Povo – TN, Italy*

[2]*C.I.Ma.I.Na. Interdisciplinary Centre for Nanostructured Materials and Interfaces, Universita' di Milano, Via Celoria 16, I-20133 Milano, Italy*

Abstract. Nowadays, there is a great interest in developing microdevices able to perform in-vitro cell assays for high throughput drug screening. In this work, we propose two modules that could be integrated in a single platform: a gold microelectrode array (MEA) for single-site cell transfection and two interdigitated electrodes (IDEs) arrays for the controlled localization of cells and the delivery of bio-chemical species. The modules have been already realized, separately tested and both have demonstrated their feasibility. The final aim is to perform single-site drug delivery in a microfluidic structure where different bio-chemical species could be addressed to different cell groups in specific areas of the chip. In perspective, an high level of integration can be obtained by combining the two miniaturized approaches through microfabrication techniques, resulting in a portable and high throughput system for fast and efficient in-vitro drug screening.

1. Introduction

Individual cell targeting is of major interest in the field of diagnosis and drug delivery. The understanding of many biological processes would greatly benefit from the ability to analyse the content of single cells. Advances in microfabrication have introduced the possibility to work easily at the single cell level. In this context the main goal is to develop an integrated microsystem that allows single cell electroporation, controlled positioning of cells, and delivery of bio-entity-loaded micro/nanoparticles. This platform will include two modules: the first one for the single cell addressed transfection based on electroporation technique [1], and the second for the electrical movement of cells and micro/nanoparticles based on dielectrophoretic (DEP) technique [2] (Fig. 1).

P. Malcovati et al. (eds.), *Sensors and Microsystems: AISEM 2009 Proceedings*, Lecture Notes in Electrical Engineering 54, DOI 10.1007/978-90-481-3606-3_63,

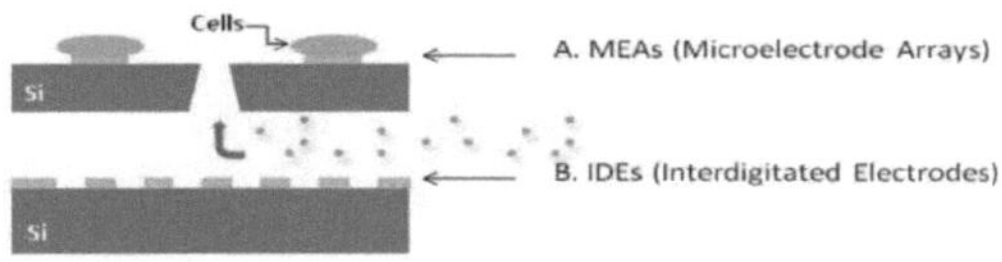

Figure 1. Schematic illustration of the integrated system: MEAs for cell electroporation (**a**) and IDEs for addressed cell drug delivery (**b**).

2. Materials and methods

The electroporation module consists of a gold cell-size MEA with several passing holes which allow the delivery of transfection solutions in desired areas of the chip (Fig. 2). The dimensions of each part are listed in Table 1.

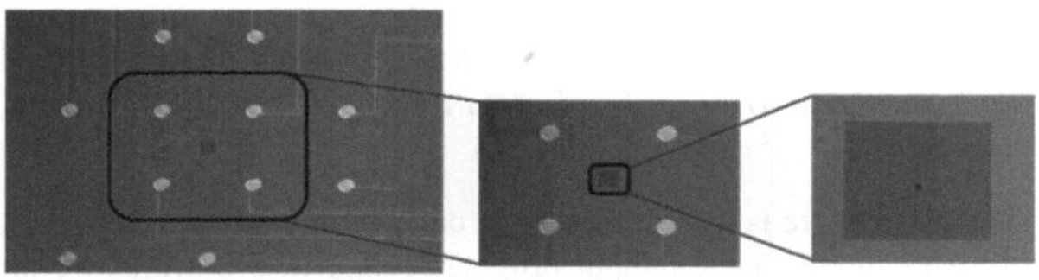

Figure 2. The microphotograph and the insets show the array of gold circular electrodes, membrane of dielectric multilayer and passing hole.

Table 1. Geometrical parameters of the MEA-based module.

Parameter	Value
Electrode diameter	50 μm
Membrane dimension	20–60 μm
Holes dimension	3–4 μm
Chip dimension	1 × 1 cm

Concerning the controlled positioning of cells and addressed transfectants delivery, the device mainly consists of two planar parallel IDEs arrays for both electric vertical (positive/negative DEP, p–n DEP) and lateral (travelling wave DEP, tw DEP) movement of micro/nanoparticles. The device layout has been investigated and realized in order to perform alternate electric fields with waveforms shifted by 180° and 90° respectively (Fig. 3).

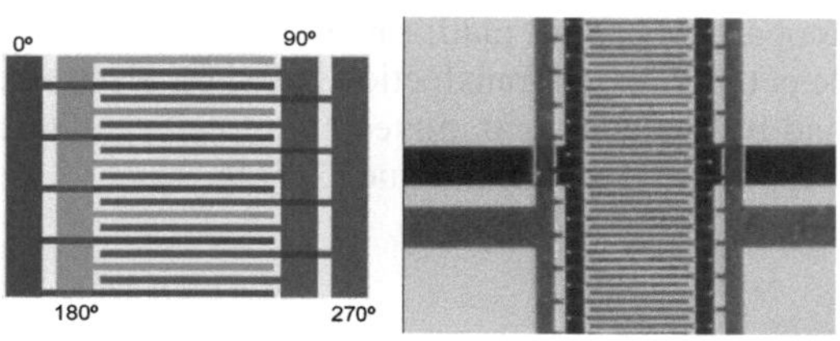

Figure 3. IDEs structure (*left*) and picture (*right*): electrode width and gap of 10 μm.

Both MEA and IDEs modules are characterized by two levels of metal structures (buried connection lines made of Al 1% Si + Ti/TiN and gold electrodes) in order to reduce the fabrication costs and the dimensions while improving the device electrical performances. Additional steps of bulk micro-machining are developed in order to realize the inlet microfluidics of the MEA-based module. Biocompatible polymers (e.g. PDMS) and quartz are used for microchannels and cells confinement respectively. Since the two devices share the same fabrication process we propose their integration in a single microsystem. The complete MEA packaging of the device is shown in Fig. 4.

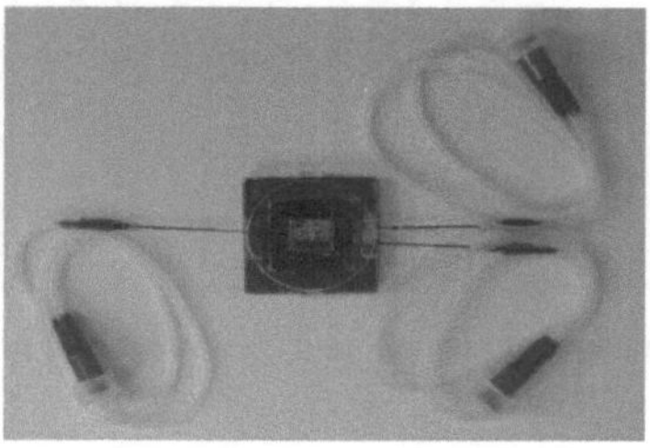

Figure 4. Packaging of the device with fluidic connections.

3. Experimental results

In order to demonstrate the feasibility of the two systems, both modules have been individually characterized.

3.1. *Microfluidics testing of MEAs*

Liquid injection has been performed in order to check the possibility to introduce transfectants in confined regions of the MEA device. Preliminary tests have involved coloured buffer solutions just to follow the liquid delivery from the injection point to the holes and verify the absence of occlusions along the microchannels.

3.2. *Device testing: polystyrene microbeads and cells dielectrophoretic experiments*

The dielectrophoretic effects of the device have been evaluated by using two types of polystyrene beads with 5 and 10 μm diameters (white and blue, respectively). Controlled p-n DEP and also separation by means of tw DEP of these different size particles have been obtained. The relative frequencies for vertical and lateral movement have been identified [3].

Moreover, three different cell lines (NIH/3T3 fibroblasts, MDCK epithelial cells and SH-SY5Y neuroblastoma cells) have been used to test also the dielectrophoretic cell movement. Relative frequencies are listed in Table 2, while an example of n-p DEP for MDCK cells is shown in Fig. 5.

Table 2. n–p DEP frequencies of NIH/3T3 (fibroblasts), MDCK (endothelial), and SHSY5Y (neuroblastoma) cells.

Cells	NIH/3T	MDCK	SHSY5Y
n-DEP	5 kHz	5 kHz	5 kHz
p-DEP	52 kHz	52 kHz	30 kHz

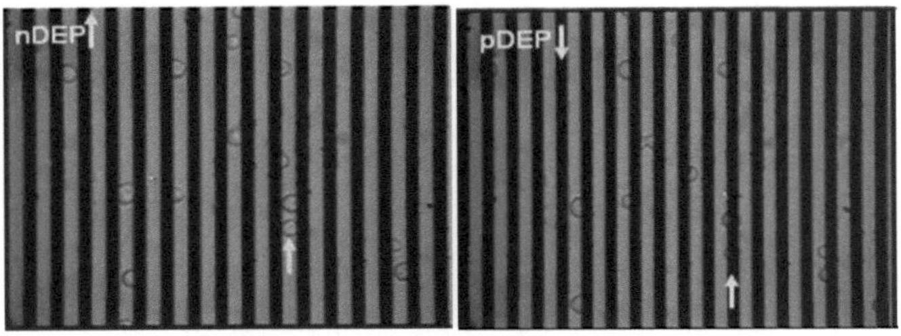

Figure 5. Example of vertical motion (n- and p-DEP) of MCDK cells.

4. Conclusions

This work presents two modules in terms of both design and microfabrication technology: a MEA for single-site cell transfection and two IDEs arrays for the controlled delivery of bio-chemical species. They have been individually characterized: both the microfluidics testing by means of liquid injection through the bottom hole and dielectrophoresis-based conveyance of microbeads/cells have been proven. Our final goal will concern the realization of a full-integrated system in order to provide a fast and efficient platform for in-vitro drug screening.

Acknowledgments This work has been supported by CARITRO Foundation Rovereto, Italy under the project: CELTIC 2006 – "Development of a integrated system based on innovative nano-microfabrication technologies for in vitro-diagnostic assays" and by AIRC under OGCG grant: "Development and integration of highthroughput technologies for the functional genomics of cancer."

References

1. S. Vassanelli, L. Bandiera, M. Borgo, G. Cellere, L. Santoni, C. Bersani, M. Salomon, M. Zaccolo, L. Lorenzelli, S. Girardi et al, "Space and time resolved gene expression experiments on cultured mammalian cells by a single-cell electroporation microarray", *New Biotechnology*, [25] pp. 55–67, 2008.
2. N. Crews, J. Darabi, P. Voglewede, F. Guo, A. Bayoumi, "An analysis of interdigitated electrode geometry for dielectrophoretic particle transport in micro-fluidics" *Sensors and Actuators B*, [125] pp. 672–679, 2007.
3. R. Cunaccia, L. Odorizzi, E. Morganti, A. Adami, C. Collini, L. Lorenzelli, E. Jacchetti, C. Lenardi, P. Milani "A nano-on-micro device for dielectrophoretic cancer cell separation and detection" *Proceedings of the 13th Italian Conference, Sensors and Microsystems* (AISEM), pp. 238–242, 2009.

A NOVEL BASED PROTEIN MICROARRAY FOR THE SIMULTANEOUS ANALYSIS OF ACTIVATED CASPASES

I. LAMBERTI[1], L. MOSIELLO[1], C. CENCIARELLI[2], A. ANTOCCIA[2] AND C. TANZARELLA[2]

[1]*ENEA CR Casaccia, Dipartimento Biotecnologie e Protezione della Salute, Via anguillarese, 301, 00123 Roma*

[2]*Dipartimento di Biologia, Università "Roma Tre", V.le G. Marconi 446, 00146 Roma*

Abstract. The aim of our work is to develop a new method to study caspases expression in apoptotic cells using an antibody-based protein microarray. Actually we performed the array for two different form of caspase-3 (cleaved and uncleaved), in the next future we plan to enlarge the microarray analysis using caspase-8 and caspase-9 antibodies in order to obtain a new analytical tool to study the expression of these markers in apoptotic cells

1. Introduction

The western blot is an analytical technique used to detect specific proteins in a given sample of tissue homogenate or cellular extracts. It uses gel electrophoresis to separate native or denatured proteins, then transferred to a nitrocellulose or PVDF membrane, and detected using antibodies specific to the target protein.

In recent years, the antibody microarray technology has made significant progress, going from proof-of-concept designs to established high-performing technology platforms capable of targeting non-fractionated complex samples, as proteoma. Antibody microarrays are an easy-to-use, cost-effective tool for high-throughput protein profiling using cell extracts, tissue lysates, and other biological samples. Apoptosis is the process of cell auto-destruction often called 'programmed cell death'. Apoptosis is a highly regulated process characterized by morphological and biochemical cellular changes, as the activation of specific cytolic proteases, called caspases. The aim of our work is to develop a new method to study caspases expression in cells induced to apoptosis using an antibody based protein microarray. According to this aim, we study the feasibility to transfer the methods of the immunological assay for caspases from classical western blot into a microarray format in order to develop a multiparametric, rapid, sensitive and inexpensive assay.

To check the feasibility of this format, we focused our studies on the most popular caspase (caspase-3, cleaved and un-cleaved forms).

P. Malcovati et al. (eds.), *Sensors and Microsystems: AISEM 2009 Proceedings*, Lecture Notes in Electrical Engineering 54, DOI 10.1007/978-90-481-3606-3_64,

2. Experimental

2.1. *Protein chip platform*

Our activities in microarray field was mainly devoted to develop protein microarray using the pulsed laser assisted deposition, PLD [2–4].

In this paper to perform microarray determination we used a commercial protein chip platform composed of a spotting machine (producing a maximum spot density of 64 spot for a 8 × 8 matrix and showing a spot diameter between 500–700 μm and a spot distance of 1 mm); for microarray reactions we used a high performance mini oven with accurate temperature control and vibration function with linear rocking movement. Finally for microarray reading we uses an optical scanner, with a 300–600 dpi resolution, equipped with image and data analysis software able to collected the results in terms of optical density for each spot. The protein chip platform is shown in Fig. 1.

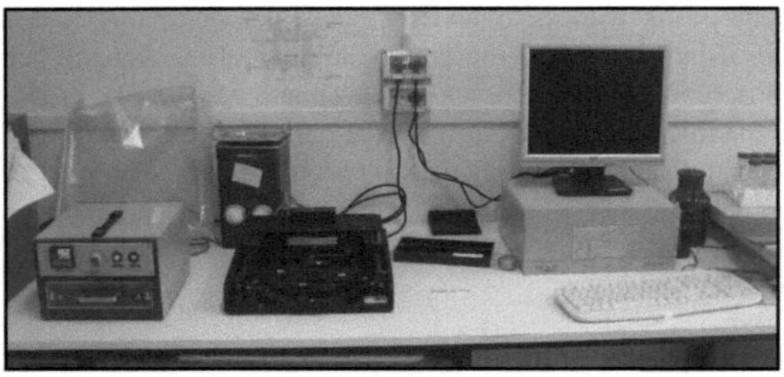

Figure 1. Protein chip platform.

2.2. *Apoptosis induction*

To validate the protein chip, H460 lung tumor cells were induced to apoptosis by treating them for 4 h with 2 μM staurosporine and performing Western blot with antibodies for activated caspase-3. In fact, the cleavage of inactive procaspase to active cleaved caspase is a recognized marker of apoptosis [5].

2.3. *Cell lysis and immunoblot*

Western blot analysis (as control of microarrays) was carried out as described in Cenciarelli et al. [1] loading 35 μg of whole cell lysate. Briefly, blotted filters were incubated with 1 μg/ml of specific antibodies recognizing caspase-3cleaved (Cell Signaling, MA, USA) and α-tubulin as loading control. Primary antibodies were detected using secondary anti-mouse antibody HRP conjugates and visualized using the enhanced chemioluminescence detection system.

2.4. *Protein spotting and microarraying*

In our method protein chip were prepared spotting crude and denaturated proteic lysate from cell H460, treated with staurosporine. The lysate was spotted at different concentrations onto a well containing a glass slide coated to copolymer of N, N-dimethylacrylamide (DMA), N,N-acryloyloxysuccinimide (NAS) and [3-(methacryloyl-oxy)-propyl] trimethoxysilyl (MAPS) and in the same way proteins obtained from untreated cells were spotted, as negative control.

This polymeric surface is demonstrated to be suitable for covalent and stable coating of biological molecules and for protein microarray technology. The proteins immobilized on this polymeric surface maintain an active conformation and are easily accessible, providing a detection limit of 54 amol/spot.

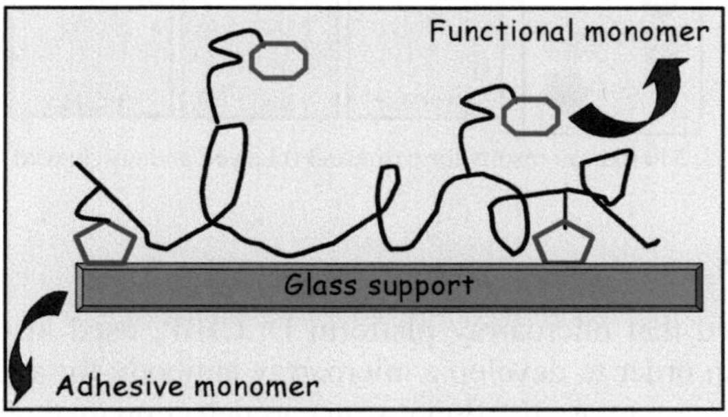

Figure 2. Glass slide coated to DMA-NAS-MAPS.

2.5. *Caspase-3 assay*

After spotting and blocking with a solution of 2% (w/v) BSA in 0.05 M Na2HPO4.2H2O, the microarrays was incubated with a solution of a murine monoclonal antibodies anti-caspase-3 (cleaved and uncleaved forms) for 1 h at 37°C. The antigen-antibody binding on the spotted microarrays was revealed adding a secondary biotinylated antibody anti-mouse-Fc specific and via the enzymatic reaction with a streptavidine-Alkaline-Phosphatase (AP) which reacts with a chromogenic substrate (nitro blue tetrazolium chloride/5-Bromo-4-chloro-3-indolyl phosphate toluidine salt, NBT/BCIP). This solution is usually used for the sensitive detection of alkaline phosphatase in immunoblotting and immunohistochemical assays and, in our method, gives a dark-blue color on the spot on which the reaction antigen-antibody is occurred. The scheme of glass slide copolymer treated, also provided by Life Line Lab (Pomezia, Italy) is reported in Fig. 2.

2.6. *Scan and data analysis*

The dark color on the microarrays was finally determinated using the microarray reader and a grey scale relative values were collected for each spot. Because the results are shown and plotted in term of ratio between "grey value for each spot/ grey value for background", the ratio decreased for spots more dark.

3. Results

In our results we observed a microarray dark spot only in sample derived from apoptotic cells and demonstrated the feasibility of microarray technology in order to characterize crude cell protein lysate for these or others tumor protein markers, comparable to the classical western blot analysis. In the Fig. 3 are shown the microarray results for caspase-3 (cleaved and uncleaved form) for the same samples spotted (crude proteins from apoptotic cells and from control cell culture). We plan in the next to enlarge our microarray analysis on others markers for apoptosis, as caspase-8 and -9 and to improve the presented method by application of a fluorescent reader for spot detection.

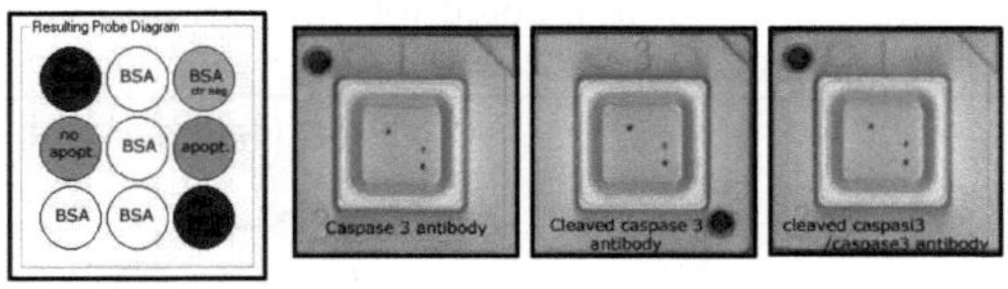

Figure 3. Microarray results for caspase-3 (cleaved and un-cleaved form).

4. Conclusions

We demonstrated that microarray platform Dr.CHIP, used in our experiments would be used in order to develop a microarray antibody for apoptosis markers, but it seem necessary to use the fluorescence as detection method. In our opinion the application of a fluorescent reader for spot detection, as a CCD camera, it would lead possible to enlarge microarray analysis on others markers for apopotosis, as caspase-8 and caspase-9 in order to obtain a multiparametric microarray assay and eliminate the need for multiple immunoprecipitation/ Western blot experiments.

Acknowledgments We are particularly grateful to dr. I. Solinas and Dr. C. Padula for assistance on analytic equipment (Dr. Chip platform) provided by Life Line Laboratory (Pomezia, Italy).

References

1. Cenciarelli, C., Tanzarella, C., Vitale, I., Pisano, C., Crateri, P., Meschini, S., Arancia, G., Antoccia, A. (2008). Apoptosis, 13(5), 659–69
2. Chiari, M., Cretich, M., Corti, A., Damin, F., Pirri, G., Longhi, R. (2005) Peptide microarrays for the characterization of antigenic regions of human chromogranin A. Proteomics, 5(14), 3600–3603
3. Cretich, M., Pirri, G., Damin, F., Solinas, I., Chiari, M. (2004) A new polymeric coating for protein microarrays. Analytical Biochemistry, 332(1), 67–74
4. Gagliardi, S., Rapone, B., Mosiello, L., Luciani, D., Gerardino, A., Morales, P. (2007) IEEE Transactions on Nanobioscience, 6(3), 242–248
5. Nuñez, G., Benedict, M.A., Hu, Y., Inohara, N. (1998). Oncogene, 17(25), 3237–3245

ELECTRORHEOLOGICAL FLUIDS BASED ON INORGANIC NANOPARTICLES FOR ROBOTIC APPLICATIONS

L. SCIBILIA[1], G. RIZZO[1], M. SORRENTI[1], G. NERI[1], P. GIORGIANNI[2], M. LEVANTI[2] AND S.S. MARCHESE

[1]*Dipartimento di Chimica Industriale e Ingegneria de Materiali, Università di Messina Contrada di Dio, 98166 S. Agata di Messina, Italy*

[2]*Signo Motus s.r.l Via Panoramica 340, 98100 Messina, Italy*

Abstract. The development of novel ER fluids based on nanostructured inorganic materials (single or mixed metal oxide, zeolites) dispersed in silicon oil is currently underway in our laboratories. Here, we describe the design of a new rheometer able to characterize the rheological properties of these ER fluids. An example of the use of ER fluids based on inorganic nanoparticles in transmitting forces for robotic applications is also reported.

1. Introduction

Electrorheological (ER) fluids are suspensions of dielectric particles in a non-conducting liquid that exhibits drastic changes in their rheological properties, including a large increases in apparent viscosity and shear stress under an applied electric field [1]. The application of an electric field in these fluids induce polarization of the suspended particles and a chain-like structure can be formed along the electric field direction in a few milliseconds [2]. Because of their controllable viscosity and short response time, ERF materials can be regarded as smart materials for active devices and actuators [3].

The recent discovery by Wen and coworkers of the giant electrorheological (GER) effect in nanoparticle-based suspensions [4] has given new impulse to the search of inorganic nanoparticles as ER fluid components. In this regard, the recent research to the development of ER fluids based on nanostructured inorganic materials (single or mixed metal oxide, zeolites) dispersed in silicon oil has been focused. The paper deals the design of a new rheometer able to characterize the rheological properties of these novel ER fluids. An example of the use of ER fluids based on inorganic nanoparticles in transmitting forces for robotic applications is also reported.

2. Rheometer design

The state of the art does not provide analytical models to predict the rheological behaviour of ER fluids. Furthermore, the existing instrumentation, although

P. Malcovati et al. (eds.), *Sensors and Microsystems: AISEM 2009 Proceedings*,
Lecture Notes in Electrical Engineering 54, DOI 10.1007/978-90-481-3606-3_65,

their excellent performances, does not concern specifically ER fluids; thus, several key parameters such as conductivity, break-down electric field, dielectric constant can not be determined [5]. For these main reasons, an innovative rheometer is under development.

2.1. *Technical specifications*

Through a careful analysis of the commercial ER fluids and future applications, the technical specifications of the rheometer, shown in Table 1, were determined.

Table 1. Technical specification of the rheometer under development.

Parameter	Value	Condition
General characteristics		
Fluid volume	18 ÷ 25 ml	
Viscosity	0.01 ÷ 100 Pa·s	Without field
Test fluid temperature range	−15 ÷ 75°C	
Resolution temperature	0.5°C	
Fluid conductivity	$1 \cdot 10^{-7}$ S/m	E_{max} = 10 kV
Break-down electric field	10 kV	$D_{elettr(MIN)} \geq$ 1mm
Polarization voltage	10 kV	
Density current absorbed by the ER fluid	125 μA/cm^2	E_{max} = 10 kV
Power supply to polarization ER fluid	30 W	
Rheometer voltage	220 V(AC)	
Shear stresses		
Rotational speed	175 r/min	
Static yield stress	270 kPa	
Shear stress	50 kPa	
Strain rate	200 s^{-1}	
Compression stresses		
Compression stress	300 kPa	
Compression speed	0.1 ÷ 2 mm/s	
Distance between electrodes	1 ÷ 10 mm	
Dielectric constant		
Frequency analysis dielectric constant	0.1 ÷ $1 \cdot 10^6$ Hz	
Sinusoidal voltage amplitude applied	0.1 ÷ 10 V	
Relativ dielectric permittivity	1 ÷ 60	
Dielectric loss factor	0.01 ÷ 20	

The system will allow the following types of measures: viscosity, static yield stress, dynamic yield stress, dynamic shear stress, compression, response time to variations in electric field, dielectric constant, conductivity, break-down electric field, temperature.

2.2. *System layout*

The structure of the rheometer is a 2 degree of freedom kinematics chain. To characterize the ER fluid it is placed inside a test chamber with a layout known as couette geometry, as shown in Fig. 1, where the fluid is heated and subjected to shear and compression stress.

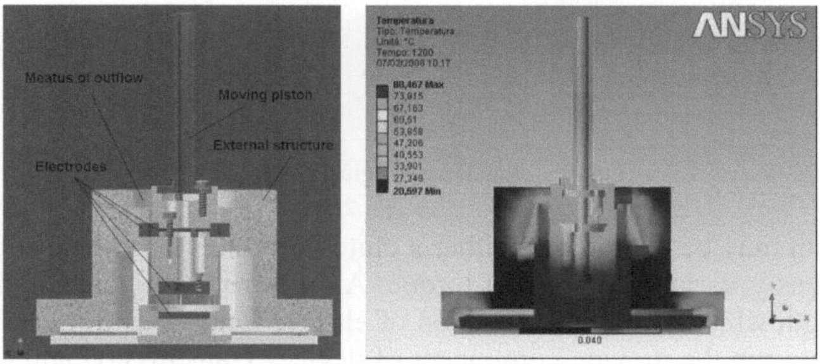

Figure 1. Couette geometry and thermal analysis of the rheometer test chamber.

To determine the heat transfer time and the temperature achieved by the materials of the couette, thermal analysis has been carried out. Thermal actuators (peltier cells) able to lead up to the maximum temperature (75°C) the couette materials in 9 min has been used.

2.3. *Management software*

The management software of the rheometer will be implemented by the use of the LabView development environment which guarantees extreme versatility, robustness and reliability in applications such as the one in question. This choice is also compatible with design choices made with regard to hardware electronic control.

3. Robotic application

The ER fluids based on inorganic nanoparticles can be used for robotic applications. For example, semi-active joint based on ER fluid can be designed to assist the walk of anthropomorphic robot. This semi-active joint can be mounted in correspondence of specific joints of the anthropomorphic robot limb (for example ankle and knee) and can modulate the transmitted torque between the connected links according to the phase of the human step. Therefore, in a robotic limb the number of electric motors can be reduce according to the required robot specific performance; some of the most complex problems of electric motors such as control circuits, high dimensions and heat transfer are overcome. An example of electrorheological fluid joint device layout, in progress in our laboratories, is shown in Fig. 2.

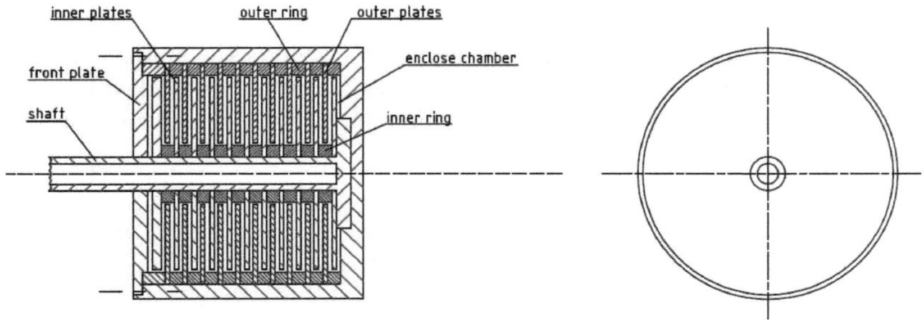

Figure 2. Electrorheological fluid joint.

This joint may be operated as either a clutch or brake to provide substantially instantaneously variable control of torque. An electrorheological fluid having a yield strength which varies as a function of electric field, provides the interactive medium between the moving component of device. Transmission and conversion of torque in device is accomplished by the gripping effect of the electrorheological fluid between inner plates and outer plates. The ability to completely disassemble and substitute either inner and outer plates solves the problem of abrasion and chemically aggressive of electrorheological fluid.

Condition responsive control of device must be provided by closed-loop feedback of a signal indicative of rotational speed of the system to the control circuit.

References

1. T. Hao, Electrorheological fluids, Adv. Mater. 13, 1847–1857, 2001.
2. J.P. Huang, Force acting on the microparticles in electrorheological solids under the application of a nonuniform ac electric field, Chem. Phys. Lett. 439, 380–383, 2004.
3. H. Böse, A. Trendler, Smart fluids and their properties – perspectives for new actuators, Proc. Euromat. 1–6, 2000.
4. W.J. Wen, X.X. Huang, S.H. Yang, K.Q. Lu, P. Sheng, The giant electrorheological effect in suspensions of nanoparticles, Nature Mater. 2, 727–730, 2003.
5. Y. Tian, Y. Meng, H. Mao, S. Wen. Electrorheological fluid under elongation, compression, and shearing, Phys. Review E 65, 031507, 2002.

WIRELESS NANOTRANSDUCERS FOR IN-VIVO MEDICAL APPLICATIONS

G. MANTINI[1], A. D'AMICO[1], C. FALCONI[1] AND Z. LIN WANG[2]
[1]*Department of Electronic Engineering, University of Tor Vergata, Via del Politecnico 1, 00133 Roma*
[2]*School of Materials Science and Engineering, Georgia Institute of Technology, Atlanta, Georgia, USA*

Abstract. Nanotechnology has the potential to allow the development of more effective and less invasive methodologies for the diagnosis and treatment of many diseases. In particular, wirelessly-linked implantable nanotransducers are nanodevices capable of transducing signals from one energy domain into another, and they also allow the interaction of internal or external macroscopic systems with implantable nano-sized devices. This ability can have a great impact on targeted drug delivery, hyperthermia, thermal ablation, energy harvesting, early diagnosis and monitoring of physiological parameters. Here we review some promising wireless nanotransducers and their possible medical applications.

1. Introduction

Minimally invasive or non-invasive alternatives to traditional, invasive methodologies are sought in almost every field of medicine. In some cases, an ideal solution is wireless: an external or internal macroscopic system should wirelessly interact with the implanted, nano-sized devices. The wireless interaction may take advantage of electromagnetic or acoustic waves (e.g. nanodevices able to transduce the received energy into heat), but also of chemical substances (e.g. nanodevices able to transduce glucose into electrical energy) in order to connect nanodevices to a macroscopic (internal or external, biological or artificial) system. The wireless interaction must be efficient and selective. Last, but not least, the nanostructures must be biosafe and biocompatible, even after long-term body exposure to the nanoparticles (especially with reference to toxicity and to the risk of undesired aggregation and obstruction of vessels).

2. In-vivo applications of wireless nanotransducers

In this section we discuss the most important in vivo applications of wireless nanotransducers.

P. Malcovati et al. (eds.), *Sensors and Microsystems: AISEM 2009 Proceedings*, Lecture Notes in Electrical Engineering 54, DOI 10.1007/978-90-481-3606-3_66,

2.1. *Wireless nanoheaters*

Cancer thermal ablation is the most investigated application of nanotechnology to medicine. Wireless nanoheaters [1] can take advantage of many different dissipative mechanisms (e.g. gold nanoshells [2], magnetic nanoparticles, wireless Joule nanoheaters [1], …). As an outstanding example, it has already been shown that gold nanoshells [2] can selectively damage cancer cells by heating, as shown in Fig. 1.

2.2. *Wireless nanotransducers for drug delivery*

Nanotechnology can revolutionize drug delivery by allowing a wirelessly controllable, selective, accurate, and non-invasive delivery of drugs. Many transduction mechanisms are possible: in general, wireless nanotransducers and therapeutics may be included in a coating which can be broken-perforated-dissolved-melt-… upon excitation, thus releasing drugs in a selective, controlled, and accurate manner. There are many possible ways to trigger drug delivery: magnetic field, ultrasound, electric field, temperature, light, mechanical forces.

2.3. *Wireless nanosensors*

Wireless nanosensors offer very promising opportunities for the non-invasive, wireless measurement of physiological parameters inside the body, which is one of the most important objectives for medicine. Research on wireless nanosensing is mainly focused on molecular imaging of cells and cellular components. Remarkably, nanoparticles functionalization can expand the imaging capabilities to almost any cellular component present in the human body.

2.4. *Wireless nanogenerators*

In general, we refer to nanotransducers that can wirelessly harvest the environment energy in order to provide electrical energy/power as wireless nanogenerators. Nanotechnology can be the key for high efficiency energy harvesting and enable the design of implantable systems, which can harvest energy from many energy sources readily available in the in vivo environment. As a preliminary attempt to convert mechanical energy into electrical energy by means of nanoscale transducers, an array of zinc oxide nanowires [5] has been used.

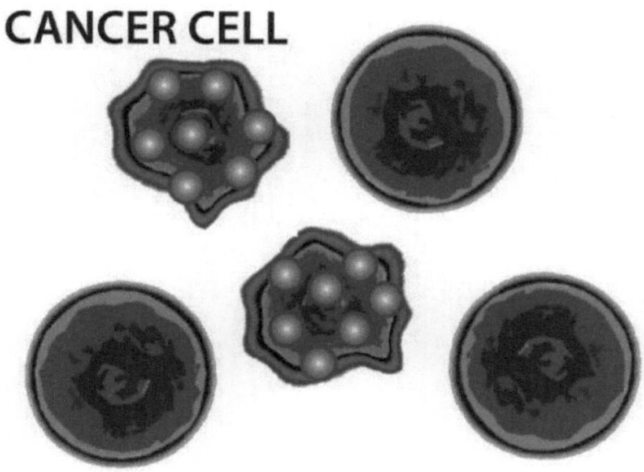

Figure 1. Active targeting allows to selectively attach gold nanoshell (or other nanostructures) on the surface of cancer cells (shown with an irregular shape).

3. Nanostructures for wireless transduction

In this section we describe a few types of wireless nanotransducers, indicating their potentialities and their limits with reference to the in vivo applications described in the previous section.

Gold nanoshells are nanoparticles beads made of a silica core coated with a thin gold shell [2]. By changing the thickness of both the core and the outer shell it is possible to accurately tune their absorption and scattering properties. Gold nanoshells are being investigated for cancer thermal ablation, cancer imaging, and wirelessly triggered drug delivery.

Carbon nanotubes [3] exhibit striking properties; in particular, carbon nanotubes are about 100 times stronger than steel with only one-sixth of its weight and have unusual thermal and electrical properties. When exposed to NIR radiation, carbon nanotubes behave as wireless nanoheaters and might therefore be used for hyperthermia, thermal ablation, and targeted drug delivery.

Quantum dots (QDs) [4] are nanocrystals with typical diameters in the range 2–10 nm, made of semiconductors; the most common QDs are made of Cadmium Selenide capped by Zinc Sulfide (CdSe/ZnS). When excited with ultraviolet light, QDs fluoresce in different neon colors depending on their size. The potential of QDs in biomedical applications has been rediscovered recently; QDs are now currently used as probes for high-resolution molecular imaging for cells and cellular components.

Recently, a variety of almost defect-free mono-crystals of zinc oxide have been fabricated [5] and some of these zinc oxide nanostructures may be used as wireless nanotransducers (a few examples are shown in Fig. 2). The excellent mechanical properties of ZnO self-assembled structures may also be useful for the fabrication of composite nanostructures for in vivo medical applications [5]. ZnO nanostructures also offer ideal biocompatibility [6].

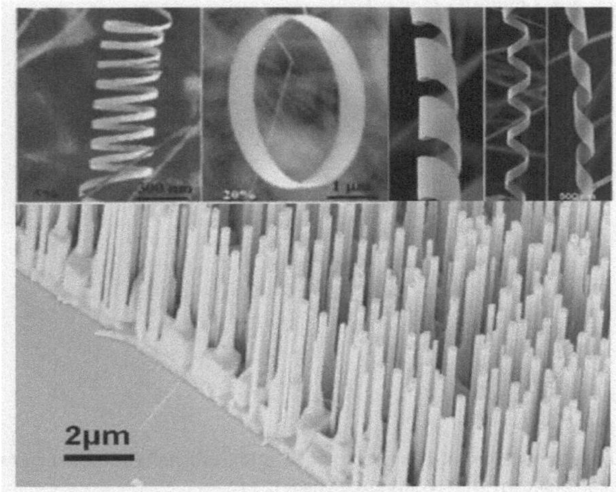

Figure 2. Zinc oxide nanostructures which can be used as wireless nanotransducers.

4. Conclusions

Nanotechnology can offer more effective and less invasive diagnostic and therapeutic alternatives to traditional approaches by allowing the fabrication of nanosized devices able to wirelessly interact with other in vivo or external devices; these wireless nanotransducers have great potential for targeted drug delivery, hyperthermia, thermal ablation, energy harvesting, and monitoring of physiological parameters. The wireless interaction between the nanostructures and the external system must be efficient and selective; the nanostructures must be safe and biocompatible.

References

1. C. Falconi, A. D'Amico, Z.L. Wang, "Wireless Joule nanoheaters", Sensors and Actuators B, 127 (2007), 54
2. C. Loo, A. Lin, L. Hirsch, M.H. Lee, J. Barton, N. Halas, J. West, R. Drezek, "Nanoshell-enabled photonics-based imaging and therapy of cancer", Technol. Cancer Res. Treat. 3 (2004), 33
3. C.J. Gannon et al, "Carbon Nanotube-enhanced Thermal Destruction of Cancer Cells in a Noninvasive Radiofrequency Field", Cancer (October 2007)
4. J.H. Kima, D. Morikisa, M. Ozkan, "Adaptation of inorganic quantum dots for stable molecular beacons", Sensors and Actuators B: Chemical, 102 (September 2004), 315
5. Z.L. Wang and J. Song, "Piezoelectric Nanogenerators Based on Zinc Oxide Nanowire Arrays", Science, Vol. 312. no. 5771 (14 April 2006), pp. 242–246
6. J. Zhou, N. Xu, Z.L. Wang, "Dissolving behavior and stability of ZnO wires in biofluids: a study on biodegradability and biocompatibility of ZnO nanostructures", Adv. Mater. 18 (2006), 2432

DEVELOPMENT OF MEMS MICROCANTILEVER DETECTORS FOR DNA SINGLE NUCLEOTIDE POLYMORPHISM DETECTION IN AUTOIMMUNE DISEASES DIAGNOSTIC

A. ADAMI[1], M. DECARLI[1], L. ODORIZZI[1], L. LORENZELLI[1], K. FINCATI[2], K. SCHICHO[3] AND H. GRUESSINGER[3]

[1]*FBK-CMM BioMEMS Foundation Bruno Kessler – Center for Materials and MicroSystems – BioMEMS Group, via Sommarive 18, 38100 Trento, Italy*
[2]*CIVEN – Coordinamento Interuniversitario Veneto per le Nanotecnologie, Via delle Industrie 9, 30175 Venezia-Marghera, Italy*
[3]*PCS Professional Clinical Software GmbH, Industriering 11, 9020 Klagenfurt, Austria*

Abstract. In this work, the development of a cantilever-based detector arrays for LOC application is presented, especially aimed to the integration in point of care systems for early diagnosis and screening of autoimmune disorders based on typing of Human Leukocyte Antigen (HLA). The design and technological implementation of single crystal cantilevers with very low thickness are presented, finalised to the evaluation of technologies and functionalisation procedures for high-density and high sensitivity detector arrays with piezoresistive readout.

1. Introduction

Nowadays, the development of Lab-On-a-Chip (LOC) technologies has opened up new perspectives in the field of personalized diagnosis and treatment, by taking advantage of reduced size, low volume requirement for samples, and rapid analysis. In particular, MEMS, microelectronics and nanobiotechnology are enabling technologies for the realization of biosensors by combining micro-systems for sample handling, signal read-out and functionalisation methodologies able to detect specific bioaffinity reactions. In this frame, LOC technologies are a powerful tool to perform a wide range of proteomic and genomic tests starting from fluid biological samples such as blood or saliva [1].

In this work, the development of a cantilever-based detector arrays for LOC application is presented, especially aimed to early diagnosis and screening of autoimmune disorders at point of care, based on typing of Human Leukocyte Antigen (HLA), Lab-on-Chip (LOC) technologies and computer based artificial intelligent algorithms. The detection of HLA sequences containing Single Nucleotide Polymorphisms (SNPs) will be performed with piezoresistive microcantilever arrays in order to evaluate the susceptibility to autoimmune diseases. This approach will also take advantage of a LOC module based on

P. Malcovati et al. (eds.), *Sensors and Microsystems: AISEM 2009 Proceedings*,
Lecture Notes in Electrical Engineering 54, DOI 10.1007/978-90-481-3606-3_67,

Polymerase Chain Reaction (PCR) for sample pre-processing and diagnostic algorithms for the realisation of an automated and portable diagnostic system. The detector module is based on arrays of silicon microcantilevers, which are typically suspended beams realised with MEMS fabrication technologies, suitable for biological and chemical sensing [2]. Sensitivity to specific analytes can be achieved by coating the beam surface with proper films, able to selectively bind a particular target molecule. A self assembled monolayer (SAM) of known ss-DNA probes will represent the functional layer [3]. The following hybridization of the complementary target sequence induces a differential stress between the top and the bottom surfaces of a cantilever, driving the deflection of the beam. This method also allows the integration of large detectors arrays, needed for multiple DNA sequences detection, where larger number of assays must be performed on the same platform with respect to other applications such as the detection of pathogens [4]. Both the design and technological implementation of single crystal cantilever with very low thickness and high sensitivity are presented, finalised to test a technology for LOC detection module for point-of-care applications.

2. Design of MEMS-based cantilever arrays

Since the sensitivity of microcantilever devices operated in standard module is proportional to the beam curvature due to surface interaction, selection of materials and device thickness is the major design and technical point. Several readout methods are available, although for point-of-care applications piezoresistive readouts take advantage of the good sensitivity, high robustness, simple readout principle and easy integration for large arrays. Both analytical and finite element analysis have been performed, focused on the evaluation of expected response with different technological approaches.

Table 1. Analytical performances evaluation achievable with different technologies.

Device type	Response @ 400 nM (ppm)
SOI	5.5
SiO_2–SOI	4.5
SiO_2–poly	3
SOI–poly	4.7

In Table 1, simulated responses to the typical concentration of 400 nM DNA with single mismatch are reported for single crystal beams with implanted resistors ("SOI" technology; thicknesses: 340 nm/Si, 200 nm/SiO_2, junction depth: 100 nm), SiO_2 beams with single crystal piezoresistors etched in the device layer of a SOI wafer ("SiO_2–SOI" technology; thickness 400 nm/SiO_2, 100 nm/Si, 200 nm/SiO_2), SiO_2 beams with poly-silicon piezoresistors ("SiO_2-poly" technology; thickness 400 nm/SiO_2, 100 nm/poly-Si, 200 nm/SiO_2:) and single crystal beams with poly-Si resistors ("SOI" technology; thicknesses 340 nm/Si, 200 nm/SiO_2,

100 nm/poly-Si resistors, 200 nm/SiO_2). The reported thicknesses are the result of an optimisation based on analytical models including the mechanical properties of the structures and an applied surface stress equivalent to DNA interaction. The optimisation has been performed in terms of detector response for each technology by taking into account technical constraint and device suitability. As shown, the highest response can be achieved with implanted resistors on single crystal Si.

The optimal design of devices has been considered by Finite Element (FE) simulations. Results of readout sensitivity for different implants types as a function of position along the beam are shown in Fig. 1, reporting the n-type implant largely outperforming p-type resistors for planar stress condition, which are the ideal conditions for long beams working in static mode.

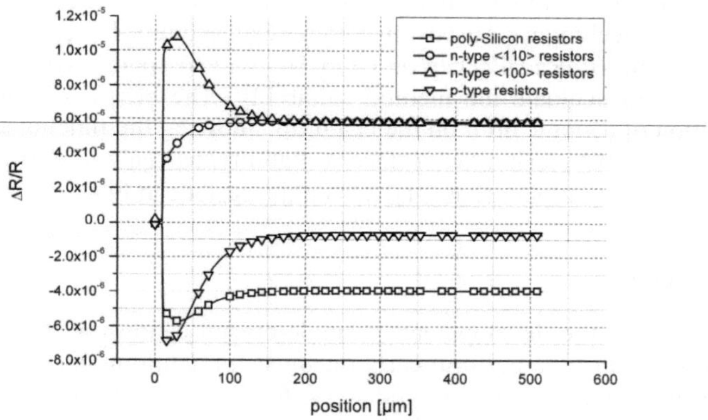

Figure 1. Simulated response of piezoresistive element to single-base mismatch at 400 nM concentration, as a function of position along the beam.

Then, the selected technological approach is based on the fabrication of microcantilevers starting from Silicon-On-Insulator (SOI) substrates, allowing an extreme reduction and tight control of the thickness of the beams and n-type implanted piezoresistors with low junction depth.

3. Fabrication and functionalisation procedures

The technological development has been initially focused on the realisation of the mechanical structure of the devices, based on 340 nm Si crystal, a PECVD SiO_2 adhesion layer and a gold layer (20 nm) for promoting the binding of DNA probes with thiol-based chemistry.

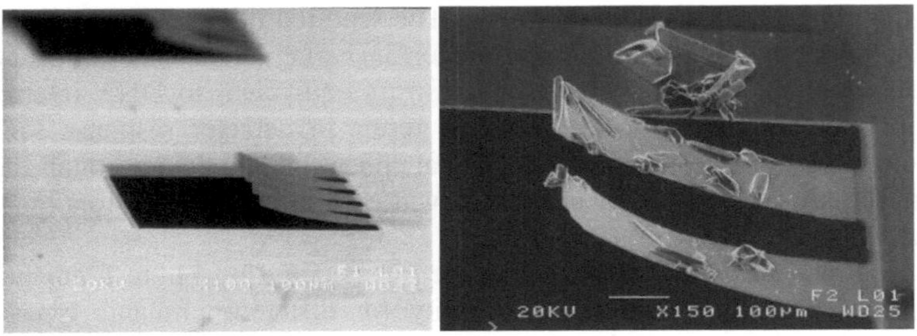

Figure 2. Array of fabricated microcantilever structures with 380 nm thickness, before and after the deposition of a sodium-phosphate buffer on the surface.

In order to provide a functionalisation procedure with high scalability to high density arrays, techniques based on commercial spotters have been selected. In particular, the *BioRad VersArray Chip Writer Pro* spotter system has been used in order to evaluate the method suitability. The best approach involves the deposition of a single drop on the beam tip, allowing the functionalisation of the complete beam by using the surface wettability, as can be inferred from the distribution of sodium-phosphate crystals on the surface in Fig. 2.

4. Conclusions

This work reports the development of single-crystal cantilever arrays for LOC applications, finalised to perform early diagnosis of autoimmune diseases based on HLA SNPs detection. Design guidelines, feasibility of mechanical structures with very low thickness, required to achieve good sensitivity and preliminary results of the procedure suitable for high density array functionalisation have been demonstrated.

Acknowledgments This work has been funded by the EC under the Integrated Project POCEMON (Grant: EU-FP7-ICT-2007-216088).

References

1. J. West, M. Becker, S. Tombrink, and A. Manz, "Micro Total Analysis Systems: Latest Achievements" – Analytical Chemistry, Vol. 80, no. 12, pp. 4403–4419, 2008.
2. C. Ziegler, "Cantilever-based biosensors", Analytical and Bioanalytical Chemistry no. 379, pp. 946–959, 2004.
3. J. Fritz, M. K. Baller, H. P. Lang, H. Rothuizen, P. Vettiger, E. Meyer, H.-J. Guntherodt, Ch. Gerber, and J. K. Gimzewski, "Translating Biomolecular Recognition into Nanomechanics", Science, Vol. 288, pp. 316–318, 2000.
4. M. G. Dobson, P. Galvin and D. E. Barton, "Emerging technologies for point-of-care genetic testing", Expert Review of Molecular Diagnostics, Vol. 7, no. 4, pp. 359–370, 2007.

A NEW APPROACH FOR CMOS FABRICATION OF MICROCANTILEVER/NANOTIP SYSTEMS FOR PROBE-STORAGE APPLICATIONS

G. BARILLARO, S. SURDO AND G.M. LAZZERINI

Dipartimento di Ingegneria dell'Informazione: Elettronica, Informatica, Telecomunicazioni, Università di Pisa, Via G. Caruso 16, 56126 Pisa, Italy

Abstract. A novel method for fabricating the mechanical part of the "Millipede", a MEMS-based scanning-probe data storage system, is reported. The proposed approach exploits a selective *p*-to-*n* HF-based electropolishing step, which allows to remove *p*-type silicon, used as a sacrificial layer, while leaving *n*-type silicon, exploited as structural material, un-etched. The approach is CMOS compatible, thus allowing to integrate both the mechanical and electronic part of the "Millipede" on the same chip. Experimental results, carried out on chips fabricated using the BCD6 process of ST Microelectronics, demonstrate the feasibility of fabricating free-standing *n*-type silicon cantilevers by selective etching of the *p*-type substrate.

1. Introduction

Several original approaches, from thermo-mechanical to holographic or magnetic, are today under investigation for the fabrication of next generation data storage devices. So far, MEMS-based scanning-probe data storage systems are among the most promising ones, in terms of ultra high density, low access time and low power consumption. In this framework, the IBM "Millipede" approach is one of the most interesting and more advanced solution [1–3]. The "Millipede" exploits nanometric-sharped tips to induce changes at the atomic scale in the data storage substrate. In particular, an electrically driven two-dimensional (2D) array of thermo-mechanical cantilever-tip systems is used to write, read and erase bits on a polymeric film placed underneath.

At the present time, one of the most challenging aspects under investigation is related to the integration of the cantilever/tip system (mechanical part) together with the circuitry (electronic part) on a single chip. In fact, in the state-of-art version of the "Millipede", the mechanical and electronic parts are fabricated on two different wafers. Moreover, the nanotip, which is the most critical component of the whole system, is fabricated by using a time-controlled reactive ion etching (RIE) step. RIE-induced variation of tip geometrical parameters, such as tip radius, angle and height, strongly affects the bit density [1].

P. Malcovati et al. (eds.), *Sensors and Microsystems: AISEM 2009 Proceedings*, Lecture Notes in Electrical Engineering 54, DOI 10.1007/978-90-481-3606-3_68,

In this work an original method for integrating both the mechanical and electronic part of the "Millipede" on the same wafer, by using CMOS-based processes, is proposed. Among the main advantages of such an approach there are the reduced fabrication costs and increased performances, as a consequence of reduced parasitic effects, access time and power consumption. Moreover, the exploitation of a time-independent etching process for the tip fabrication also reduces spreading of tip parameters, thus allowing to increase both bit density and reliability. Preliminary experimental results, performed on chips fabricated by using a CMOS-based commercial process, are also reported to validate such an approach.

2. A novel idea

The main idea of the proposed approach is the exploitation of a selective *p*-to-*n* etching (HF-based electropolishing) which allows to remove *p*-type silicon layers, while leaving *n*-type layers un-etched. Following this approach, the mechanical part of the "Millipede" is defined by controlling both *p*-type and *n*-type dopants implantation and diffusion steps, in a *p*-type substrate [4]. The *p*-doped silicon is then used as a sacrificial layer, whereas the *n*-doped silicon is exploited as a structural material for tip and cantilever fabrication.

The main technological steps for the fabrication of the microcantilever/nanotip system are sketched in Fig. 1.

Starting from a *p*-type silicon substrate, an *n*-type implant is used to define the cantilever and its anchor (Fig. 1a). A *p*-type implant is then carried out to define the tip, on top of the *n*-type cantilever, by using a suitable circular mask (Fig. 1b). The geometrical tip characteristics (radius, angle, height) are defined by finely tuning the *p*-type dopant diffusion (Fig. 1c). Finally, the *p*-type material is selectively removed by means of an electrochemical etching step (HF-based electropolishing), while leaving *n*-type silicon regions un-etched. The etching time is long enough to remove the *p*-type silicon until free-standing cantilevers, with a tip on top, are obtained (Fig. 1d). Since silicon electropolishing in HF-based electrolytes is activated by holes [5], it is possible to remove the *p*-type silicon with respect to the *n*-type material with a high selectivity, thus allowing to fabricate the free-standing cantilever/tip system with a time-independent

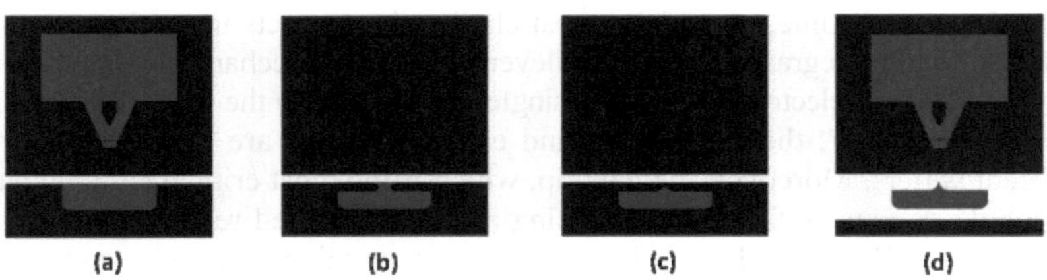

Figure 1. Fabrication process steps. The starting material is a *p*-type wafer (dark layer): (**a**) *n*-type implant and diffusion (light layer); (**b**) *p*-type implant; (**c**) controlled *p* dopant diffusion; (**d**) etching of the *p*-type material to obtain a free-standing *n*-type cantilever/tip system.

process. It is to note that the cantilever is still connected to the substrate through its anchor, which has dimensions much larger than the cantilever width. Furthermore, electropolishing step in aqueous HF solution has been already demonstrated as fully-compatible with microelectronic commercial processes, thus assuring the integration of both the micromechanical and electronic parts of the "Millipede" on the same chip, using CMOS-based processes [6, 7].

3. Experimental results

In order to validate the here above proposed approach, a few preliminary experiments were carried out on chips fabricated by means of the BCD6 process of ST Microelectronics. The starting material was a p^+-type silicon substrate with a p^--type epitaxial layer on top. Several mechanical structures with different width, length and geometry were defined by exploiting n^--type implantation and diffusion steps in the p^- material. Small dots with higher doping (n^+ regions) were also formed over the n^- regions. Finally, several oxide layers were deposited on the silicon substrate, for chip passivation. After the BCD6 sequence was completed, the samples were rinsed in a HF(48%)/Ethanol (1:1 by vol.) solution for 30 min in order to remove the oxides and expose the silicon surface for the next electropolishing step. This step was performed using a 5% HF (by weight) aqueous solution, with an anodic voltage of 1.6 V, for several, different etching times. These conditions were found to guarantee the removal of the *p*-type material, while leaving the *n*-type silicon un-etched.

Figure 2 shows SEM views of *n*-type structures suspended from the *p*-type substrate, obtained by selective electropolishing of the *p*-type material. The sample was anodized at 1.6 V for 480 s. As expected, the *n*-type material was un-etched, while the *p*-type silicon was fully removed. This clearly proves the high selectivity of the HF-based electropolishing between *p*-type and the *n*-type silicon, thus validating the proposed approach for fabricating mechanical structures. Moreover, it is to note that n^+ dots on top of *n*-regions were also left un-etched, thus enhancing the flexibility of the selective *p*-to-*n* electropolishing approach for surface micromachining applications.

Figure 2. SEM views of free standing *n*-type structures suspended over the *p*-Si substrate, fabricated by means of selective *p*-to-*n* electropolishing in HF-based solutions.

4. Conclusions

In this work a novel method for fabricating the mechanical cantilever/tip system of the IBM "Millipede" probe-storage device has been proposed. Experimental results, carried out on chips fabricated by using the BCD6 process of ST Microelecronics, validate the proposed method and open new horizons for CMOS-compatible fabrication of mechanical structures. Among the advantages of such an approach there are: (1) definition of mechanical structures by means of conventional implantation and diffusion steps, which guarantee an accurate control of their geometrical parameters (thickness, width, shape), (2) time-independent etching process, due to the high selectivity between *p*-type silicon etching with respect to *n*-type, (3) actual compatibility of the whole fabrication process with modern CMOS technologies, which allows to integrate mechanical structures together with control circuitry, within the same chip.

Acknowledgments Authors wish to thank personnel at STMicroelectronics for the chip fabrication, in the framework of the Università di Pisa-STMicroelectronics Joint R&D Center.

References

1. P. Vettiger, M. Despont, U. Drechsler, U. Dürig, W. Häberle, M. I. Lutwyche, H. E. Rothuizen, R. Stutz, R. Widmer and G. K. Binnig, "The Millipede – More than one thousand tips for future AFM data storage", IBM J. Res. & Dev., vol. 44(3), pp. 323–340, 2000.
2. A. Pantazi, A. Sebastian, T. A. Antonakopoulos, P. Bachtold, A. R. Bonaccio, J. Bonan, G. Cherubini, M. Despont, R. A. DiPietro, U. Drechsler, U. Durig, B. Gotsmann, W. Haberle, C. Hagleitner, J. L. Hedrick, D. Jubin, A. Knoll, M. A. Lantz, J. Pentarakis, H. Pozidis, R. C. Pratt, H. Rothuizen, R. Stutz, M. Varsamou, D. Wiesmann and E. Eleftheriou, "Probe-based ultrahigh-density storage technology", IBM J. Res. & Dev., vol. 52(4/5), pp. 493–511, 2008.
3. M. Despont, U. Drechsler, R. Yu, H. B. Pogge and P. Vettiger, "Wafer scale microdevice transfer/interconnect: from a new integration method to its aplication in an AFM based data storage system" Transducers '03- The 12th International Conference on Solid State Sensors, Actuators and Microsystems, Boston, pp. 1907–1910, June 8–12, 2003.
4. G. Barillaro, A. Diligenti, C. Riva, R. Campedelli and S. Losa, "Process for manufacturing microelectromechanical interaction systems for storage medium", PCT-06CA240, International Patent Application (2007).
5. V. Lehmann, "Electrochemistry of Silicon", WILEY-VCH Weinheim, 2002.
6. T. L. Ritzdorf, G. J. Wilson, P. R. McHugh, D. J. Woodruff, K. M. Hanson and D. Fulton, "Design and modelling of equipment used in electrochemical processes for microelectronics" IBM J. Res. & Dev., Vol. 48(1), pp. 65–77, 2005.
7. G. Barillaro, P. Bruschi, F. Pieri and L. M. Strambini, "CMOS-compatible fabrication of porous silicon gas sensors and their readout electronics on the same chip", Phys. Stat. Sol. (a) 204, pp. 1423–1428, 2007.

CHARACTERIZATION AND TESTING OF A DOUBLE AXIS SCANNING MICROMIRROR

F. BATTINI[1], E. VOLPI[1], E. MARCHETTI[1], T. CECCHINI[1], F. SECHI[1], L. FANUCCI[1], M. DE MARINIS[2] AND U. HOFMANN[3]

[1]*Department of Information Engineer, University of Pisa, via Caruso, 16 I-56122, Pisa*
[2]*SensorDynamics AG, via Giuntini, 63, I-56023, Navacchio (Pisa), Italy*
[3]*Fraunhofer Institute for Silicon Technology ISIT, Itzehoe, Germany*

Abstract. Testing and characterization of Micro-Electro-Mechanical Systems (MEMS) and Micro-Opto-Electro-Mechanical Systems (MOEMS) can be very challenging due to the multi-domain nature of these devices. Nowadays high volume, high-cost, and accurate measuring systems are necessary to characterize and test MEMS and MOEMS especially to examine motions, deflections and resonance frequencies that are the distinguishing characteristics of these systems. This paper presents an environment for MEMS and MOEMS testing and characterization. The environment is based on ISIF (Intelligent Sensor InterFace) platform developed by SensorDynamics AG in collaboration with the University of Pisa. The environment has been used to perform several measurements on a scanning micromirror developed by the Fraunhofer Institute for Silicon Technology (ISIT). The testing environment has been validated comparing measurement results with results obtained by FEM simulation performed with Comsol Multiphysics™. Finally, these results have been used to create an electrical equivalent model of the micromirror.

1. Introduction

Micro-Electro-Mechanical Systems (MEMS) devices are employed in many areas, such as communications, automotive, signal sensing and space technology. The continuous progress in MEMS fabrication technologies has led to the development of Micro-Opto-Electro-Mechanical Systems (MOEMS) that represent the outstanding state of the art of the projection technology. The increasing success of these systems stems from their low power consumption, low manufacturing cost, miniaturization and the possibility of integration with electronic read-out circuits.

On the other hand due to the intrinsic multi-domain nature of such systems, testing and characterization of MEMS and MOEMS system comprehensive of their conditioning electronic whether in the development or production phase can be very challenging compared to pure electrical test. A variety of custom systems relying on interferometry and vibrometers have been developed for deflection measurement but they require significant amount of development time and cost [1–3].

P. Malcovati et al. (eds.), *Sensors and Microsystems: AISEM 2009 Proceedings*, Lecture Notes in Electrical Engineering 54, DOI 10.1007/978-90-481-3606-3_69,

In this paper a fast-developing characterization and testing environment for MEMS and MOEMS based on the ISIF platform is presented. The test environment can be customized thanks to the deep programmability of the ISIF platform.

2. ISIF platform

The ISIF (Intelligent Sensor InterFace) platform developed by SensorDynamics AG in collaboration with the University of Pisa can be used as a complete solution for sensor signal conditioning as well as an actuation electronic driver for MEMS and MOEMS [4]. Thanks to its deep programmability and high flexibility, the ISIF platform can be easily connected to the real sensor and different architectures, data paths and signal processing can be quickly experimented and evaluated on the field allowing a fast development of a working interface prototype. The platform is composed by an analog and a digital section (Fig. 1). The analog section features four input channels for signal acquisition and analog conditioning composed by amplifiers and filters. After the input channel an ADC performs the digital conversion. Other analog IPs (DAC, current references, offset compensations and regulators) complete the analog section. All IPs are fully programmable by a set of registers linked together by a JTAG-like serial register chain.

The digital section is composed by the LEON 32 bit RISC DSP processor, and other IPs for digital signal processing (a four-channel demodulator, a modulator and a Numerical Controlled Oscillator). The IPs can be hardware interconnected and accessed by software. An input/output analog test bus is also provided to supply stimuli and to probe output signals for each block.

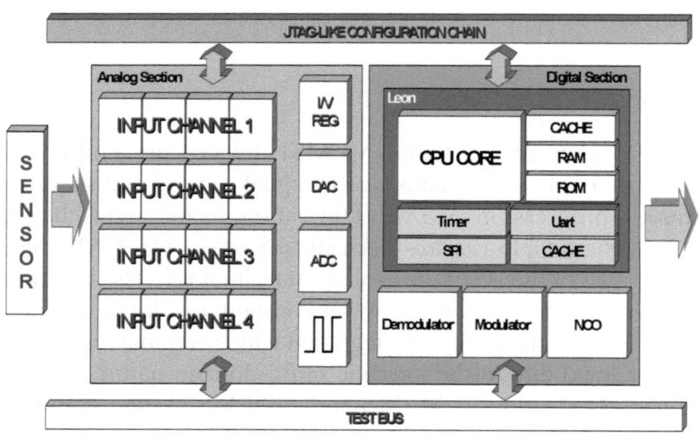

Figure 1. ISIF platform block diagram.

3. Fast characterization and test approach

A fast characterization and test approach for MEMS and MOEMS has been developed. The approach is based on the use of ISIF and the following flow:

- Design space exploration
- Preliminary configuration of the ISIF and connection to the MEMS
- First results, cross-check with the expected results, run-time minor changes of the configuration and settings of the IPs
- Tests on final configuration
- Comparison and validation with simulations (if possible)
- Analysis of data and results

4. Case study: micromirror characterization

The case study used to validate the characterization environment is the test of the micromirror prototype developed by ISIT [5]. The testing environment is able to perform a closed-loop driving of the micromirror. The system is able to lock the resonating frequency of the micromirror using a PLL implemented by the digital part of the ISIF and to control the amplitude of the deflection. This approach is very important because the micromirror must work in resonance conditions in order to obtain large deflection angles with lower voltage signals.

The testing environment, together with Finite Element Method (FEM) simulations, has led us to obtain an electrical equivalent circuit of the micromirror (Fig. 2) validated by measures on the field and allowed to perform a dynamic measurement of the sensing capacitance of the micromirror. It is worth to notice that the availability of an electrical equivalent model of the sensor is very important towards the design of higher performances read-out and driving stages. Figure 3 shows the block diagram of the system that performs the closed loop driving of the micromirror. The micromirror is shown on the left. The analog section actuates the micromirror (DAC and High Voltage driver) and measures the value of the micromirror's sensing capacitance. The digital section implements a digital PLL used to lock the resonance frequency of the micromirror.

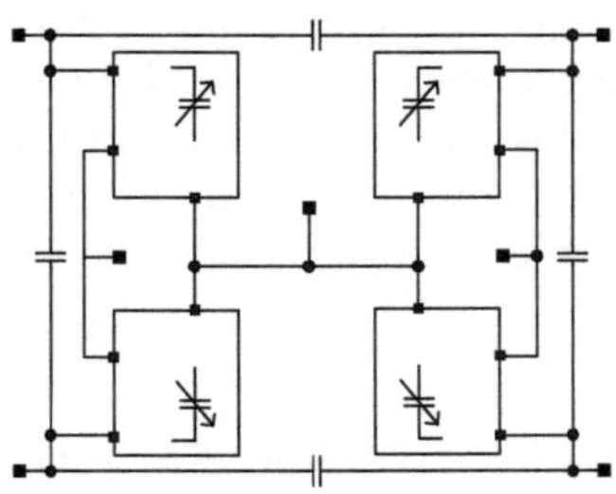

Figure 2. Electrical equivalent model of one axis of the micromirror. Each axis of the micromirror has been modeled as a variable capacitance whose value depends on the deflection angle and on the physical design of the micromirror. *sin_minus* and *sin_plus* are two counter-phase sine waves. *vbias_slow* is a constant voltage. The sensing is made using the SA terminals.

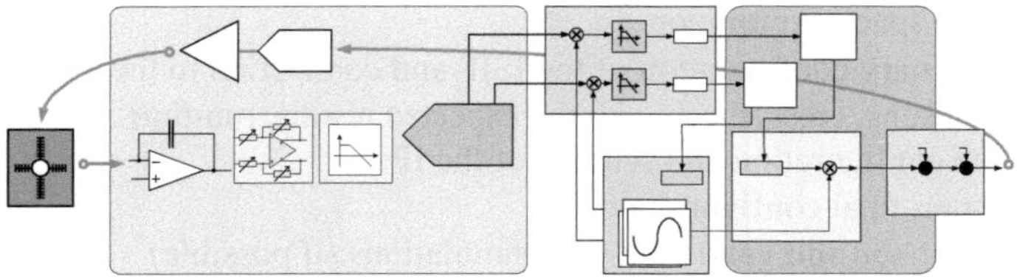

Figure 3. Block diagram of the system performing the closed loop driving on one axis of the mirror.

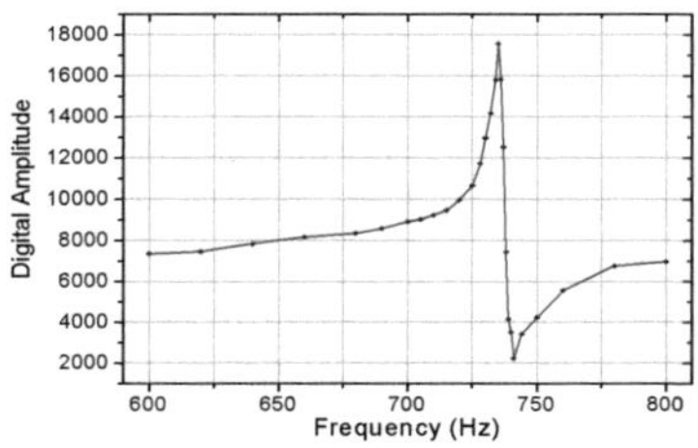

Figure 4. Result of the dynamic sensing capacitance measure. The mirror has been actuated with a 90 V peak to peak sine wave. For frequencies lower than 735 Hz (resonance frequency of the mirror), the signals due to parasitic and sensing capacitances are in phase. For higher frequencies, the sensing capacitance signal is delayed by π, while the parasitic capacitance signal is not delayed, hence the output signal amplitude changes.

5. Conclusions

A fast developing test and characterization system for MEMS and MOEMS has been presented. The system exploits the ISIF platform developed by Sensor Dynamics AG. The platform, together with a High Voltage hardware section and a graphical user interface, allowed us to create a lab test environment which can easily and rapidly explore different types of architecture for MEMS characterization. As an application of our environment, the characterization and test of a double axis scanning micromirror has been performed. After the static and dynamic capacitance measurements, a closed loop driving has been developed. An example of the results is shown in Fig. 4. FEM simulations have been performed in order to validate this type of approach and to create an electrical model of the micromirror which allowed us to compare and validate our results.

References

1. MSV 300 – Micro Scanning Vibrometer, “Polytec PI product information”, Tustin, CA, USA.
2. R. A. Lawton et al., “MEMS characterization using scanning laser vibrometer”, Proc. SPIE 1999, Symp. Microelectronic Manufacturing, 1999.
3. Reliability, Testing and Characterization of MEMS/MOEMS, Proc. of SPIE, www.spie.org.
4. F. D’Ascoli, M. Tonarelli, M. Melani, M. De Marinis, A. Giambastiani, L. Fanucci, “Intelligent sensor interface for automotive applications”, ICECS 2005, Gammarth (Tunisia), Dec 2005.
5. U. Hofmann, M. Oldsen, H. Quenzer, J. Jane, M. Heller, M. Weiss, G. Fakas, L. Ratzmann, E. Marchetti, F. D’Ascoli, M. Melani, L. Bacciarelli, E. Volpi, F. Battini, L. Mostardini, F. Sechi, M. De Marinis, B. Wagner, “Wafer-level vacuum packaged micro-scanning mirrors for compact laser projection displays”, SPIE Photonics West 2008, San Jose, California USA, vol. 6887, pp. 100–114, 2008.

A HIGH-VOLTAGE PWM CURRENT DRIVER FOR HOT-WIRE ANEMOMETERS

E. VOLPI[1], L. FANUCCI[1], F. D'ASCOLI[2] AND E. PARDI[2]
[1]*Department of Information Engineering, University of Pisa, via Caruso, 16, I-56122, Pisa*
[2]*SensorDynamics AG, via Giuntini, 63, I-56023, Navacchio (Pisa), Italy*

Abstract. The design of a high-voltage PWM current driver for hot-wire anemometers is presented. The circuit is able to provide a PWM output current whose value can be fixed by means of a negative feedback loop. The circuit will be used, together with other analog and digital IPs, to perform a constant temperature fluid speed measure based on a hot-wire anemometer. The proposed solution has been designed and simulated in a 0.35 µm Bipolar-CMOS-DMOS technology.

1. Introduction

The measurement of fluid speed is a key issue in many industrial and laboratory applications (e.g. in process control for dosing reagents and combustibles) but now it is extending in more domestic applications like control of air conditioning of closed environments and monitoring of potable water. However commercially available sensors are usually quite expensive especially when high reliability is required. Integrated flow sensors represent a very interesting solution especially for the possibility to have low manufacturing costs, high performances and to be integrated with the read-out circuits in the same package or even in the same die. There are several examples of measurement methodologies for flow detection in literature. The more common methodologies are based on: (1) measurement of pressure variation between two points along the fluid path; (2) measurement of the speed of a mechanical turbine wheel; (3) measurement of the energy transferred between the fluid and the flow sensor such as heat transfer. Some flow sensors based on the last principle are the hot-wire anemometers [1, 2]. Hot-wire anemometers can be operated either fixing the total power dissipation on the sensor (constant power measure) or fixing the sensor temperature (and resistance) (constant temperature measure). Some implementations of constant temperature measurement systems based on hot-wire anemometer have been reported in literature [3, 4] but they usually require external drivers with high-current capabilities that are difficult to be achieved in a fully integrated solution.

P. Malcovati et al. (eds.), *Sensors and Microsystems: AISEM 2009 Proceedings*,
Lecture Notes in Electrical Engineering 54, DOI 10.1007/978-90-481-3606-3_70,

The proposed solution is a high-voltage PWM current driver integrated in a BCD 0.35 μm technology whose output current value is fixed by means of a negative feedback loop. The circuit is designed to drive a hot-wire anemometer (MAF sensor) developed by FhG/ISiT (Fig. 1) [5]. The MAF sensor includes a heater (R_h of 50.0 ± 0.5 Ω) exposed to the fluid and a second resistor (R_t of 50.0 ± 0.5 Ω) exploited to measure the ambient (fluid) temperature.

The PWM current driver is used inside the control loop of Fig. 2. R_h and R_t are the MAF sensor resistors. The PWM current driver and the other blocks are integrated on the same chip except for the two resistors R_{he} and R_{te} used to complete the Wheatstone bridge and the capacitor C. The control loop is used to fix the sensor temperature and perform a constant temperature measure. In order to keep the sensor temperature constant, a Proportional-Integral controller changes the duty cycle of the PWM current and so the power provided to the sensor depending on the fluid rate (v). The current pulse width contains the information about the fluid rate (v). H(s) block evaluates the fluid rate (v).

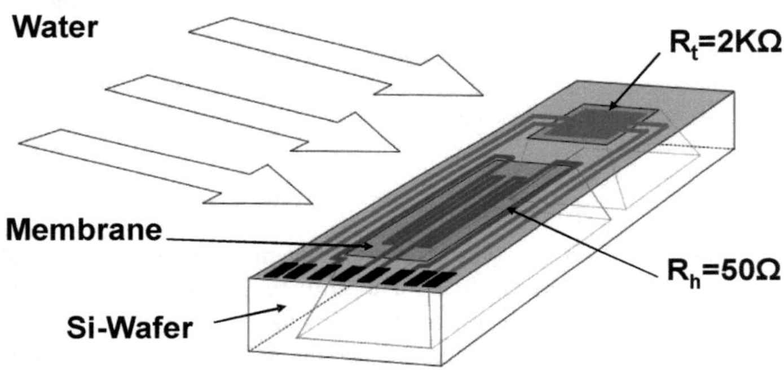

Figure 1. Schematic diagram of the MAF sensor: R_h are the fluid heaters, R_t are used to measure the fluid temperature.

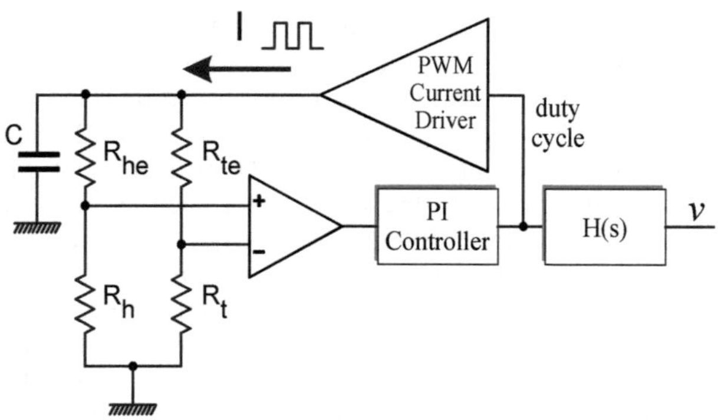

Figure 2. Block diagram of the system used to drive the flow sensor (resistors R_h and R_t).

2. Circuit description

The simplified schematic of the driver is shown in Fig. 3: PDM are high-voltage PDMOS devices available in the BCD 0.35 μm technology, NM are low-voltage NMOS devices. The proposed circuit is designed to work with a supply voltage (VBATT) between 4 V and 16 V and is intended to generate a PWM current up to 100 KHz.

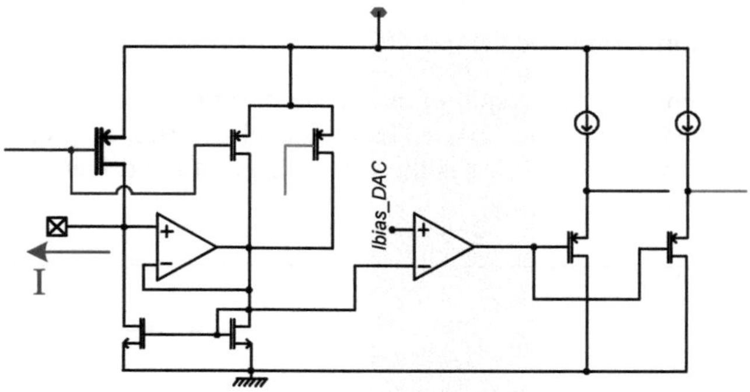

Figure 3. Simplified schematic of the PWM driver.

The output current value can be selected between 50 mA and 130 mA using a current DAC. The circuit can be divided in three main sections: the sensing section, the cascode OPA section (OPA2) and the buffer section. The sensing section is used to monitor the current of the driver. The sensing is performed exploiting a smaller replica (PDM2) of the output driver (PDM1). The gate of PDM1 and PDM2 are short circuited and the proposed design solution realized with MOSs NM1/2 and OPA1 forces PDM1 and PDM2 to work with the same V_{DS} voltage. This architecture is necessary due to the low matching performances of high-voltage devices. As a consequence the ratio between I_{PDM1} and I_{PDM2} is equal to the ratio between the aspect ratios of PDM1 and PDM2. It is worth to notice that without the voltage buffer, the difference between V_{DS1} and V_{DS2} might be quite high since the output node voltage can ideally varies between 0 and VBATT depending on the load and the driver output current selected. The aspect ratio of PDM1/2 and NM1/2 are made in order to have $I_{NM2} << I_{PDM1}$ (about 1/1,000 of I_{PDM1}) so that $I_{load} \approx I_{PDM1}$. The cascode OPA section performs the comparison between I_{NM1} and a bandgap derived and programmable current I_{bias_DAC}. The OPA2 guarantees a high AC gain and it closes the negative feedback loop on the gate of the voltage buffer PDM4 (buffer section). PDM3 and PDM5 have the same role of PDM2 and PDM4 respectively during the driver turn-off phase. This solution prevents current spikes of the output current and reduces the settling time of the feedback loop when the driver is turned-on or off. The turn-off and turn-on of the current driver is implemented turning-on and off the two current sources I_{bias_sf1} and I_{bias_sf2}. Low-voltage devices cannot exceed 3.3 V (equal to the supply voltage V_{DD} of NM1/2 and OPA2) with their V_{DS}, so they must be correctly protected. NDMOS devices with their gate connected to V_{DD} are placed on the drain of NM1/2 (not shown in Fig. 3 for simplicity). These protections keep the drain of NM1/2 a V_{GS} (about 800 mV) down to V_{DD}. In this way the exceeding voltage is kept by the NDMOS transistors that are able to sustain high V_{DS}.

3. Result of simulations

In Fig. 4 is reported the result of a transient simulation in these working conditions: PWM frequency = 100 KHz, duty cycle = 50%, I_{load} set to 120 mA, 100 Ω load, VBATT = 16 V. V1 is the gate voltage of the driver and I_{load} is the driver output current.

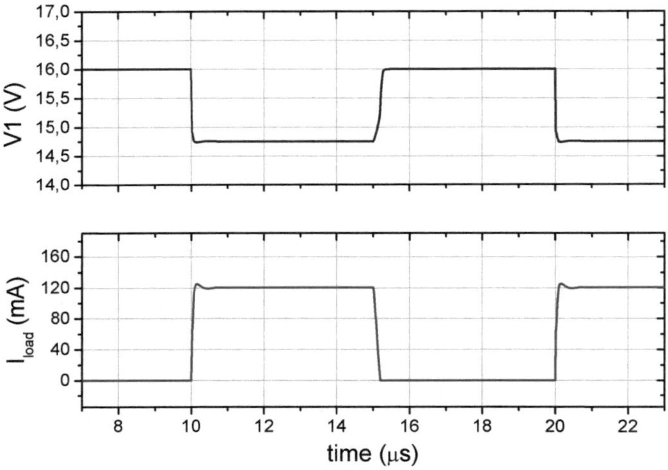

Figure 4. Transient signals of PDM1 gate voltage and of the driver output current.

4. Conclusions

A high-voltage driver integrated in a 0.35 μm BCD technology has been presented. The circuit will be used to drive a hot-wire anemometer with an output PWM current whose value can be fixed by means of a negative feedback loop. The circuit has been simulated to verify the effectiveness of the proposed topology. The circuit is also provided with a power-down circuitry in order to reduce the power consumption almost to zero (except for leakage currents) when the driver is not used.

References

1. B.W. van Oudheusden, "Silicon thermal flow sensors", Sensors and Actuators A: Physical, vol. 30, pp. 5–26, 1992.
2. M. Elvenspoek, "Thermal flow micro sensor", Semiconductor Conf., Proc. of CAS'99, 1999.
3. A. Oliveira et al., "An Anemometer with PWM Excitation", Proceedings of IEEE IECON, 21st International Conference, 2005
4. P. Bruschi et al., "A flow sensor for liquids based on a single temperature sensor operated in pulsed mode", Sensors and Actuators A: Physical, vol. 110, pp. 269–275, 2004.
5. M. Melani et al., "Hot Wire Anemometric MEMS Sensor for Water Flow Monitoring", Design, Automation, and Test in Europe, Munich, Germany, pp. 10–14, March 2008.

A MEMS PIEZORESISTIVE INCLINATION SENSOR WITH CMOS ASIC FRONT-END INTERFACE

S. DALOLA, V. FERRARI AND D. MARIOLI

Dipartimento di Elettronica per l'Automazione, Università degli Studi di Brescia Via Branze 38 – 25123 Brescia, Italy

Abstract. In this paper a system for inclination measurement is presented. It consists of a MEMS piezoresistive accelerometer manufactured in silicon bulk micromachining and an application specific integrated circuit (ASIC) specifically designed for resistive-bridge sensors. The interface circuit excites the sensor with a constant current and converts both the unbalance and the overall resistance of the sensor bridge respectively into the frequency and the duty cycle of a rectangular-wave output signal. DC current excitation of the sensor makes the configuration unaffected by wire resistance and parasitic capacitances, therefore the sensor can be placed remotely from the electronics without suffering accuracy degradation.

1. Introduction

Typically inclination sensors are based on a very sensitive accelerometer used with an input lower than 1 g [1]. Various types of silicon micromachined accelerometers with both piezoresistive and capacitive principles have been developed in recent years [1, 2]. The capacitive accelerometers have higher sensitivity and lower temperature coefficients but their signal could be difficult to measure due to the parasitic capacitance of the connecting wires. Micromachined piezoresistive accelerometers show simple sensor structure and fabrication process, high sensitivity, good linearity and low level of cross-axis sensitivity. Piezoresistive sensors, adequately conditioned, can be suitable for use in applications where the sensor must operate remotely from the electronics, such as in critical environmental conditions, without suffering accuracy degradation due to lead wire resistance and reduce the effect of stray and cable capacitances. In this paper an application specific integrated circuit (ASIC) for resistive-bridge sensors [3] is proposed to excite a MEMS piezoresistive accelerometer [4], configured as an inclinometer.

2. Sensor description

The sensor is a multi-axial accelerometer with piezoresistive detection, manufactured in silicon bulk micromachining with *Multi-Project-Wafer SensoNor*

P. Malcovati et al. (eds.), *Sensors and Microsystems: AISEM 2009 Proceedings*, Lecture Notes in Electrical Engineering 54, DOI 10.1007/978-90-481-3606-3_71,

technology. As shown in Fig. 1, the device consists of a seismic mass (2,550 × 2,550 × 400 μm) linked to the silicon substrate by means of four suspension beams (550 × 438 × 23 μm) placed symmetrically at the sides of the mass. Two piezoresistors are implanted at both ends of each beam in order to detect the deformation of the beams due to the acceleration applied to the seismic mass.

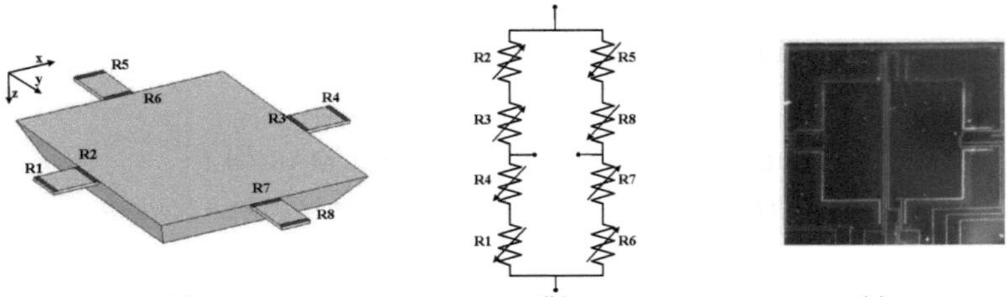

(a) (b) (c)

Figure 1. Schematic diagram (**a**), piezoresistors configuration (**b**), and top-view picture (**c**) of the sensor.

The eight piezoresistors are only partially connected internally to the sensor, while selected nodes are made accessible through external pads. The set of piezoresistors can be either configured in two separate Wheatstone bridges for the detection of the *x* and *y* components of the acceleration in the chip plane, or in a single full bridge for the detection of the *z* component normal to the plane. For inclination detection, the piezoresistors were configured in the *z* bridge in order to measure acceleration along the *z* axis. The symmetry of the structure virtually gives no cross-axis sensitivity [4].

3. ASIC interface description

The ASIC interface, manufactured in the *Alcatel* 0.7 μm standard CMOS technology, was designed for resistive-bridge sensors. The circuit, whose block diagram is shown in Fig. 2, is based on a relaxation oscillator in which the frequency *f* of the rectangular-wave output is related to the fractional bridge unbalance, and the duty cycle *dc* depends on the overall bridge resistance, which is typically related to temperature. In this way, two independent pieces of information are simultaneously and independently carried on the same output signal [3].

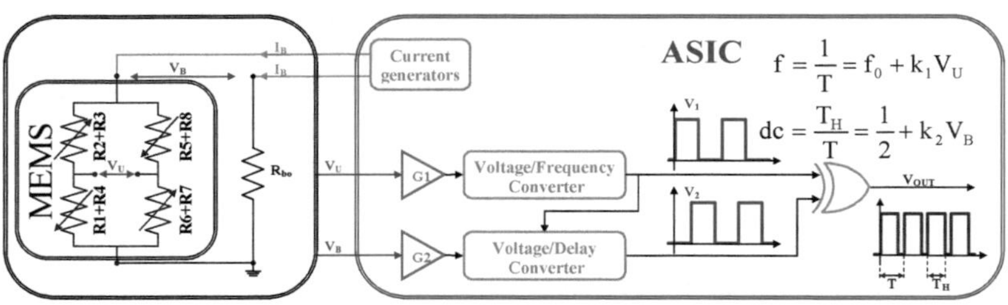

Figure 2. Schematic diagram of the MEMS inclinometer connected to the ASIC interface.

A multiple-output current generator supplies two identical constant DC currents I_B: one drives the sensor bridge, while the other flows into an external reference resistor R_{b0}. The bridge output voltage V_U sets the frequency of the square-wave voltage V_1, while the square-wave voltage V_2 is phase-shifted with respect to V_1 by an amount dependent on the voltage V_B, proportional to the difference between the overall bridge resistance and the reference value R_{b0}. The rectangular-wave output V_{OUT} is obtained by the EX-OR of V_1 and V_2. In this way the frequency shift and duty cycle of V_{OUT} are proportional to the bridge unbalance and overall resistance, respectively. The quasi-digital output increases the noise immunity and allows to interface to digital systems directly. The ratio-metric topology gives a first-order independence from power supply voltage, since fluctuations of power supply voltage affect both the thresholds of the comparators and the integration current of the same factor.

4. Experimental results

The sensor was directly connected to the ASIC interface for the characterization. With the auxiliary external components chosen, the ASIC central frequency was set at about 4,600 Hz, and the frequency sensitivity at 5.6 Hz/(μV/V). The experimental sensitivity of the sensor for *z* axis is of about 29 (μV/V)/g at room temperature, therefore the nominal sensitivity of the sensor plus electronics results of about 162 Hz/g. The sensitivity of the inclinometer at room temperature was measured by mounting the sensor on a rotating fixture, as shown in Fig. 3. The sensor was tilted of an angle θ between the chip plane and the gravity vector ranging from −90° to +90°, i.e. a static acceleration between −g and g was applied. The frequency of the output waveform V_{OUT} was measured by a frequency counter (Philips PM6680).

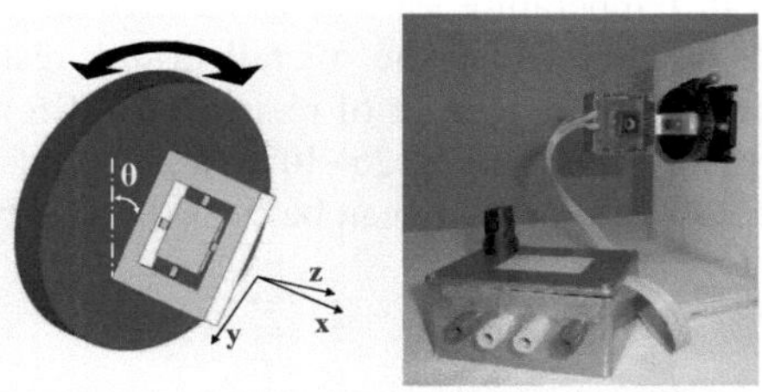

(a) (b)

Figure 3. Schematic diagram (**a**) and picture (**b**) of the measurement system.

As shown in Fig. 4a, the measured frequency shows a good fit to the expected sinusoidal trend, since the equivalent acceleration projected on the sensitive axis *z* of the device is proportional to $g{\cdot}sin(\theta)$. By plotting the data versus the equivalent applied *z*-axis acceleration, a linear result is obtained as shown in Fig. 4b.

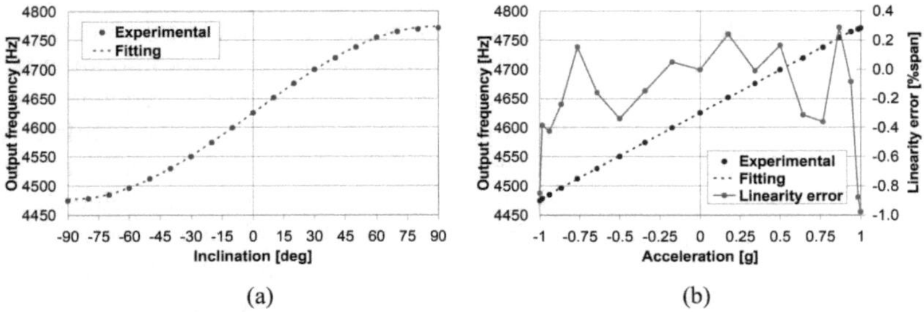

Figure 4. Measured output frequency versus z-axis inclination (**a**) and versus z-axis acceleration (**b**).

The resulting sensitivity, expressed as the output frequency versus the *z*-axis acceleration, is about 148 Hz/g, in good agreement with the theoretical expectations. The resolution of the system was evaluated by measuring the output frequency at constant inclination ($\theta = 0°$) and the standard deviation over 30 samples population with frequency-counter gate time of 1 s. The obtained resolution is about 0.26 Hz that corresponds to an equivalent inclination of 0.1°. In order to evaluate the behaviour of the sensor at different operating temperatures, the accelerometer was heated by means of a resistive heater, attached to the bottom face of the sensor substrate, while the ASIC was kept at distance and maintained at room temperature. The sensor temperature was monitored with a Pt100 resistive sensor. By measuring the fractional bridge unbalance of the sensor versus angle at different temperatures in the range 25–65°C, the sensitivity as a function of temperature was determined, as shown in Fig. 5a. The sensor temperature can be determined by measuring the bridge resistance, related to the output duty cycle. The overall sensor bridge resistance R_B and the duty cycle *dc* of the output waveform V_{OUT} were measured simultaneously as a function of the sensor temperature for a constant inclination ($\theta = 0°$). As shown in Fig. 5b the duty cycle follows the overall bridge resistance behaviour, as expected. The temperature coefficient of resistance (TCR) results of about 190 ppm/°C and 680 ppm/°C in ranges 25–40°C and 40–65°C respectively. The thermal dependence of the sensitivity can be compensated by the read-out unit.

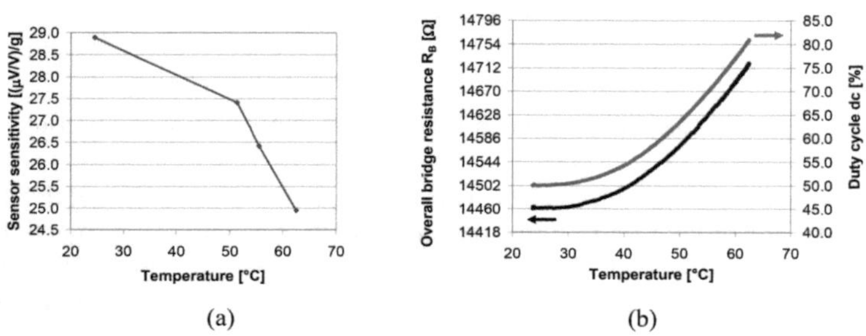

Figure 5. Measured sensor sensitivity (**a**), bridge resistance and output duty cycle (**b**) versus temperature.

5. Conclusions

A system for the measurement of static inclination, based on a MEMS piezoresistive accelerometer and a CMOS ASIC front-end interface has been proposed and characterized. The system is suitable for operation of the sensor remotely from the interface, such as in a high-temperature environment. In good agreement with theoretical predictions, the system shows a measured sensitivity of about 148 Hz/g and a resolution of about 0.1° at room temperature.

References

1. J. Courteaud, P. Combette , N. Crespy, G. Cathebras, A. Giani, "Thermal simulation and experimental results of a micromachined thermal inclinometer", Sensors and Actuators A, [vol. 141], pp. 307–313, 2008.
2. R. P. Van Kampen, R. F. Wolffenbuttel, "Modeling the mechanical behavior of bulk-micromachined silicon accelerometers", Sensors and Actuators A, [vol. 68], pp. 137–150, 1998.
3. V. Ferrari, A. Ghisla, Z. S. Kovács Vajna, D. Marioli, A. Taroni, "ASIC front-end interface with frequency and duty cycle output for resistive-bridge sensors", Sensors and Actuators A, [vol. 138], pp. 112–119, 2007.
4. V. Ferrari, A. Ghisla, D. Marioli, A. Taroni, "MEMS accelerometer with multiaxial response by dynamic reconfiguration of piezoresistive bridges", Proceedings of the Eurosensors XX Conference, Göteborg (Sweden), 17–20 September 2006.

ACTIVELY CONTROLLED POWER CONVERSION TECHNIQUES FOR PIEZOELECTRIC ENERGY HARVESTING APPLICATIONS

A. ROMANI[1], C. TAMBURINI[1], A. GOLFARELLI[1], R. CODELUPPI[1], M. DINI[1], E. SANGIORGI[1], M. TARTAGNI[1] AND R.P. PAGANELLI[2]

[1]*School of Engineering II, Campus of Cesena, University of Bologna, Via Venezia, 52 – 47023 Cesena, Italy*

[2]*CNR-IEIIT, at DEIS-University of Bologna, Viale Pepoli, 3/2 – 40123 Bologna, Italy*

Abstract. The current advances in ultra-low power design let foresee great opportunities in energy harvesting platforms for self-powered systems. This paper presents two conversion schemes based on active control for harvesting energy with a higher efficiency than traditional passive approaches. A prototype has been developed and the approaches have been validated for piezoelectric energy harvesters with both measurements in realistic conditions (i.e. irregular vibrations) and mixed-signal circuital simulations. The proposed converters may increase harvested power of at least 25% and up to three times with respect to a passive rectifier. The harvested power is about 40 μW in presence of weak vibrations (a_{RMS} = 1.18 m/s^2) obtained in a train passenger car in motion with a 28 × 6 × 0.5 mm^3 cantilever made of PZT-A4E with a 20 g mass attached at the free end.

1. Introduction and experimental setup

Power supplies of many electronic devices still mainly rely on electrochemical cells with limited stored charge and often unpractical to replace. However, the environment is an intrinsic source of low-density highly-available energy, in either steady or intermittent and irregular forms [1], such as for example mechanical vibrations. The current advances in ultra-low power design let foresee great opportunities in harvesting this kind of energy for supplying electronic systems. As the available power densities usually range below 100 μW/cm^2, much care has to be put into the design of efficient power converters. This paper will describe how solutions based on active control of power conversion boost up the harvested energy.

P. Malcovati et al. (eds.), *Sensors and Microsystems: AISEM 2009 Proceedings*, Lecture Notes in Electrical Engineering 54, DOI 10.1007/978-90-481-3606-3_72,

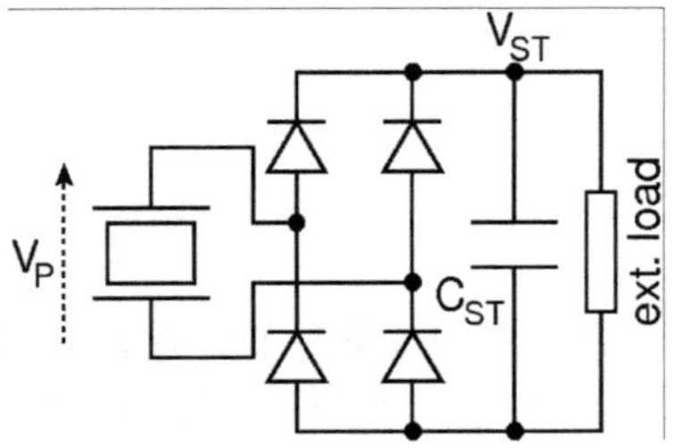

Figure 1. A passive diode rectifier bridge for power harvesting from piezoelectric transducers.

Without lack of generality, a piezoelectric mass-cantilever harvester will be considered as a case study. A conventional diode rectifier bridge (Fig. 1) will be assumed as a reference for comparing different types of power conversion. In order to provide reliable power estimates, the amplitude of input vibrations was recorded in specific cases of interest with a Kionix KXP-74 accelerometer. In Fig. 2 the accelerometer was bound to the wall of a train passenger car and acquired data during motion at about 100 km/h. These data were used as input for:

1. Circuital simulations (including an equivalent electro-mechanical circuit of the piezo transducer) for reliably estimating the behavior of conversion circuits.
2. A mechanical shaker allowing to jointly test piezo transducers and conversion circuits.

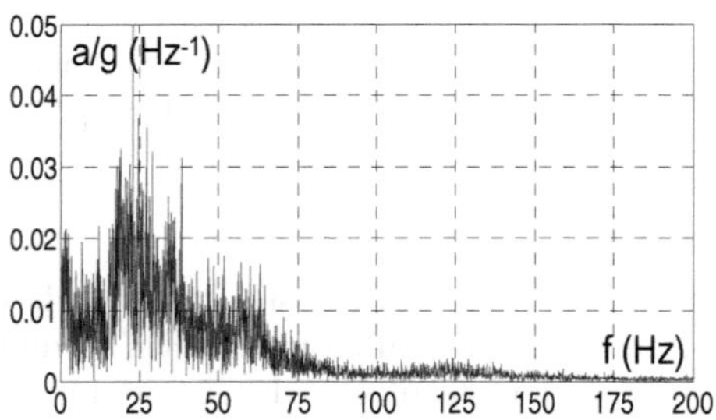

Figure 2. Amplitude spectrum of vibrations in a train passenger car at 100 km/h ($a_{RMS} = 2.37 m/s^2$).

2. Description of the proposed circuits

With a traditional diode rectifier bridge (Fig. 1), as long as $|V_P(t)| - 2V_\gamma < V_{ST}$, no energy is transferred from the piezoelectric source despite its motion (i.e. vibrations are performing mechanical work). Moreover, the maximum power transfer is achieved only for a specific load current which strongly depends on the input vibration pattern. This circuit is not efficient in real systems, where the load operates discontinuously and vibrations are irregular: it is extremely difficult to bias the rectifier bridge with its optimum current.

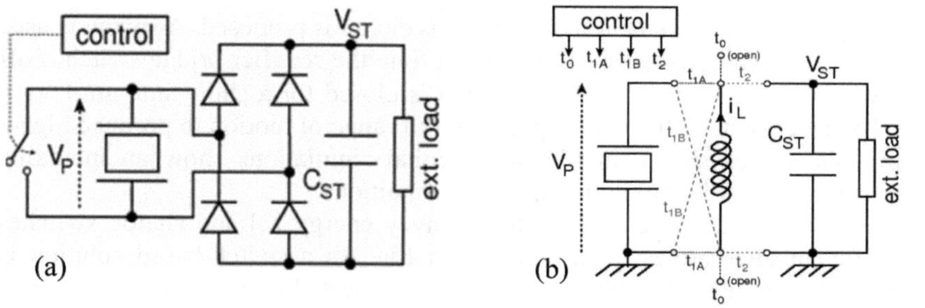

Figure 3. (**a**) The proposed improvement to the rectifier bridge for energy harvesting. (**b**) The proposed switch-based power converter.

An active control which operates according to the state of the piezo transducer may improve harvesting efficiency. This can be done in real-time because of the low frequency of vibrations. Two schemes will be proposed herein.

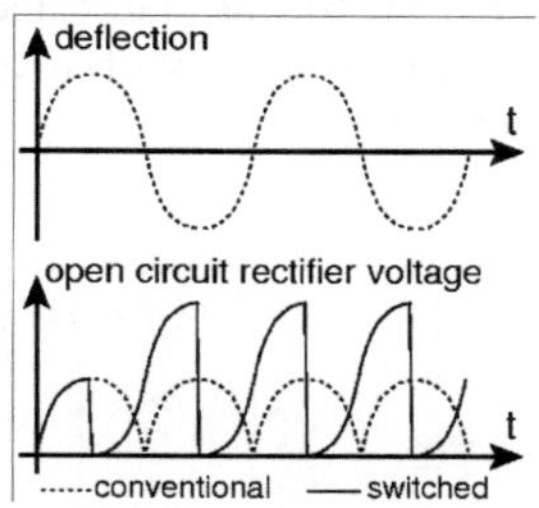

Figure 4. Operating principle of the improved rectifier: higher voltages are produced. The depicted waveforms correspond to no load and no storage capacitor.

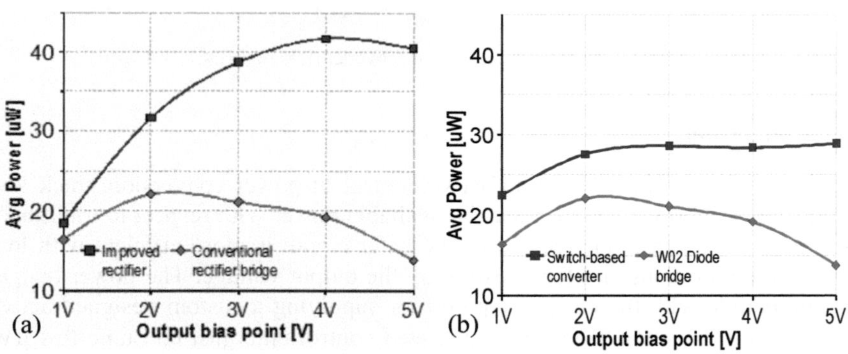

Figure 5. (**a**) Simulation results for available output power for the improved rectifier (Fig. 3a) compared to the conventional diode bridge (Fig. 1); NSR0130P2 diode models were used in both cases; (**b**) Measured output power for the prototype of switch-based converter (Fig. 3b), compared to measurements of a W02 diode bridge (Fig. 1); vibrations were reproduced with a shaker system. All simulated and experimental data in (**a**) and (**b**) refer to a commercial PZT-5H 28 × 6 mm^2 piezo transducer with a 20 g tip mass and the same input vibrations shown in Fig. 2. A switched resistor was used as an external load to modulate the output current and determine the output bias point.

In Fig. 3a an improvement of the previous circuit is proposed. A switch is used to reset the piezo transducer voltage. Each time the rectifier bridge switches off (i.e. when $|V_P(t)| - 2V_\gamma < V_{ST}$) the switch is closed for a short time until V_P = 0V. The reset switch allows to exploit the full range of motion to produce higher peak voltages, as shown in Fig. 4. Circuital simulations show an increased harvested power (Fig. 5a) in the same test conditions.

Moreover, inductors can efficiently convey energy [2] and analog switches are efficient alternatives to diodes [3]. In Fig. 3b a switch-based solution is proposed. The piezo element is normally disconnected (configuration t_0). When $V_P > 0$ and a local maximum [$V_P < 0$ and a local minimum] is detected on V_P, configuration t_{1B} [configuration t_{1A}] is applied and C_p is discharged by an increasing current through L. When V_P = 0V (and i_L is at its maximum value), configuration t_2 is applied and i_L flows towards C_{ST}. When i_L = 0A, all switches are opened again (configuration t_0). Transients last no longer than a quarter of period of oscillations and precise timings are required for energy efficiency. This circuit was implemented (Fig. 6): the harvested power is almost constant and in the worst case it is 25% higher than with a conventional rectifier bridge in the same test conditions (Fig. 5b).

Figure 6. Prototype of the switch-based converter (L = 10 mH, C_{ST} = 4.7μF).

3. Conclusions

The proposed circuits, based on active control of power conversion, track low frequency vibrations and increase the available power with respect to traditional solutions based on diode rectifiers. The latter circuit is more efficient with low amplitude vibrations and for boosting up the output voltage. The power budget is compatible with the energy required for supplying a custom designed active control unit: very low frequency integrated control units just consume few μW [4]. The above schemes are suitable for storing energy in either capacitors or electrochemical cells (e.g. button batteries) and for being implemented in standard CMOS technologies.

Acknowledgments This work was supported by Eurotech Group as part of the research project "Self-Powered Portable and Wireless Electronic Systems" carried out jointly with University of Bologna. The presented technology is patented.

References

1. S. Roundy, P. K. Wright, J. Rabaey, "A study of low-level vibrations as a power source for wireless sensor nodes", Computer Communications, vol. 26, no. 11, pp. 1131–1144, 2003
2. E. Lefeuvre, A. Badel, C. Richard, L. Petit, D. Guyomar, "A comparison between several vibration-powered piezoelectric generators for standalone systems", Sensors and Actuators A, vol. 126, no. 2, pp. 405–416, 2005
3. A. Romani, C. Tamburini, R. P. Paganelli, A. Golfarelli, R. Codeluppi, E. Sangiorgi, M. Tartagni, "Dynamic switching conversion for piezoelectric energy harvesting systems", in Proc. of IEEE Sensors 2008, pp. 689–692
4. I. Doms, P. Merken, R. P. Mertens, C. Van Hoof, "Capacitive power-management circuit for micropower thermoelectric generators with a 2.1μW controller", in 2008 IEEE ISSCC Digest of Technical Papers, pp. 300, 615

FEM ANALYSIS OF PIEZOELECTRIC NANOSTRUCTURES FOR ENERGY HARVESTING

G. MANTINI[1], A. D'AMICO[1], C. FALCONI[1] AND Z. LIN WANG[2]

[1]*Department of Electronic Engineering, University of Tor Vergata, Via del Politecnico 1, 00133 Roma*

[2]*School of Materials Science and Engineering, Georgia Institute of Technology, Atlanta, Georgia, USA*

Abstract. Piezoelectric zinc oxide nanostructures have been recently used to convert mechanical energy into electrical energy. Here we give reasons for previously reported experimental results and provide insight for the design of high-efficiency piezoelectric nanogenerators. Moreover, we investigate some of the most important 2-contacts piezo-electric nanowires by means of Finite Element Method (FEM) simulations.

1. Introduction

Recently, zinc oxide nanowires have been used for transducing vibrations into electric energy; interestingly, it is possible to integrate an enormous number (e.g. billions per square centimeter) of piezoelectric nanotransducers on a single substrate [1].

In order to effectively harvest mechanical energy, many practical issues must be considered: how the piezoelectric potential can be simultaneously generated in a large number of nanowires; how the electric energy can be extracted and applied to an external load; how the nanostructures and the microsystem can be fabricated, assembled, and packaged.

Obviously, the critical challenge for energy harvesting is high efficiency. Various issues must be solved in order to power a microsystem by means of piezoelectric nanostructures; moreover, the overall efficiency of the complete system will, in general, depend on many parameters. However, in this work, without regard to technological or practical issues, we exclusively focus on the conversion of a given static force into a voltage based on theoretical calculation. With this approach, we consider both previously reported and novel piezoelectric nanostructures for energy harvesting and systematically compare their output potentials when a given static force is applied; both theoretical considerations and FEM (finite element method) simulations allow the identification of the optimal nanostructures (from the point of view of the output voltage for a given input force).

P. Malcovati et al. (eds.), *Sensors and Microsystems: AISEM 2009 Proceedings*, Lecture Notes in Electrical Engineering 54, DOI 10.1007/978-90-481-3606-3_73,

2. FEM calculations

For all the FEM simulations, for simplicity, without loss of generality, we consider zinc oxide as the piezoelectric material; moreover, we neglect its electrical conductivity and we restrict our attention to static simulations. Our focus will be the estimation of the output potential under various deformation configurations and with different positions of the contacts. These assumptions can be tolerated as our goal is to qualitatively compare the potentialities of different nanostructures and provide insight for the design of high efficiency piezoelectric nanostructures, rather than quantitatively predict the piezoelectric potential output; besides, our results can be a good start for the development of more sophisticated models.

2.1. *Top-bottom piezoelectric nanowire*

Similarly to [2], a nanowire with cylindrical geometry (50 nm diameter and 600 nm length) has been laterally deflected by a lateral bending force of 80 nN; the force has been applied to the top surface of the nanowire in order to avoid punctual deformations [2], which results, in practice, in the application of a surface force approximately equal to 40×10^6 N/m^2. The bottom surface was affixed and entirely grounded. The nanowire has been considered as an ideal dielectric medium (i.e. its electrical conductivity is zero). Figure 1a shows the potential distribution viewed from side (this simulation is in excellent agreement with [2] and, in particular, shows the parallel plate capacitor model of piezoelectric potential).

2.2. *Bottom-bottom piezoelectric nanowire*

A bottom-bottom square-base nanowire has been considered and bent by the same force as before. The two identical gold contacts have been placed laterally at the base of the NW; the bottom-left contact has been grounded. All the other parameters were the same as in the previous calculations.

Figure 1b shows the potential distribution; the output potential between the left and right contacts is about 6 V. In fact, in this configuration, the regions subject to the highest strain (i.e. the bottom part of the nanowire) are not covered by conductive films. FEM simulations consistently confirm that nanowires with total bottom contact are not optimal from the point of view of the output potential generated by lateral bending; clearly, in some cases, technological issues might favor the use of sub-optimal nanostructures.

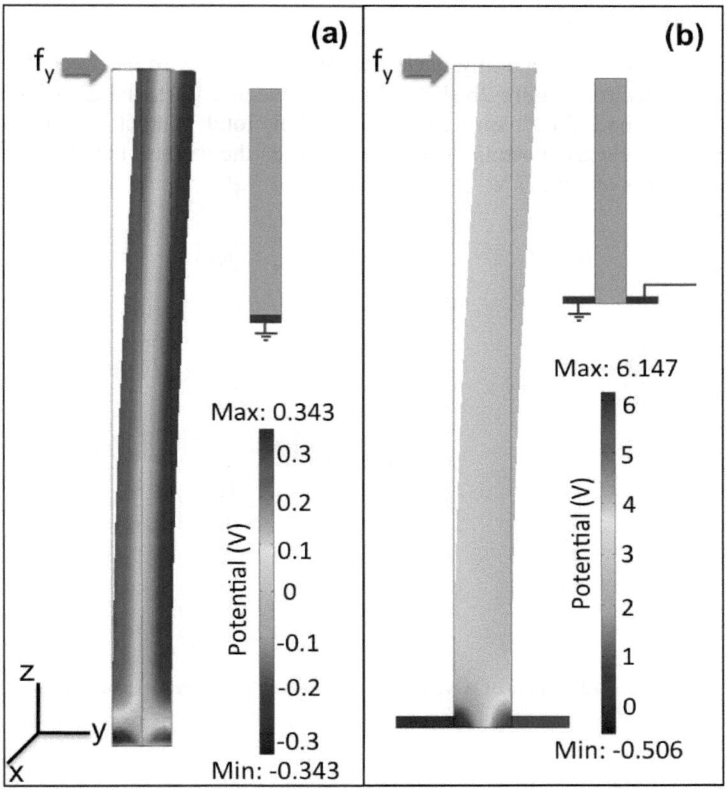

Figure 1. Piezoelectric potential distribution of bent nanowires. (**a**) top-bottom nanowire; (**b**) bottom-bottom nanowire with two lateral gold contacts.

3. Vertical compression

All the previous simulations assume a lateral bending with force applied at the free tip of the nanowire; it is, however, possible to apply a vertical compressive force (i.e. parallel to the axis of the nanowire). Contrary to lateral bending forces, in this case, obviously, the electric potential along the central axis of the nanowire is not constant; for this reason, a high output potential can be generated even when both the base and the tip are totally covered by conductive materials, a potentially critical advantage for practical implementations. Figure 2a shows the potential in a ZnO nanowire when a force is applied in the z direction; the NW is compressed and a negative potential of about 2.5 V is generated between the top surface of the nanowire and the grounded base.

4. Lateral stretching

As presented in [3], a laterally packaged piezoelectric wire has been developed as power generator. Figure 2b shows the piezoelectric potential distribution of a laterally stretched ZnO nanowire with the left, total contact of the nanowire grounded. The electric potential generated between the top and bottom (grounded) metal contact is about 3.7 V.

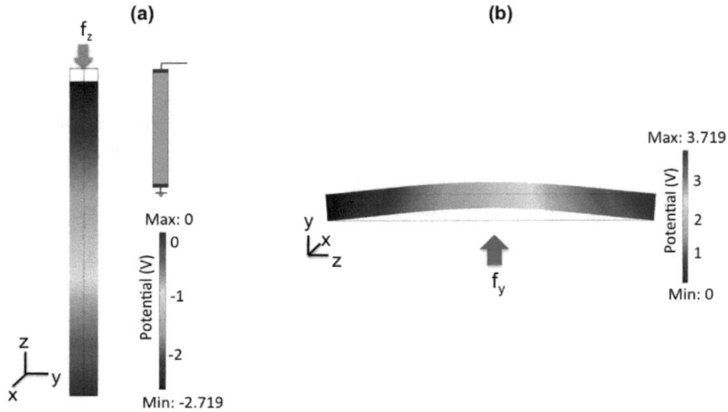

Figure 2. Potential distribution of a vertically compressed nanowire (**a**) and laterally stretched nanowire (**b**).

5. Conclusions

In summary, we have shown that the piezoelectric potential mechanically induced in a deformed nanostructure can critically depend on the position of the metal contacts and on the way the input force is applied. The results are confirmed by FEM simulations and can provide insight for the development of high-efficiency piezoelectric nanostructures for energy harvesting.

References

1. Z.L. Wang, J. Song, "Direct-current nanogenerator driven by ultrasonic waves", Science, [312], pp. 102–105, April, 2006.
2. Y. Gao, Z.L. Wang, "Electrostatic potential in a bent piezoelectric nanowire. The fundamental theory of nanogenerators and nanopiezotronics", Nano Lett., vol. 7 (8), pp. 2499–2505, July, 2007.
3. R. Yang, Y. Qin. L. Dai, Z.L. Wang, "Power generation with laterally packaged piezoelectric fine wires", Nature Nanotechnology, vol. 4, pp. 34–39, November, 2008.

PIEZO-POLYMER-FET DEVICES BASED TACTILE SENSORS FOR HUMANOID ROBOTS

RAVINDER S. DAHIYA[1], GIORGIO METTA[1], MAURIZIO VALLE[2], LEANDRO LORENZELLI[3] AND ANDREA ADAMI[3]

[1]*Robotics, Brain and Cognitive Sciences Department, Italian Institute of Technology, Via Morego 30, Genoa, 16163, Italy*
[2]*DIBE – University of Genoa, Via Opera Pia 11 A, Genoa, 16145, Italy*
[3]*Bio-MEMS – Fondazione Bruno Kessler, Via Sommarive 18, Trento, 38050, Italy*

Abstract. This work presents novel POSFET (Piezoelectric Oxide Semiconductor Field Effect Transistor) tactile sensing devices and the approach, for developing high density tactile sensing arrays for the fingertips of humanoid robot. The POSFET tactile sensing device is obtained by spin coating, and processing in situ, a piezoelectric polymers (PVDF-TrFE) film on the gate area of a MOS device. The feasibility study of the approach was earlier done by developing piezoelectric polymer-MEA (microelectrode array) based test chips – each comprising of 32 'taxels' (tactile elements) epoxy-adhered with a piezoelectric polymer (PVDF-TrFE) film. The successful implementation of POSFET touch sensing devices will be followed by their use in tactile sensing arrays.

1. Introduction

Touch sensing, needed for tasks like manipulation and exploration, is an important component of a robotic platform [1]. Movement of robots from the structured environment of manufacturing plants to the daily life has also added new tasks like safe interaction in a human-populated environment. The way robots interact with the environment is an important issue, as real-world objects exhibit rich physical interaction behaviors which depend on how heavy and hard the object is when hold, how its surface feels when touched, how it deforms on contact and how it moves when pushed. Such interaction behaviors can be better understood by touching or physically interacting with them – as humans do.

Over the years, touch sensing technology has improved and many sensors, using different materials and transduction methods, viz: Resistive/Piezoresistive, Quantum Tunneling, Capacitive, Optical, Ultrasonic, Magnetic, Piezoelectric etc. have been developed [2, 3]. Most of these sensors are generally big in size and thus unsuitable for body sites like robot's fingertips where large numbers of sensors with high density are needed. As in human fingertips, it is desirable to have fast and high density and tactile arrays for exploration and fine manipulation. This requires a large number of taxels (25–100 with ~1 mm spatial resolution)

P. Malcovati et al. (eds.), *Sensors and Microsystems: AISEM 2009 Proceedings*, Lecture Notes in Electrical Engineering 54, DOI 10.1007/978-90-481-3606-3_74,

in a small space (~1 × 1 cm) and having fast response (of the order of few milliseconds) [4]. Considering the limited available space on the robot finger, miniaturization seems to be the solution and is also adopted m this work – which is targeted towards the humanoid robot lcub [5].

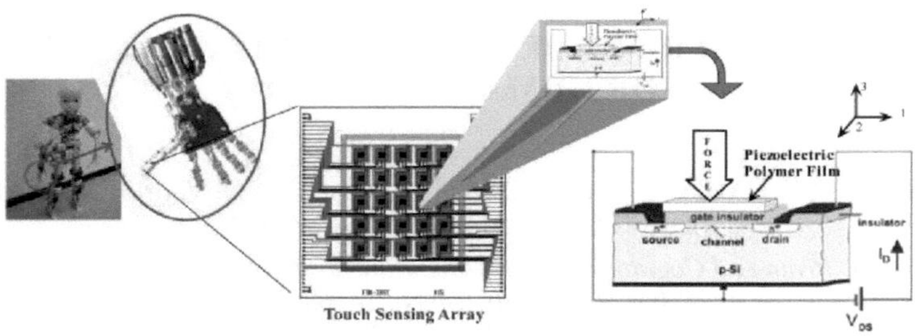

Figure 1. The approach for developing touch sensing arrays for the fingertips of robot.

This work presents novel POSFET (Piezoelectric Oxide Semiconductor Field Effect Transistor) tactile sensing devices and the approach [1], shown in Fig. 1, for developing high density tactile sensing arrays for the fingertips of humanoid robot. The POSFET touch sensing devices are fabricated by spin coating thin (~2.5 μm) film of piezoelectric polymer, PVDF-TrFE (Polyvinylidene Fluoride-Trifluoroethylene), directly on to the gate area of MOS transistor [6]. The POSFET device, thus, represents an integral "sensotronic" unit comprising of transducer and the transistor and capable of 'sensing and partially processing at same site'. Besides this, the marriage of sensing material and the electronics would improve signal to noise ratio and hence the force sensitivity.

2. POSFET touch sensors – concept and implementation

A piezoelectric film working in the sensing mode generates a charge/voltage when mechanical force/stress is applied. This charge/voltage is proportional to the applied force/stress [7] and they are related as:

$$Q_t = d_{33} \times F \qquad (1)$$

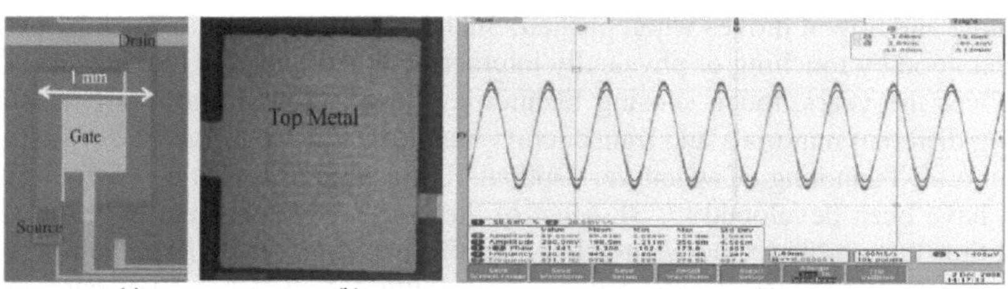

(a) (b) (c)

Figure 2. Enlarged picture of fabricated POSFET devices before (**a**) and after (**b**) deposition and processing of PVDF-TrFE piezoelectric polymer film, (**c**) The response of POSFET tactile sensor when normal dynamic force (1 N, 930 Hz sinusoidal) is applied on top of the sensor.

where, Q, is the charge developed on a taxel due to applied force, F and d_{33} is the piezoelectric constant. By depositing the piezoelectric polymer on the gate area of MOSFET, as shown in Fig. 1, such a charge/voltage of the piezoelectric polymer can be used to modulate the charge in the induced channel of MOSFET. Thus, POSFET touch sensing devices directly convert force into voltage, which can be amplified by POSFET itself and further processed by an electronic circuitry that may be present on the same chip.

Somewhat similar approach for the development touch sensors has been reported in [1, 8]. The extended gates, epoxy-adhered with PVDF (Polyvinylidene Fluoride) and PVDF-TrFE films are used in these works. The extended gate approach brings the sensor and conditioning electronics closer and hence the overall response is better than that of conventional approach – in which, the sensor and conditioning electronics are placed at a distance. However, the extended gates also introduce a large substrate capacitance, which in turn, significantly attenuates the charge/voltage available at gate terminals of MOS transistors and benefits offered by closely located sensor and electronics, are not fully exploited [9, 10]. POSFET touch sensing devices, on other hand, are free from such problems.

The POSFET touch sensing device, implemented on silicon substrate is shown in Fig. 2. Detailed fabrication process of POSFET touch sensing devices is described elsewhere [6]; however, the key points are presented here. The MOS device is fabricated by using the n-MOS technological module of a non standard 4 μm Al gate p-well ISFET/CMOS technology [11]. In order to match the spatial acuity of receptors in human skin, POSFET touch sensing devices are designed to have size of 1 × 1 mm. A better spatial acuity can be obtained by reducing the size of transistor. To obtain large transconductance, n-MOS devices are designed to have W = 7,500 μm and L = 12 μm. A 2.5 μm thick PVDF-TrFE piezoelectric polymer film is then spin coated from solution and processed on the gate, following the steps explained elsewhere [12]. The performance of POSFET touch sensor is evaluated by connecting it in a source-follower configuration and applying a dynamic normal force of 1N at various frequencies in the range 2 Hz–1 KHz. Figure 2c shows the output of POSFET device when sinusoidal force at 930 Hz is applied.

3. Conclusions

POSFET touch sensing devices, presented in this work, show a linear response over range of forces. In addition to sensing and processing at same site and the improved performance, POSFET devices as an integral "sensotronic" unit offer practical advantages like reduction in number of wires – which is a key issue in robotics. The performance, utility and local processing capability of POSFET touch sensors can be further improved by including complex circuitry and following a SOC/SiP approach.

Acknowledgments This work is supported in part by the European Commission Project RobotCub (IST-FP6-004370) and in part by the Italian Ministry of Education University and Research under the project PRIN 2007 "Tactile Sensing System for Humanoid Robots using Piezo-polymer-FET devices".

References

1. R. S. Dahiya, M. Valle, G. Metta, and L. Lorenzelli, "Tactile Sensing Arrays for Humanoid Robots using Piezo-Polymer-FET Devices," in AISEM 2008, 13th Annual Conference on Sensors and Microsystems, Rome, Italy, 2008.
2. R. S. Dahiya, G. Metta, M. Valle, and G. Sandini, "Tactile Sensing: From Humans to Humanoids," IEEE Transactions on Robotics, pp 1-20, 2009, DOI: 10.1109/TRO.2009.2033627
3. M. R. Cutkosky, R. D. Howe, and W. Provancher, "Force and Tactile Sensors," in Springer Handbook of Robotics, B. Siciliano and K. Oussama, Eds. Berlin Heidelberg: Springer 2008, pp. 455–476.
4. R. S. Dahiya, M. Valle, and G. Metta, "Touch Sensing by Piezoelectric Polymer Oxide Semiconductor Field Effect Transistor (POSFET) Devices," in 2nd IEEE International Workshop on Electron Devices and Semiconductor Technology, Mumbai, India, 2009, DOI: 10.1109/TRO.2009.2033627
5. G. Sandini, G. Metta, and D. Vernon, "RobotCub: An Open Framework for Research in Embodied Cognition," in IEEE-RAS/RSJ International Conference on Humanoid Robots, Humanoids 2004, Santa Monica, Los Angeles, CA, USA, 2004.
6. R. S. Dahiya, M. Valle, G. Metta, L. Lorenzelli, and A. Adami, "Design and Fabrication of POSFET Devices for Tactile Sensing, "in TRANSDUCERS 2009, The 15th IEEE International Conference. on Solid-State Sensors, Actuators and Microsystems, Denver, USA, 2009, DOI: 10.1109/TRO.2009.2033627
7. R. S. Dahiya, M. Valle, and L. Lorenzelli, "SPICE Model of Lossy Piezoelectric Polymers," IEEE Transaction, on Ultrasonics, Ferroelectrics and Frequency Control, vol. 56, pp. 387–396, 2009.
8. R. S. Dahiya, G. Metta, and M. Valle, "Development of Fingertip Tactile Sensing Chips for Humanoid Robots," in 5th IEEE International Conference. on Mechatronics, Malaga Spain, pp. 1–6, 2009.
9. R. S. Dahiya, M. Valle, G. Metta, and L. Lorenzelli, "POSFET Based Tactile Sensor Arrays," in IEEE ICECS'07, The 14th International Conference. on Electronics, Circuits and Systems, Marrakech, Morocco, pp. 1075–1078, 2007.
10. R. S. Dahiya, M. Valle, G. Metta, and L. Lorenzelli, "Bio-inspired Tactile Sensing Arrays," in Bioengineered and Bioinspired Systems, SPIE Europe Conference on Microtechnologies for new Millennium, Dresden, Germany, pp. 1–9, 2009.
11. S. Martinoia, N. Rosso, M. Grattarola, L. Lorenzelli, B. Margesin, and M. Zen, "Development of ISFET array-based Microsystems for bioelectrochemical measurements of cell populations'" Biosensors & Bioelectronics, vol. 16, pp. 1043–1050, 2001.
12. R. S. Dahiya, M. Valle, L. Lorenzelli, G. Metta, and S. Pedrotti, "Deposition Processing and Characterization of PVDF-TrFE Thin Films for Sensing Applications," in IEEE Sensors 2008, Lecce, Italy, pp. 490–493, 2008.

INTEGRATED OPTOFLUIDIC MACH-ZEHNDER INTERFEROMETER

R. BERNINI[1], G. TESTA[1], L. ZENI[2] AND P. M. SARRO[3]
[1]*IREA-CNR, Via Diocleziano 328, 80124 Napoli, Italy*
[2]*DII, Seconda Università di Napoli, Via Roma 29, 81031 Aversa, Italy*
[3]*ECTM-DIMES, TUDelft NL-2600 GB Delft, The Netherlands*

Abstract. In this paper, we propose an innovative configuration for a Mach-Zehnder interferometer (MZI), entirely composed by liquid core waveguides. It is an optofluidic asymmetric MZI based on antiresonant reflecting optical waveguides (ARROWs) with liquid core. An experimental characterization of the proposed device has been carried out and the experimental results confirm the correct behaviour of the device.

1. Introduction

Optofluidics is a new emerging field in which integrated complex optical systems are realized with fluids [1]. It presents an innovative approach for the implementation of new optical devices in which the fluids become the fundamental part, allowing for the realization of adaptive optical systems characterized by a high degree of reconfigurability and tunability. In this context, we propose a novel MZ configuration based on liquid core ARROW waveguides. Such a configuration takes the advantage of a high interaction between the probing light and the sample as they are both confined in the same waveguiding channels [2, 3]. Several integrated Mach-Zehnder interferometers (MZIs) have been realized, producing hundreds of publications in the literature every year [4, 5]. Such interferometers are extensively employed in different area ranging from telecommunication and sensing applications. In the context of optical biosensor, such interferometers are particularly attractive since they are capable of detect very low concentration of biochemical substances.

Typically, MZIs are based on conventional solid core waveguides in which a phase shift is induced in one arm by a variation of effective refractive index of the waveguide.

The sensing performance of these devices is limited to the partial sensitivity [6], defined as the rate of change of the effective refractive index of the guided mode as the liquids index varies. The maximum partial sensing sensitivity reported for MZI based on SiON technology is about 0.23 [6, 7]. This value increases by increasing the interaction between the evanescent field and the sample under analysis. In our MZI configuration, since the mode field is almost

P. Malcovati et al. (eds.), *Sensors and Microsystems: AISEM 2009 Proceedings*, Lecture Notes in Electrical Engineering 54, DOI 10.1007/978-90-481-3606-3_75,

completely confined into the liquid core of the waveguide, the interaction of the sample and field is strongly enhanced as compared to the solid core MZ in which only the evanescent part of the field is sensitive to the surrounding liquid medium. Hence, in our case, the partial sensitivity can approaches to one.

2. Operating principles

Our MZ configuration is based on liquid core ARROW waveguides. In an ARROW waveguide the light confinement is achieved through a high quality Fabry-Perot reflector formed by two cladding layers deposited on the side walls of the core. This waveguide has been realized with the standard silicon technology. It is composed by two halves joined together; both halves are composed by two cladding layers of silicon dioxide (n_2 = 1.457) and silicon nitride (n_1 = 2.227) deposited on a silicon substrate, with thicknesses $d_1 \approx$ 266 nm and $d_2 \approx$ 266 nm, respectively. The thicknesses of such layers are chosen to assure a low-loss propagation of the structure for a liquid core (n_c = 1.330) at the wavelength λ = 633 nm.

A schematic drawing of the proposed asymmetric integrated Mach-Zehnder interferometer is shown in Fig. 1.

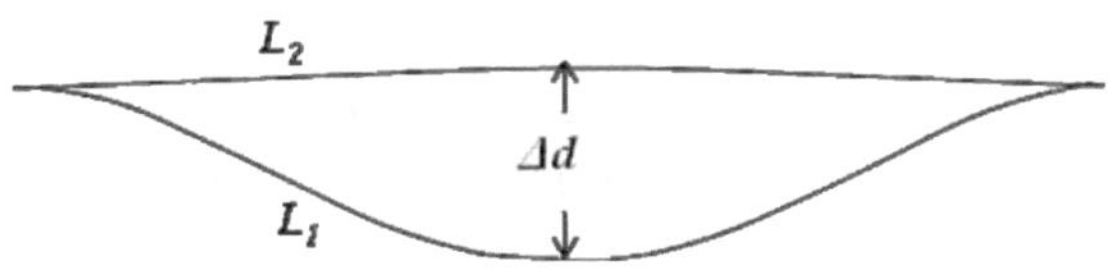

Figure 1. Schematic drawing of the asymmetric Mach-Zehnder interferometer.

The light is coupled into the first straight waveguide and then split by an y-junction into the left and right arms, both shaped to reduce the bend losses; in order to provide two different optical paths, and thus the phase delay between beams, the two branches have been designed with different geometrical length (L_1, L_2), giving rise to an unbalanced configuration, as shown in Fig. 1. Hence, for a fixed input wavelength λ, the phase difference accumulated at the output is

$$\Delta\varphi = \frac{2\pi}{\lambda} n_{eff} (L_1 - L_2) , \tag{1}$$

where n_{eff} is the effective refractive index of the guided mode. The transmitted optical intensity at the output in a MZ interferometer is

$$I \propto 1 + V \cos(\Delta\varphi) , \tag{2}$$

where the phase difference $\Delta\varphi$ is given by Eq. (1), V = ($I_{max} - I_{min}$)/($I_{max} + I_{min}$) $= 2\sqrt{k}/(1+k)$ is the visibility of the interferometer and $k = I_1/I_2$ is the beam intensity ratio between the two arms of the device.

3. Experimental results

Two MZ configurations have been realized varying the distance between the two arms. In the first configuration (MZ1) the separation between the left and right branches is $\Delta d_1 = 310$ μm giving rise to a difference $\Delta L_1 = (L_1 - L_2) = 17.05$ μm, in the second one (MZ2) is $\Delta d_2 = 510$ μm, $\Delta L_2 = 47.30$ μm. In both cases the total length of the devices is 1.5 cm and the required core volume is about 1.5 nl.

In Fig. 2a, b are shown the scanning electron microscope (SEM) pictures of the y-junctions of the MZ1 and MZ2 interferometers. The images illustrate the lower wafer before adding the upper wafer on top of it.

Figure 2. SEM images of the y- junctions of the (**a**) MZ1, (**b**) MZ2 interferometers.

In order to characterize the devices, the light from an unpolarized white lamp was end-coupled into the waveguide core through an optical fiber with a 50 μm core diameter. The transmitted light was collected with an optical fiber (50 μm core) connected to a CCD spectrometer (range 550–750 nm). The open-ended hollow waveguide core was filled with methanol ($n_c = 1.325$) by capillary effect due to the very small dimension of the device. The normalized spectra transmitted from the MZ1 and MZ2 are reported in Fig. 3a, b. From this figure, we recognize the typical spectral behavior of an antiresonant reflecting optical waveguide with broad antiresonant peak centered around the design value (λ = 633nm) and the interference pattern of the Mach-Zehnder interferometer. From the measured transmitted spectrum we have evaluated a free-spectral range of $FSR_1 \approx 16.5$ nm for the MZ1 and of $FSR_2 \approx 5.8$ nm for the MZ2. These values are in good agreement with the expected free spectral range of $FSR_1 \approx 16$ nm and $FSR_2 \approx 6$ nm around 610 nm. The measured visibilities are $V_1 = 0.453$ for the MZ1 and $V_2 = 0.375$ for the MZ2. This MZ interferometer can have interesting application especially in the sensing field. If the liquid core refractive index changes, a variation of the optical output power can be detected at fixed wavelength. The sensing device exploits a large refractive index tunability; in fact, the guiding condition in the hollow core is satisfied for a quite large index range $n_c = 1.32 \div 1.45$, that corresponds to the detection range of the sensing device. It has been estimated that an intensity variation of 10^{-2} corresponds to a refractive index change $\Delta n = 5.4 \times 10^{-3}$ RIU and $\Delta n = 1.6 \times 10^{-5}$ RIU for MZ1 and MZ2 respectively, at $\lambda = 633$ nm and around $n_c = 1.33$.

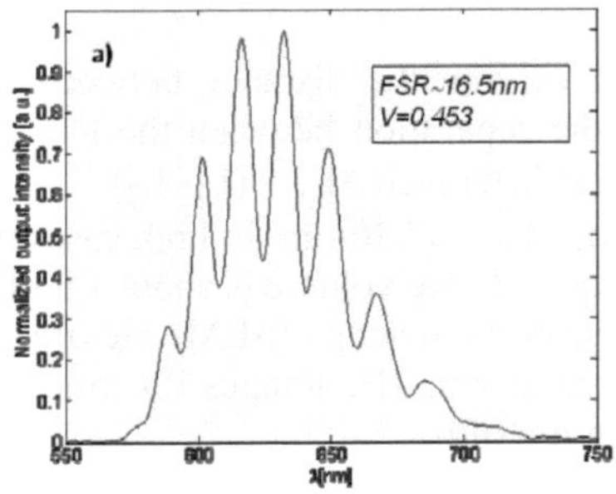

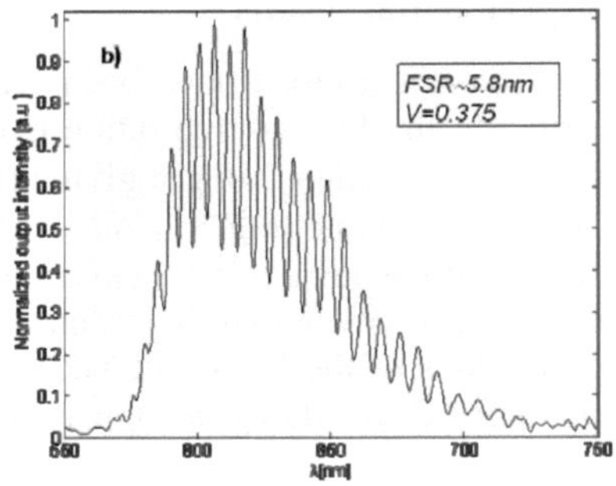

Figure 3. Normalized transmitted intensity from (**a**) MZ1, (**b**) MZ2.

4. Conclusions

A novel integrated optofluidic MZ interferometer has been proposed and experimentally characterized by measuring the transmitted spectrum. The experimental results show that interferometers with a good visibility can be achieved in good agreement with the theoretical results, proving the feasibility of an integrated optofluidic Mach-Zehnder interferometer.

References

1. D. Psaltis, S. R. Quale, and C Yang, "Developing optofluidic technology through the fusion of microfluidics and optics, Nature, [vol. 442], pp. 381–386, 2007.
2. C. Monat, P. Domachuk, and B. J. Eggleton, "Integrated optofluidics: A new river of light", Nature Photonics, [vol. 1], pp. 106–114, 2007.
3. R. Bernini, G. Testa, L. Zeni, and P. M. Sarro, "Integrated optofluidic Mach-Zehnder interferometer based on liquid core waveguide" Appl. Phys. Lett. [vol. 93], pp. 011106–3, 2008.
4. R. G. Heideman and P. V. Lambeck, "Remote opto-chemical sensing with extreme sensitivity: design, fabrication and performance of a pigtailed integrated optical phase-modulated Mach–Zehnder interferometer system", Sens. Actuators B [vol. 61], pp. 100–127, 1999.
5. F. Prieto, B. Sepúlveda, A. Calle, A. Llobera, C. Domínguez, A. Abad, A.Montoya, and L. M. Lechuga, "An integrated optical interferometric nanodevice based on silicon technology for biosensor applications", Nanotechnology [vol. 14], pp. 907–912, 2003.
6. K. Tiefenthaler and W. Lukosz, "Sensitivity of grating couplers as integrated-optical chemical sensors", J. Opt. Soc. Am. B [vol. 6], pp. 209–220, 1989.
7. Paul V Lambeck, "Integrated optical sensors for the chemical domain," Meas. Sci. Technol. [vol. 17], pp. R93–R116, 2006.

INTELLIGENT WIRELESS E-NOSE FOR POWER SAVVY DISTRIBUTED CHEMICAL SENSING

S. DE VITO, E. MASSERA, G. BURRASCA, A. DE GIROLAMO AND G. DI FRANCIA

ENEA Research Center, P.le E. Fermi, 1 80055 Portici (NA), Italy

Abstract. In this work, we present the preliminary results for a wireless electronic nose platform embedding local sensor fusion component for the analysis of gas mixture in the framework of indoor pollution monitoring. This approach allow for significant reduction of power consumption by exploiting sensor censorship algorithms i.e. avoiding the transmission of uninformative contents to data sink. At the same time local situation assessment capabilities will allow the mote to perform adaptive, local reaction strategies e.g. duty cycle modifications, actuation etc. Performance are encouraging both on discrimination and estimation problem, we believe that a significant performance enhancement can be obtained by using a tapped delay neural network architecture.

1. Introduction

Air quality monitoring scenarios such as indoor volatiles monitoring, waste management plants odour monitoring, city air pollution monitoring, are thought to greatly benefit from the use of multiple, distributed intelligent sensing units. Pervasive monitoring can, in fact, overcome the intrinsic fluid-dynamic complexities that hamper the outcome of single point of measure architectures in this scenarios [1]. For efficient indoor air quality monitoring, we propose the use of a group of battery powered, networked sensing nodes each one equipped with a small array of chemical sensors, i.e. a wireless network of electronic nose units cooperating for extracting an olfactive “image” of the sensed environment. During the last 2 years, the availability of commercial motes platforms have suggested their possible use in networked, wireless chemical sensing scenarios (see [2, 3]). Our group is developing a first low power w-nose platform, called TinyNose, based on TeloB motes and equipped with room temperature operating sensors [4]. Recently two mesh networked Tinynose prototypes have been tested for cooperative detection of gases in an ad-hoc test environment [5].

Actually, most of the power consumption of a wireless sensing unit is needed for transmitting/receiving purposes; on the other hand, most of the time, data transmission is not needed (e.g. no pollution at all or no significative pollutant concentration is detected by sensors). In this cases, the capability to locally infer decision on the basis of sensor responses can extend the network operative life

P. Malcovati et al. (eds.), *Sensors and Microsystems: AISEM 2009 Proceedings*, Lecture Notes in Electrical Engineering 54, DOI 10.1007/978-90-481-3606-3_76,

from few weeks to years. Local intelligence can furthermore enhance homogeneity in different motes relying on sensors exhibiting different response characteristics.

2. Experimental

In this work, we present preliminary results of embedding local sensor fusion capabilities on the TinyNose prototype targeting both gas detection and discrimination, trying to express at least a rough gas concentration estimation. Prototypes have been equipped with four room temperature operating sensors base on dispersion of a carbon black filler in different non conductive polymeric matrices [4]. The sensor array of each of the developed w-nose, was exposed while connected to the sensing mote, to two cycles of each four different concentrations of Acetic Acid (225, 450, 900, 1,800 ppm) and Ethanol (200, 500, 1,000, 2,000 ppm) in a controlled atmosphere chamber setup. For this preliminary experiment, steady state response was sampled and the (R-R0/R0) values were used for off-line training of two three-layer back propagation neural networks (BPNNs), one for each Tinynose prototype. BPNNs represent an optimal choice, because of their low footprint as regard as both memory and computational needs. The network architectures were designed and trained to detect and discriminate the presence of Acetic Acid or Ethanol and estimate their concentrations level using instantaneous sensor array responses. An optimal design would have required the use of three networks, one for the discrimination task and the other two for concentration estimation purposes at the cost of a significant increase in the total footprint.

A neural network component equipped with the resulting weights and input normalizing vectors was then developed in NesC language, embedded in the on-board measuring sw architecture and coupled to the normalized instantaneous sensors array readout.

For the discrimination task, thresholds have been preliminary set on the network classification output, so that the unit can locally discriminate the presence of one of the selected analytes. Discrimination output is also shown locally by using two TelosB LEDs. Thanks to this component, the single unit can now decide whether to transmit or not the sampled data to the datasink or eventually implement local reaction strategies such as modify the duty cycle of the other units to respond more quickly to the environmental changes.

The NN equipped single units have been tested in the same controlled chamber setup, sensors have been sampled each 15 s and both sensors responses and local estimations have been transmitted to datasink. There they were recorded for evaluation purposes. In Figs. 1 and 2, the response of the embedded sensor fusion component is compared with the expected output extracting the absorption phase from the entire exposure cycle.

3. Results and Conclusions

Results for both dynamic and steady state response are pictured in Figs. 1a–2b. Performances appear encouraging both on discrimination and estimation problem, however we believe that a significant performance enhancement can be obtained by using a tapped delay neural network architecture. The performance-computational cost trade off of tapped delay architectures is currently under investigation.

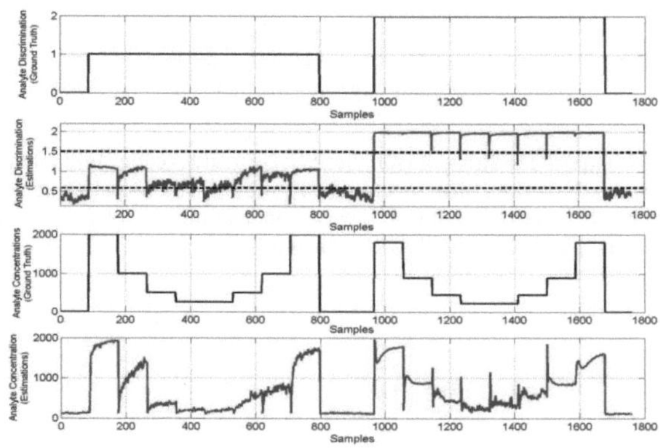

Figure 1. (**a**) Unit 1 is exposed to Ethanol (samples 90–800) and Acetic Acid (samples 967–1,676). The mote 1 is capable to correctly discriminate the analytes. Errors can be spotted when detecting Ethanol at lowest concentrations due to low sensors response. As regards as concentration estimation, the unit seems to be able to discriminate at lest three different concentration levels (low, medium, high) for both analytes.

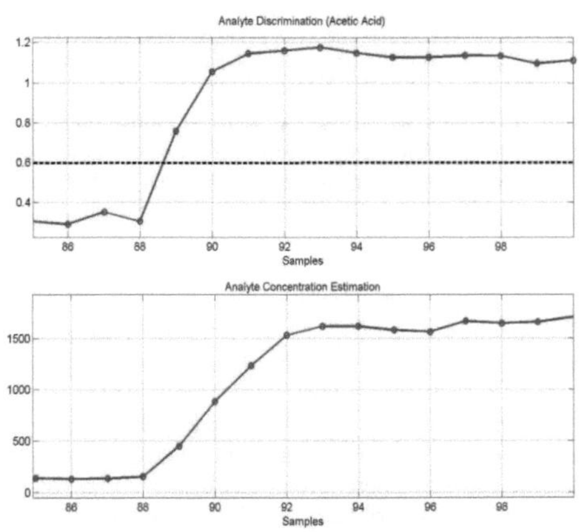

Figure 1. (**b**) Unit 1 NN responses to Ethanol (2,000 ppm) in the initial exposure phase. Results account for a fast response considering that measurement chamber filling transient is estimated in 2 min. Lower threshold for Acetic Acid detection is reached at first sample after analyte onset. However, the use of dynamic features could result in a further dynamic performance boost.

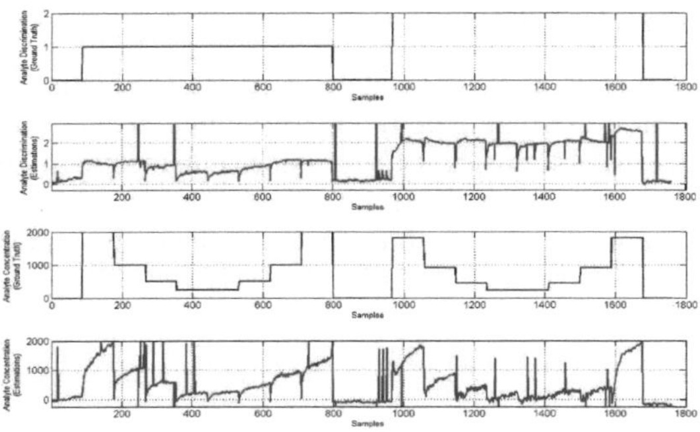

Figure 2. (**a**) Unit 2 is exposed to Ethanol (samples 90–800) and Acetic Acid (samples 967–1,676). Mote 2 exhibit a spiky behaviour due to electrical problems, nonetheless it proves capable to correctly discriminate the analytes even at lowest concentrations. As regards as concentration estimation, unit 2 seems to perform slightly worse than unit 1.

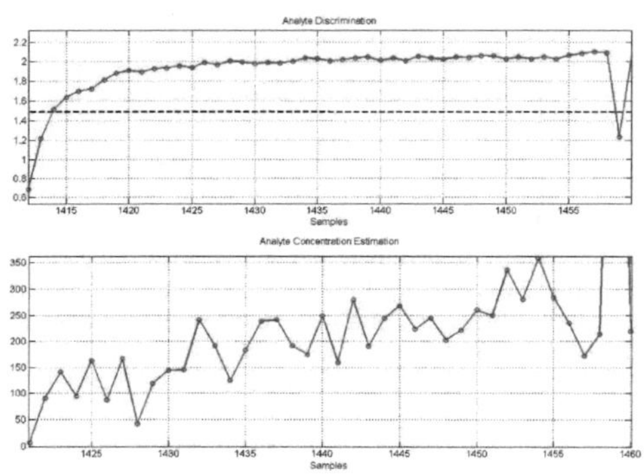

Figure 2. (**b**) Unit 2 NN responses to Acetic Acid (450 ppm). While discrimination dynamic performance are sufficient, concentration estimation reach target values very slowly, partly due to slow sensors responses at lowest concentrations, we expect this problem to be corrected by the use of a tapped delay architecture.

References

1. T. Nakamoto, "Chemical Sensing in Spatial/Temporal Domain", Chem. Rev., 108 (2), 2008, 680–704
2. C.A. Grimes et al., "A sentinel sensor network for hydrogen sensing", Sensors, 3, 2003, 69–82.
3. R. Shepherd et al., "Monitoring chemical plumes in an environmental sensing chamber with a wireless chemical sensor network", Sens. Act. B, 121, 2007, 142–149.
4. S. De Vito et al., "Enabling distributed VOC sensing applications Toward Tinynose, a polymeric wireless e-nose ", Proc. of XII Italian Conf. on Sensors and Microsystems, Naples, Feb. 2007, Eds. World Scientific.
4. S. De Vito et al., "TinyNose: Developing a wireless e-nose platform for distributed air quality monitoring applications", Proc. of IEEE Sensors Conference 2008, in press, available from authors on request.
5. S. De Vito et al., "Gas concentration estimation in ternary mixtures with room temperature operating sensors using tapped delay architectures", Sens. Act. B Chem., 124, 2007, 309–316.

SMART RFID-LABEL FOR MONITORING THE PRESERVATION CONDITIONS OF FOOD

D. CARTASEGNA, A. CITO, F. CONSO, A. DONIDA, M. GRASSI, L. MALVASI, G. RESCIO AND P. MALCOVATI

Department of Electrical Engineering, University of Pavia, Via Ferrata 1, 27100 Pavia, Italy

Abstract. This paper presents an integrated smart label for tracing food information and monitoring its preservation conditions. The system includes humidity, temperature, and light intensity sensors with the respective interface circuits, an A/D converter, and a 13.56-MHz RFID transponder for transmitting and receiving data, as well as for gathering from an external reader the energy for recharging the on-board microbattery and powering the transmitter.

1. Introduction

In the context of attention for food safety, the implementation of microelectronic systems for tracing food products through low-cost miniaturized devices is quite important. These devices shall be able to identify the product, track its path along the complete food distribution chain, and monitor the environmental conditions to which the product is subject over time, in order to verify that the preservation prescriptions are respected [1]. The first step toward the implementation of such tracing systems is the development of a smart RFID label, capable of storing the data concerning the product, monitoring the environmental conditions, and transmit the collected information upon request.

A block diagram of the proposed smart RFID label is shown in Fig. 1. The integrated microsystem includes sensors (temperature, humidity, and light intensity) and very low-power circuits (sensor interface circuits, energy management, and conversion circuit, A/D converter, and wireless transceiver). The most important specifications of the proposed smart label are summarized in Table 1.

2. Sensors and interface circuits

Each of the three considered sensors is connected to its specific interface circuit, which delivers a voltage output suitable for A/D conversion. One single A/D converter is used for all the sensors, which are therefore read-out sequentially, through a multiplexer. When are not used, both the sensors and the A/D converter are switched-off in order to reduce the power consumption. The temperature

P. Malcovati et al. (eds.), *Sensors and Microsystems: AISEM 2009 Proceedings*, Lecture Notes in Electrical Engineering 54, DOI 10.1007/978-90-481-3606-3_77,

sensor (Fig. 2a) exploits the difference between the base-emitter voltages of two bipolar transistors with different emitter areas, to produce a voltage proportional to the absolute temperature T.

The bandgap reference used (Fig. 2b), is based on the same basic principle of the temperature sensor, with the addition of two resistors.

Table 1. Specification of the proposed RFID smart label.

Humidity sensor	2.1.1.
Range, resolution	10% ÷ 90% RH, 1% RH
Maximum current consumption	20 μA
Temperature sensor	2.1.2.
Range, resolution	−40°C ÷ 80°C, 0.2°C
Maximum current consumption	20 μA
Light intensity sensor	2.1.3.
Range, resolution	1,000 W/m^2, 2 W/m^2
	2.1.4.
Maximum current consumption	30 μA
A/D converter	2.1.5.
Input range peak-peak differential	1 V
Resolution	10 bits
Maximum current consumption	20 μA
RFID transceiver	2.1.6.
Operating frequency	13.56 MHz
Standard	ISO 15693
Maximum current consumption	150 μA
Rectifying antenna	2.1.7.
Output current, voltage @ recharging	3.3 V, 100 μA
Output current, voltage @ transmission	1.8 V, 150 μA
Power management	2.1.8.
Regulated output voltage	1.8 V
Output current	200 μA

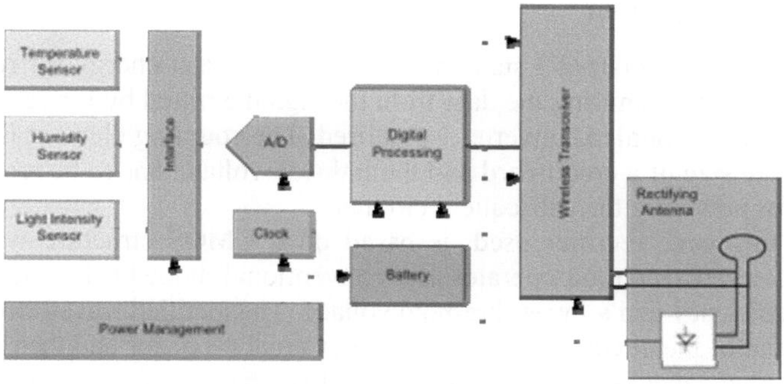

Figure 1. Block diagram of the proposed smart RFID label.

The humidity sensor is realized using an interdigitated capacitor coated with polyimide, whose dielectric permittivity changes with the relative humidity (RH). The capacitance variation is read-out using the switched-capacitor (SC) circuit (Fig. 3).

The light intensity sensor consists of an integrated reverse-biased photodiode, which delivers a current proportional to the incident light intensity.

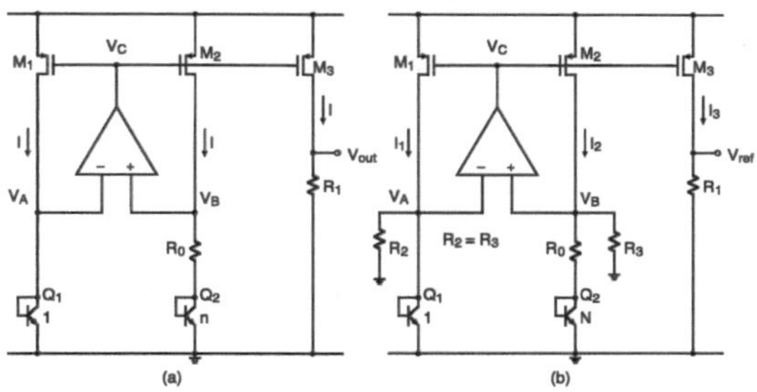

Figure 2. Temperature sensor and band-gap reference.

The developed A/D converter is based on the incremental architecture, implemented with the SC technique.

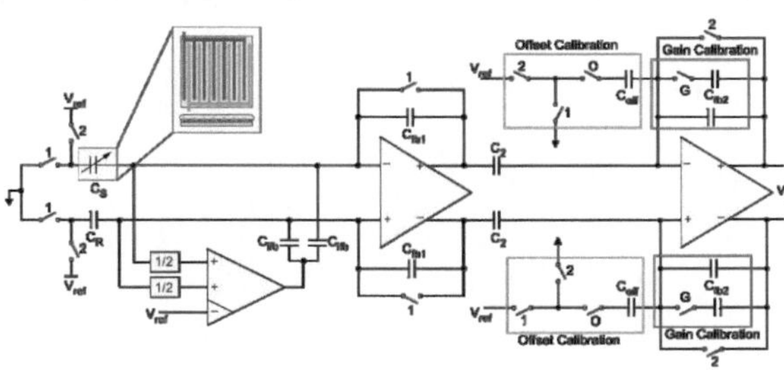

Figure 3. Schematic of the humidity sensor interface circuit.

3. Power management

According to the ISO-15693 standard, in order to extract energy for recharging the battery and transmitting the data from the signal emitted by the reader, a RF-to-DC magnetic-coupled converter is required. The coupling element is a simple coil with the size of a credit-card and the induced voltage has to be rectified and adapted for powering the subsequent circuits.

The full-wave rectifier used, is based on a CMOS structure with cross-coupled gates. This circuit operates as a conventional diode bridge, but achieves a higher efficiency and a lower threshold voltage. The rectified waveform obtained at the output of the circuit is filtered with an external capacitor and then is filtered by an external capacitor and delivered to the two charge pumps, which, using the same structure (Fig. 4), realize both the increasing and regulating functions to the rectified voltage.

In order to supply the different circuits we introduced a Low-Drop-Out Voltage Regulator that maintains the required output voltage at the value of 1.8 V, independently of the battery voltage.

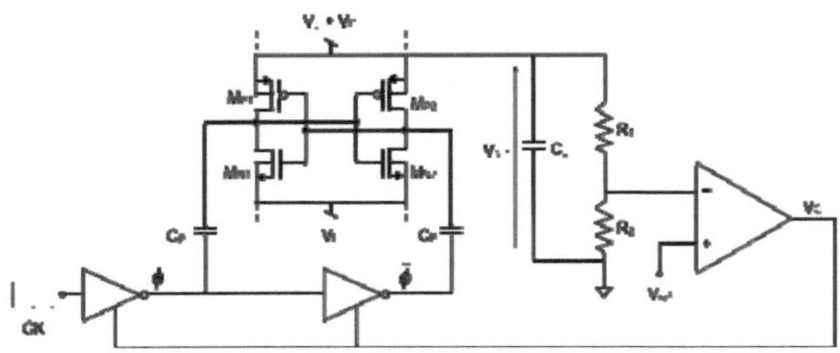

Figure 4. Schematic of the charge pump.

4. Wireless transceiver

The reader and the tag are never transmitting data concurrently, since they are using the same carrier, also used to transfer energy from the reader to the tag. The transmission of data from the tag to the reader is performed by ohmic load

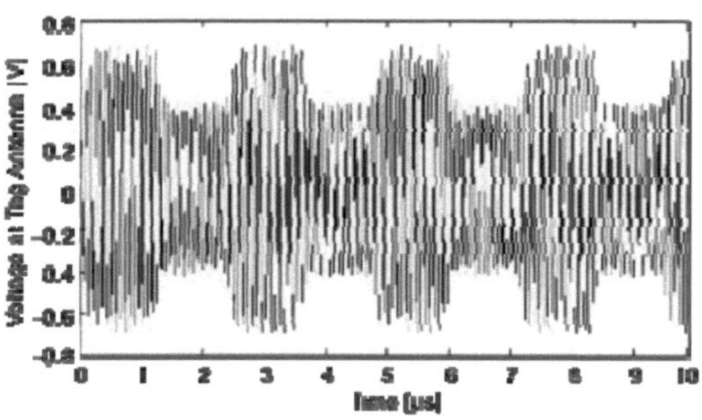

Figure 5. Simulation of the voltage at the tag antenna during data transmission (load modulation) with the minimum magnetic field.

modulation of the tag antenna (Fig. 5). The transmission from the reader to the tag is performed by ASK. The receiver in the tag consists of an envelop detector, followed by a comparator and a finite-state machine.

5. Simulation results and conclusions

The proposed smart RFID label has been designed using a 0.18-μm CMOS technology. The different blocks, as well as their interactions, have been simulated at transistor-level. All of the blocks achieve the required specifications, both in terms of performance and power consumption, thus demonstrating the feasibility of a smart RFID label for monitoring the preservation conditions of food.

References

1. T. Kelepouris, K. Pramatari, and G. Doukidis, "RFID-enabled traceability in the food supply chain," Industrial Management and Data Systems, vol. 107, pp. 183–200, Feb. 2007.

IMPROVING PIANO MUSIC TRANSCRIPTION BY ELMAN DYNAMIC NEURAL NETWORKS

G. COSTANTINI[1], M. TODISCO[2] AND M. CAROTA[3]

[1]*Department of Electronic Engineering, University of Rome "Tor Vergata" via del Politecnico, 1 – 00133 Rome, Italy*

[1]*Institute of acoustics "O. M. Corbino" via del Fosso del Cavaliere, 100 – 00133 Rome, Italy*

[2]*Department of Electronic Engineering, University of Rome "Tor Vergata" via del Politecnico, 1 – 00133 Rome, Italy*

[3]*Department of Electronic Engineering, University of Rome "Tor Vergata" via del Politecnico, 1 – 00133 Rome, Italy*

Abstract. In this paper, we present two methods based on neural networks for the automatic transcription of polyphonic piano music. The input to these methods consists in live piano music acquired by a microphone, while the pitch of all the notes in the corresponding score forms the output. The aim of this work is to compare the accuracy achieved using a feed-forward neural network, such as the MLP (MultiLayer Perceptron), with that supplied by a recurrent neural network, such as the ENN (Elman Neural Network). Signal processing techniques based on the CQT (Constant-Q Transform) are used in order to create a time-frequency representation of the input signals. The processing phases involve non-negative matrix factorization (NMF) for onset detection. Since large scale tests were required, the whole process (synthesis of audio data generated starting from MIDI files, comparison of the results with the original score) has been automated. Test, validation and training sets have been generated with reference to three different musical styles respectively represented by J. S. Bach's inventions, F. Chopin's nocturnes and C. Debussy's preludes.

1. Introduction

The target of this work dealt with the problem of extracting musical content (i.e., a symbolic representation of musical notes) from audio data, particularly with reference to polyphonic piano music. Actually, the harmonic components of notes that simultaneously occur in polyphonic music significantly obfuscate automated transcription. The first algorithms developed by Moorer [1] used comb filters and autocorrelation in order to perform transcription of very restricted duets. The most important work in this research field is the Sonic project [2] developed by Marolt.

In this paper, we propose two supervised classification methods that infer the correct note labels based only on training with labeled examples. These methods

P. Malcovati et al. (eds.), *Sensors and Microsystems: AISEM 2009 Proceedings*, Lecture Notes in Electrical Engineering 54, DOI 10.1007/978-90-481-3606-3_78,

performs polyphonic transcription via a system of MultiLayer Perceptron (MLP) [3] classifiers and a system of Elman Neural Network (ENN) [3] classifiers that have been trained starting from spectral features obtained by means of the well known Constant-Q Transform (CQT) [4]. In order to investigate the influence of musical styles on classification systems, we used three different MIDI files containing works by three composers lived in different ages: J.S. Bach (three part inventions), F. Chopin (nocturnes) and C. Debussy (preludes). The processing phase starts in correspondence to a note onset. The onset detection algorithm, proposed in [5], is based on a suitable binary time-frequency representation of the audio signal followed by Non-Negative Matrix Factorization (NMF) [6].

The paper is organized as follows: in section 2 the generation of audio data set will be described, section 3 illustrates the spectral features; section 4 will be devoted to the description of the note classification sensor interface; finally, the results concerning the transcription of the notes in the score will be shown.

2. Audio data set

In order to investigate the influence of musical styles on classification systems, we used three different MIDI files containing works by three composers lived in different ages: J.S. Bach (three part inventions), F. Chopin (nocturnes) and C. Debussy (preludes).

Table 1. Training-validation audio data set.

Bach's & Chopin's Pieces	
Major Tonality	
Training	Validation
Bach's 3 part inv in Cmajor	Bach's 3 part inv In DM→CM
Chopin's Nocturn in AbM→CM	Chopin's Nocturn in GM→CM

Bach's & Chopin's Pieces	
Minor Tonality	
Training	Validation
Bach's 3 part inv in Aminor	Bach's 3 part inv in Em→Am
Chopin's Nocturn in Bbm→ Am	Chopin's Nocturn in Cm→ Am

Debussy's Pieces	
Training	Validation
Debussy's Prelude n. 1	Debussy's Prelude n. 4

We have trained the neural networks with five different training-validation data sets: two major tonalities and two minor tonalities, taken from Bach and Chopin repertoires, and one taken from Debussy's pieces (Table 1).

3. Spectral features

The processing phase starts in correspondence to a note onset. Firstly, the attack time of the note is discarded (in case of the piano, the longest attack time is equal to about 32 ms). Then, after a Hanning windowing, as single CQT of the following 64 ms of the audio note event is calculated. Figure 1 shows the complete process.

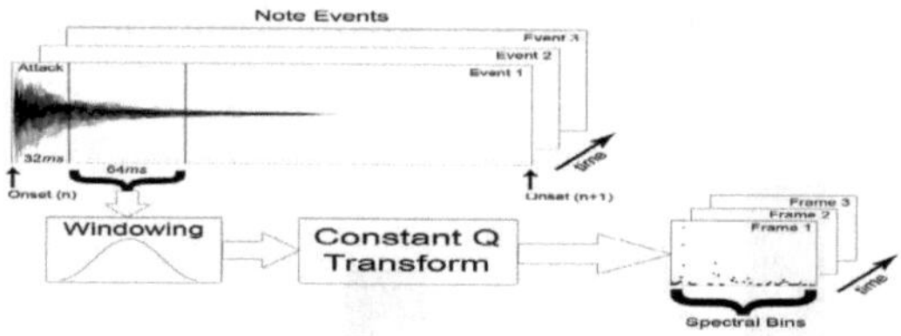

Figure 1. Complete spectral features extraction process.

4. Note classification sensor interface

Our note classification sensor interface is constituted by three components: a microphone, a signal pre-processing based on our onset detection algorithm and extraction of CQT events as explained in previous session and separate one-versus-all (OVA) feed-forward Neural Network and Elman Neural Network binary classifiers were trained on the spectral features of the most important degrees in the major and minor scales. The input to this sensor interface consists in live piano music acquired by a microphone, while the pitch of all the notes in the corresponding score forms the output.

The following figure shows the note classification sensor interface (Fig. 2).

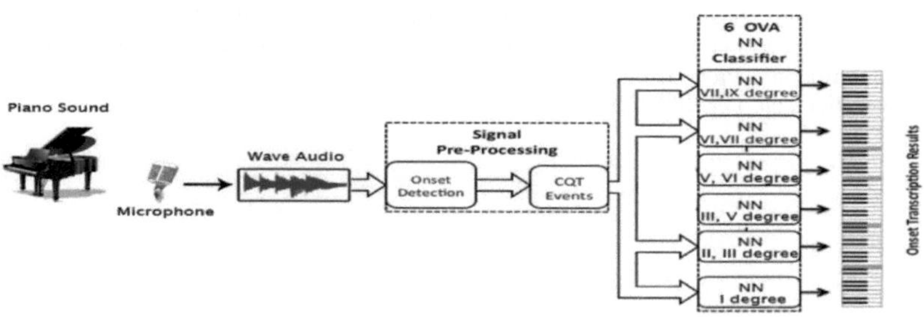

Figure 2. Note classification sensor interface.

5. Conclusion and discussion

We tested the algorithm on all Bach's Inventions in three parts, all Chopin's Nocturnes and all Debussy's Preludes. The average accuracy, in terms of fmeasure, was 83% and 76% found in ENN and MLP classifiers respectively.

Figure 3 shows transcription accuracy difference between ENN *fmeasure* and MLP *fmeasure* decreasing with respect musical styles from Bach to Chopin to Debussy. The success of dynamic neural network on Bach's Inventions and on

Chopin's Nocturnes is due to the composition method, based on classical harmony, adopted by these two authors. On the contrary, in Debussy's compositions, all the passage from a musical note to another is equally likely. In this case, the performances of the dynamic classifier and the static classifier are comparable.

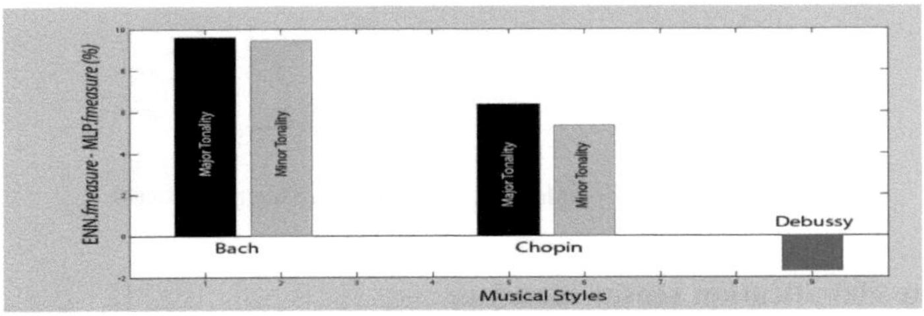

Figure 3. Transcription accuracy difference.

References

1. J. A. Moorer, "On the transcription of musical sound by computer", Computer Music Journal, 1(4): 32–38, 1977
2. M. Marolt, "A connectionist approach to automatic transcription of polyphonic piano music," IEEE Transactions on Multimedia, vol. 6, no. 3, pp. 439–449, 2004.
3. J. Hertz, A. Krogh and R.G. Palmer, Introduction to the theory of neural computation, Addison-Wesley, Reading, MA, 1991.
4. J. C. Brown, "Calculation of a constant Q spectral transform", Journal of the Acoustical Society of America, vol. 89, no. 1, pp. 425–434, 1991.
5. G. Costantini, R. Perfetti and M. Todisco, "Event based transcription system for polyphonic piano music", Signal Process. (2009), doi:10.1016/j.sigpro.2009.03.024
6. P. Smaragdis and J. C. Brown, "Non-negative matrix factorization for polyphonic music transcription", Proc. IEEE Workshop of applications of signal processing to audio and acoustics", pp. 177–180, October 2003.

A MULTISENSOR SYSTEM FOR HIGH RELIABILITY PEOPLE FALL DETECTION IN HOME ENVIRONMENT

M. GRASSI[1], A. LOMBARDI[1], G. RESCIO[1], P. MALCOVATI[1], A. LEONE[2], G. DIRACO[2], C. DISTANTE[2], P. SICILIANO[2], M. MALFATTI[3], L. GONZO[3], V. LIBAL[4], J. HUANG[4] AND G. POTAMIANOS[4]

[1]*Department of Electrical Engineering, University of Pavia, Italy*
[2]*Institute for Microelectronics and Microsystems, CNR-IMM of Lecce, Italy*
[3]*Integrated Optical Sensor Group, FBK-irst, Povo (TN), Italy*
[4]*IBM T. J. Watson Research Center, Yorktown Heights (NY), USA*

Abstract. This paper presents a hardware and software system for reliable fall detection in the home environment, with particular focus on the protection and assistance to older people. The prototype includes three different sensors: a 3D time-of-flight range camera, a wearable MEMS accelerometer and a microphone. These devices are connected with custom interface circuits to a central PC that collects and processes the information with a multi-threading approach. For each of the three sensors, an optimized algorithm for fall-detection has been developed and benchmarked on a collected multimodal database.

1. Introduction

In the last few years, research has been increasingly focusing on studying systems for tracking and recognizing people and their behavior, appearance and activities. In particular, home assistance and protection is becoming a main topic in indoor sensor systems. In this work, emerging sensing technologies are exploited in order to detect possible falls of older people in their own home environment, delivering an alarm flag to emergency operators or relatives if necessary, with the target of pursuing the maximum possible reliability and efficiency, thus minimizing false alarms rate and obviously avoiding undetected falls. For this reason, various projects and consortia have been formed and funded under the coordination of the European Community, including the Netcarity project [1], within which the work described in this paper has been carried out.

2. System architecture

The developed architecture, whose block diagram is shown in Fig. 1, includes three different sensors: a 3D time-of-flight range camera, a wearable wireless MEMS accelerometer and a commercial microphone. These devices are connected with ad-hoc interface circuits to a central host embedded PC (e_PC), as shown in Fig. 2,

P. Malcovati et al. (eds.), *Sensors and Microsystems: AISEM 2009 Proceedings*, Lecture Notes in Electrical Engineering 54, DOI 10.1007/978-90-481-3606-3_79,

that receives and processes the information employing a multi threading approach. The information provided by the 3D camera allows us to describe the environment quantitatively, in terms of appearance and depth images at QCIF resolution. In addition, a wearable three-axis accelerometer, by means of a ZigBee wireless module, delivers acceleration components to the e_PC, which recognizes specific patterns related to falls. Finally, acoustic scene analysis is performed using a commercial Shure microphone plugged into the 16-bit audio card of the e_PC.

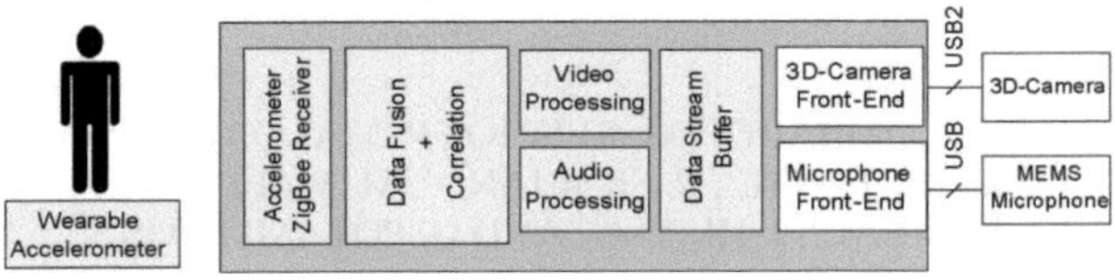

Figure 1. Block diagram of the developed system.

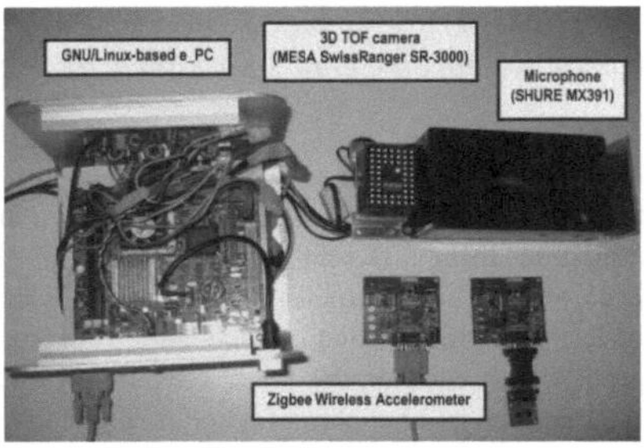

Figure 2. Photograph of the developed system.

In order to score maximum reliability and efficiency, the data delivered by each sensor is first processed by separate algorithms, and eventually, data fusion over a given time window will be performed subsequently by the e_PC. Algorithms for fall detection, optimized for each of the three kind of sensors, have been developed, using real data gathered in a data collection campaign following guidelines provided in [2], with 13 different actors, that produced more than 450 events including approximately 210 falls.

3. Fall detection algorithms

For processing 3D-vision information, the developed system employs a people detection and tracking algorithm, to provide quantitative information about the person in the scene and to determine the distance of its center-of-mass from the floor. Moreover, an ad-hoc software allows us to detect a fall event by thresholding the center of mass distance to the floor (0.4 m constitute a good choice in our experimental setup to avoid false alarms). Figure 3 shows the typical pattern of the centroid distance when fall events occur (in circles), while Table 1 reports the efficiency and reliability for developed 3D vision algorithms.

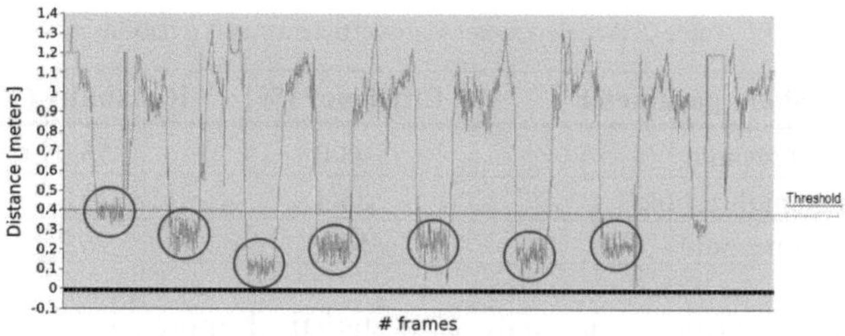

Figure 3. Pattern of the centroid height in 3D vision.

Table 1. 3D-vision alarm efficiency and reliability.

Threshold for centroid height	Efficiency (%)	Reliability (%)
0.3 m	51.1	99.2
0.4 m	80.0	97.3
0.5 m	81.3	89.3

For processing the acceleration data, we developed an algorithm, based on the information from each of the accelerometer axes of sensitivity. We verified the algorithm over more than 20 experimental acquisition sessions, containing about 10 fall events each. An example of the output waveform of the accelerometer is reported in Fig. 4 (X axis). It is possible to notice that a fall event is characterized by an acceleration peak, followed by some relatively constant values, and again followed in the experimental sessions by a peak due to the rise of the actor. The output bit-stream of each of the axes of the accelerometer is computed separately.

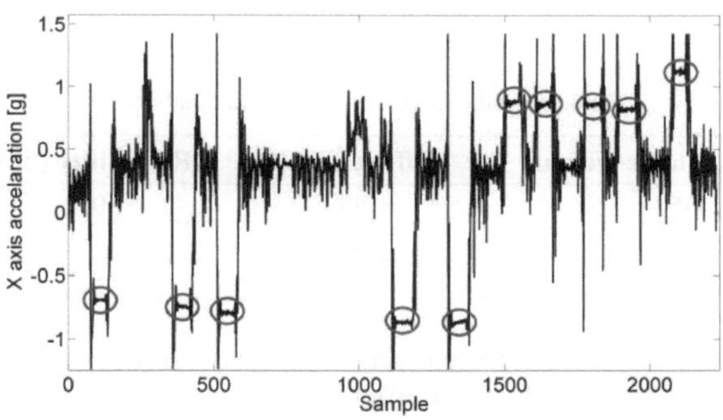

Figure 4. X-axis acceleration data during a session.

A digital block compares the absolute difference values between every consecutive four samples to a threshold. When the sum is higher than the given threshold, it means that the person has fallen. Table 2 reports tested efficiency and reliability for developed wearable wireless accelerometer device at different threshold levels.

Table 2. Accelerometer alarm efficiency and reliability.

Threshold parameter	Efficiency (%)	Reliability (%)
Low threshold	98.0	56.0
Medium threshold	88.4	79.3
High threshold	50.1	96.2

Finally, for acoustic scene analysis, a statistical approach based on hidden Markov models (HMMs) has been employed. These HMMs are analogous to whole-word speech models, with a separate model used for each acoustic event class of interest. Following training, recognition of the acoustic events employs an HMM network, similar in fashion to speech recognition decoding. An acoustic event is considered to be correctly detected, when its temporal center lies within the reference timestamps or vice versa. In Fig. 5 an example of an audio waveform including a non-fall and a fall event is reported, while Table 3 summarizes the system performance under different decoding parameters, where the acoustic weight specifies how much the acoustic scores contribute to the final score.

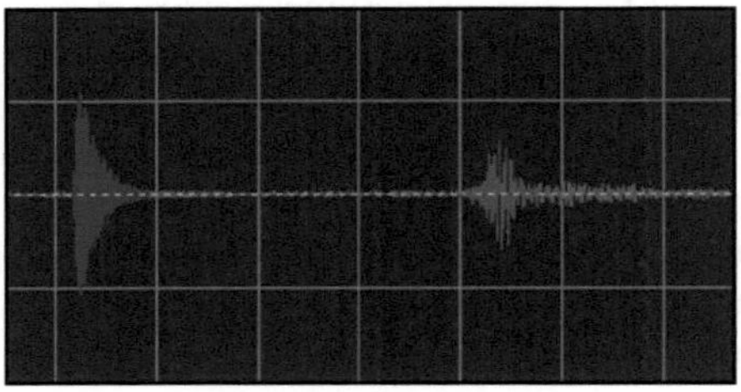

Figure 5. Example of acoustic signal.

Table 3. Audio alarm efficiency and reliability.

Acoustic weight	Efficiency (%)	Reliability (%)
0.005	59.6	59.1
0.025	80.9	42.2
0.25	83.0	35.2

Data fusion of the three different sensors is being developed with the aim of getting near to 100% reliability and efficiency.

References

1. http://www.netcarity.org
2. N. Noury, A. Fleury, P. Rumeau, A. K. Bourke, G. Ó. Laighin, V. Rialle, J. E. Lundy, "Fall detection – Principles and methods", *Proceedings of the 29th International Conference of the EMBS*, pp. 1663–1666, IEEE, 2007.

APPLICATIONS

WESNEP: A WIRELESS ENVIRONMENTAL SENSOR NETWORK FOR PERMAFROST STUDIES

A. CRISTIANI[1], G.M. BERTOLOTTI[1], G. BELTRAMI[1], R. GANDOLFI[1], R. LOMBARDI[1], R. SEPPI[2] AND F. ZUCCA[2]

[1]*Dipartimento di Informatica e Sistemistica, Università degli Studi di Pavia*
Via Ferrata 1 – 27100 Pavia
[2]*Dipartimento di Scienze della Terra, Università degli Studi di Pavia*
Via Ferrata 1 – 27100 Pavia

Abstract. The aim of this paper is to give an overview of WESNEP, an environmental wireless sensor network which is currently being developed at the University of Pavia in order to study alpine permafrost. After a brief introduction on environmental sensor networks and the definition of permafrost, the motivation of WESNEP project and the architecture of the network are described, finally the main benefits expected from the project are presented.

1. Introduction

1.1. *Wireless sensor networks*

According to [1] an Environmental Sensor Network (ESN) "comprises an array of sensor nodes and a communications system which allows their data to reach a server (...). The sensor nodes gather data autonomously and a data network is usually used to pass data to one or more base stations, which forward it to a Sensor Network Server (SNS)".

The sensor nodes can be fixed or mobile, and network's range depends on the environment being sensed: in Large Scale Single Function Networks single purpose nodes are used to cover a wide geographical area, while in Localised Multifunction Sensor Networks a small area is monitored in more detail, and wireless systems are often used.

A Wireless Sensor Network (WSN) is generally composed by distributed nodes with attached sensors that send data to each other and to a base station through a local radio link (e.g. UHF). The information is then transferred by the base station to a remote station, by means of mobile communication (e.g. UMTS, GPRS), or internet, or other data transfer systems [2].

The first WSNs were born for military applications (mainly for battlefield surveillance), but are now used in many civilian application areas, such as environment and habitat monitoring, healthcare applications, home automation, and traffic control [3].

P. Malcovati et al. (eds.), *Sensors and Microsystems: AISEM 2009 Proceedings*, Lecture Notes in Electrical Engineering 54, DOI 10.1007/978-90-481-3606-3_80,

1.2. *Permafrost*

Permafrost (i.e. ground that remain at or below 0°C for at least 2 consecutive years) is widespread in alpine environments at altitudes above 2,400–2,600 m a.s.l. (Fig. 1). Changes in the ground thermal regime are very likely occurring on permafrost areas due to climate warming, as evidenced by recent large rock falls, which occurred in many alpine areas at altitudes up to 3,500 m. These rock falls were attributed to the permafrost degradation and the current warming of permafrost has recently been confirmed by the thermal profiles of several alpine boreholes. The presence and thermal state of permafrost is indirectly revealed by numerous parameters. Among these, the ground temperature regime recorded at the surface and at different depths is of particular interest.

Figure 1. A typical alpine permafrost environment.

2. The WESNEP project

2.1. *Motivation*

Up to now at Earth Science Department, University of Pavia researches on alpine permafrost have been carried out using three different types of data logger (DL): UTL1 (Universal Temperature Logger, extensively presented in [4]), HOBO Temp (by Onset), Tinytag Plus 2 (by Gemini Data Loggers). All these DLs require a manual download of acquired data, through serial or USB connections to a notebook, with evident troubles especially when monitored areas are rugged and difficult to reach. Furthermore, the increasing number of areas to be monitored has resulted in an increasing time spent for collecting data. For these reasons, the implementation of a WSN (Wireless Sensor Network) for measuring the ground temperature in real time has started. WESNEP (Wireless Environmental Sensor Network for Permafrost studies) could open new opportunities in improving our full understanding of the processes connecting ground thermal regime and the environmental variables. In particular, it could be possible to perform real-time measurements of heat fluxes at the surface and at different depths in the ground, relating their evolution with the variations of the external environmental parameters, such as air temperature, precipitations, snow cover. Furthermore, the consequences on the ground heat transfer processes of particular events like snowpack melting, air circulation, and water infiltration during precipitations could be accurately investigated.

2.2. *Architecture*

WESNEP is based on a group of sensor nodes (SNs) that are positioned along the area to be monitored. Each SN is composed by a microcontroller (ADUC7027, by Analog Devices), an RF module (Tiny-Pro 868 MHz, by One RF Technology) a temperature probe (a modified BFI Optilas PTC probe), and a battery pack (Fig. 2, left side). The RF modules support a mesh network protocol (Mesh Lite protocol, by One RF Technology). Within this type of network, all devices can communicate with each other using a robust address deployment protocol and a flexible hierarchy.

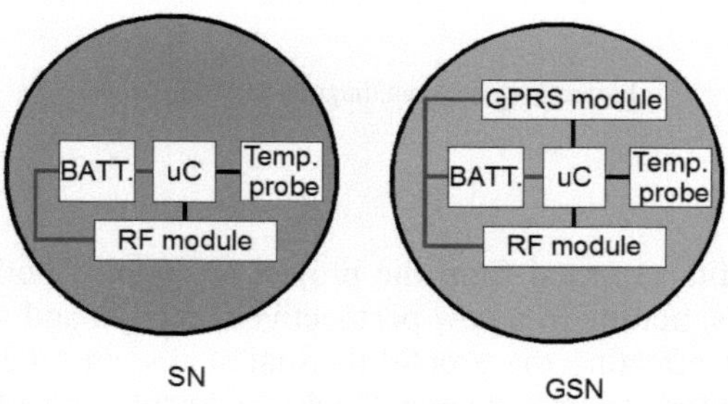

Figure 2. Block diagram of a Sensor Node (SN) and a Gateway Sensor Node (GSN) in WESPNET.

The network topology can be set either manually or using a fully automatic engine, with auto-associate and auto-repair functionalities. Moreover, the Mesh Lite protocol supports a low power mode, giving a good life length to battery powered applications. Among net's nodes, there is one which acts as a gateway, collecting data from the other nodes and sending them to a remote station through a GPRS connection. Other than the RF module, in fact, the Gateway Sensor Node (GSN) is equipped with a Telit GM862-GPS module (Fig. 2, right side). The GSN must be placed in a location which is covered by GPRS cells (Fig. 3). In areas where GPRS coverage is not available, a different type of GSN will be used, a Satellite Gateway Sensor Node (SGSN), i.e. a SN equipped with a satellite module.

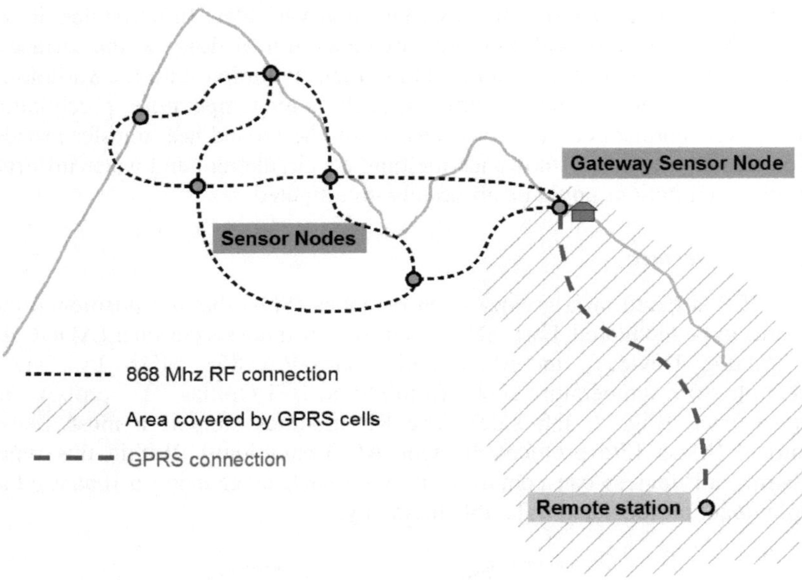

Figure 3. Functional diagram of WESPNET.

3. Conclusions

The major benefits expected from the project are related both to an effective temporal analysis, putting in a new perspective detection and monitoring of the complex process effecting the ground thermal status, and the added-value for risk and natural resource management thanks to potential coupling of WESNEP with other measurements systems (e.g. satellite sensors, field measurements).

References

1. J.K. Hart, K. Martinez, "Environmental sensor networks: a revolution in the earth system science?", Earth-Science Reviews, [vol. 78], pp. 177–191, 2006.
2. A. Hasler, I. Talzi, J. Beutel, C. Tschudin, S. Gruber, "Wireless sensor networks in permafrost research: concept, requirements, implementation, and challenges", Proc. of 9th International Conference on Permafrost, pp. 669–674, 2008.
3. S. Hadim, N. Mohamed, "Middleware: middleware challenges and approaches for wireless sensor networks", IEEE Distributed Systems Online, [vol. 7], no. 3, pp. 1–23, March, 2006.
4. M. Hoelzle, M. Wegmann, B. Krummenacher, "Miniature temperature data loggers for mapping and monitoring of permafrost in high mountain areas: first experience from the Swiss Alps", Permafrost Periglac. Process, [vol. 10], pp. 113–124, 1999.

A MULTI-PURPOSE WIRELESS SENSOR NETWORK BASED ON ZIGBEE TECHNOLOGY

G.M. BERTOLOTTI, G. BELTRAMI, A. CRISTIANI, R. GANDOLFI AND R. LOMBARDI

Dipartimento di Informatica e Sistemistica, Università degli Studi di Pavia
Via Ferrata 1 – 27100 Pavia

Abstract. Among a wide offer of wireless technologies, ZigBee is one of the most actractive for connecting low-volume devices, such as sensors. This paper aims at presenting a project which is being carried on at Microcomputers and Biomedical Devices laboratory (Department of Computer Science, Faculty of Engineering, University of Pavia, Italy). The goal of the project is the development of a prototype of ZigBee sensors network for temperature monitoring. The idea is to evaluate the capability of ZigBee technology in order to build wireless sensors networks for environmental monitoring. The use of a single chip which integrates a microcontroller and a ZigBee transceiver gives the possibility of developing wireless devices with small dimensions, low power consumption, and a good computing capability.

1. Introduction

Thanks to the improving of wireless technology, an increasing number of organizations are using wireless sensors for a wide range of applications, such as home monitoring, environmental quality monitoring, healthcare, building, industrial and agricultural systems. The success of a wireless sensors network strongly depends on the wireless connectivity infrastructure, which must be robust, energy efficient and economical. For this reason, users are increasingly working with ZigBee technology. While other wireless technologies, such as Bluetooth, are designed to connect high-volume, packet-based devices (laptops, phones, and major peripherals), using more energy and being quite expensive, ZigBee connects low-volume devices, such as sensors, and works with simpler, less expensive, easy-to-use technology [1].

2. Hardware set-up

This paper deals with the development of a wireless sensor based on the ZigBee technology. The device consists in two parts: a mainboard which hosts a Jennic JN5139M00 (or JN5139M02) ZigBee module, and a daughterboard which hosts the transducer and its conditioning circuit. The boards can be connected through

P. Malcovati et al. (eds.), *Sensors and Microsystems: AISEM 2009 Proceedings*, Lecture Notes in Electrical Engineering 54, DOI 10.1007/978-90-481-3606-3_81,

12-pins Molex PCB connectors (the socket being placed on the mainboard, the header on the daughterboard).

The JN5139Mxx modules are based on a JN5139 wireless microcontroller, which is a low power, low cost wireless microcontroller suitable for IEEE802.15.4 and ZigBee applications. The main characteristics and features of JN5139 wireless microcontroller are summarized in Table 1.

The JN5139M00 has a ceramic antenna on board and it is suitable for applications requiring radio connections of 1 km max; otherwise, the JN5139M02 is equipped with a power amplifier and a connector for an external antenna, providing a range of 4 km max [2].

Table 1. Main characteristics and features of JN5139 wireless microcontroller.

JN5139 wireless microcontroller
– 32-bit RISC processor
– 2.4 GHz IEEE802.15.4 integrated transceiver
– 192 kB of ROM
– 96 kB of RAM
– 4-inputs 12-bit ADC
– 2-inputs 11-bit DACs
– 2 comparators
– 2 application timer/counters
– 3 system timers
– 2 UARTs (Universal asynchronous receiver-transmitters)
– 1 SPI (Serial peripheral interface) port
– 2-wire serial interface
– 21 GPIO (General purpose input/output) pins

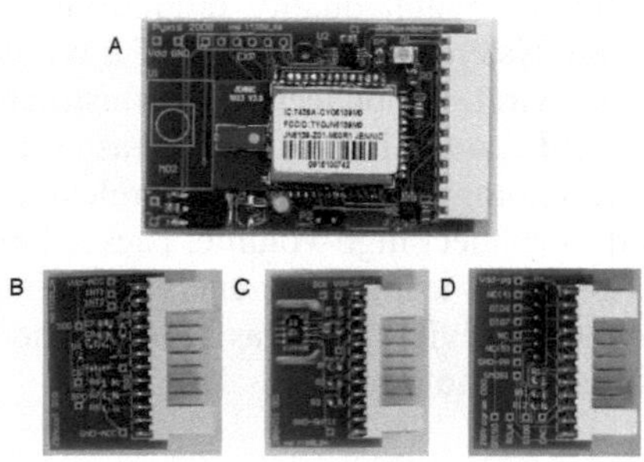

Figure 1. Wireless sensor node. Mainboard (**a**), and daugterboards: temperature and humidity sensor board (**b**), accelerometer board (**c**), programming board (**d**).

At this time, three daughterboards have been developed: one hosts a temperature and relative humidity digital sensor (SHT11, by Sensirion), the second one is provided with a digital 3-axis accelerometer (LIS302DL, by ST Microelectronics), the last is equipped with all the circuits needed to configure the wireless sensor with a Personal Computer and it actually acts as a "programming board" (Fig. 1).

Furthermore, an additional board hosting a JN5139Mxx module and providing a USB connection has been developed, resulting in a sort of ZigBee to USB adapter (Fig. 2). This can be use to allow a notebook or a Personal Computer being a node of the ZigBee sensors network. Thus, all the data acquired by the sensors can be collected and elaborated by the PC.

Figure 2. ZigBee – USB adapter.

3. Network prototype set-up

To test reliability and robustness of our devices, we are planning to set up a prototype of a wireless temperature-humidity sensors network on our Department, placing some nodes inside offices and laboratories, and a couple of nodes outside the building.

Figure 3 shows a map of Floor D – Department of Computer Science, Faculty of Engineering, University of Pavia – and the proposed topology of the network. In this example, there are eight nodes: the coordinator of the ZigBee network (the computer provided with the ZigBee-USB adapter, the latter indicated by a red circle); four routers, represented by the blue spots, and three end devices (the green circles).

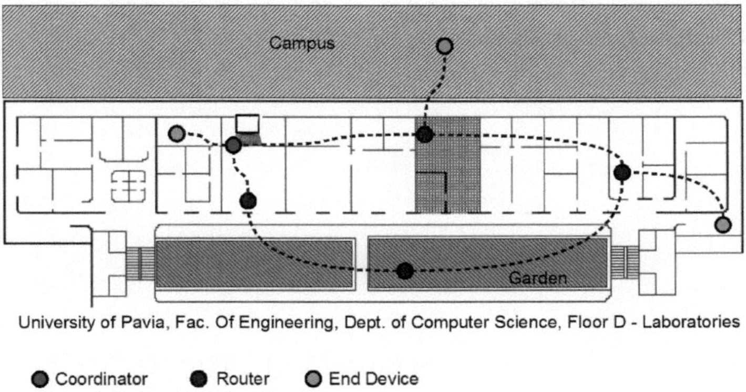

Figure 3. Scheme of a ZigBee temperature-humidity sensor network prototype.

The ZigBee coordinator is the most capable device, it forms the root of the network tree and might bridge to other networks. In general, there is exactly one

ZigBee coordinator in each network since it is the device that started the network originally. The ZigBee Router can pass data from other devices. The ZigBee End Device contains just enough functionality to talk to the parent node (either the coordinator or a router) and it cannot relay data from other devices.

This relationship allows the node to be asleep a significant amount of the time thereby giving long battery life [3].

A PC-based LabVIEW application has been developed, to provide the user the capability of configuring the net, and collecting and storing data coming from all sensor nodes.

4. Conclusions

Tests on a small network prototype are currently being conducted at Microcomputers and Biomedical Devices laboratory (Department of Computer Science, University of Pavia). The aim of these tests is to evaluate the performances of the system, with special attention to power consumption.

Next step will be the deployment of the network, as previously described, and a data acquisition over some months. We are also planning the deployment of a wireless accelerometers network to monitor vibrations of a building.

Acknowledgments The work has been supported by Pyxis srl, an academic spin-off company of the University of Pavia.

References

1. D. Geer, "User make a beeline for ZigBee sensor technology", Computer, [vol. 33], pp. 16–19, December, 2005.
2. "JN5139-001 and JN5139-Z01, IEEE802.15.4 and ZigBee Wireless Microcontrollers", [Online document], Available at http://www.jennic.com/files/support_files/JN-DS-JN5139-1v6.pdf.
3. "ZigBee Specification", [Online document], Available at http://www.zigbee.org/en/spec_download/zigbee_downloads.asp.

A WIRELESS SENSORS SYSTEM FOR SPORT STUDIES

G. M. BERTOLOTTI, G. BELTRAMI, A. CRISTIANI, R. GANDOLFI, R. LOMBARDI

Dipartimento di Informatica e Sistemistica, Università degli Studi di Pavia
Via Ferrata 1 – 27100 Pavia

Abstract. A system based on wireless sensors for biomechanical and sport applications is presented. It is currently used to study the interaction between athletes and sport surfaces, such as soccer turves and athletic tracks, with special focus on ankle vibrations. Thanks to a Bluetooth connection, the wireless sensors can communicate with a PC, where data are visualized and elaborated in real time. Thus, the subject under test can move freely, since there are no cable constraints.

1. Introduction

The Laboratory of Microcomputers and Biomedical Devices (Department of Computer Science, University of Pavia) is currently involved in various projects regarding biomechanical studies of interaction between athletes and sport surfaces, such as soccer turves and athletic tracks. The main goal is to achieve information about athlete's lower limbs accelerations and vibrations during sport tasks performed on different kinds of surfaces.

To do such type of measurements, two wireless sensors and a software application have been developed.

2. Hardware set-up

Each wireless sensor is composed by a ADUC842 microcontroller by Analog Devices, a Parani ESD110 Bluetooth module, a ADXL321 2-axis accelerometer by Analog Devices, a force sense resistor (FSR) by Interlink Electronics, and a pack of 3 AA 1.5 V batteries (Figs. 1 and 2).

The AduC842 integrates a self-calibrating multichannel ADC, a dual DAC, and a 20 MHz 8-bit MCU on a single chip. ADuC842 uses a 32 kHz crystal with an on-chip PLL generating a programmable core clock up to 16.78 MHz. The microcontroller is an optimized 8,052 core offering up to 20 MIPS peak performance. Three different memory options are available offering up to 62 kBytes of non-volatile Flash/EE program memory. Four kBytes of non-volatile Flash/EE data memory, 256 bytes RAM, and 2 kBytes of extended RAM are also integrated on-chip [1].

P. Malcovati et al. (eds.), *Sensors and Microsystems: AISEM 2009 Proceedings*,
Lecture Notes in Electrical Engineering 54, DOI 10.1007/978-90-481-3606-3_82,

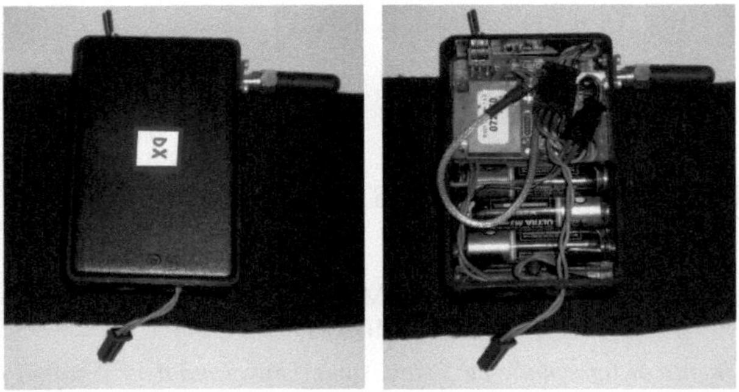

Figure 1. The wireless sensor: external view (*left*); inside view (*right*).

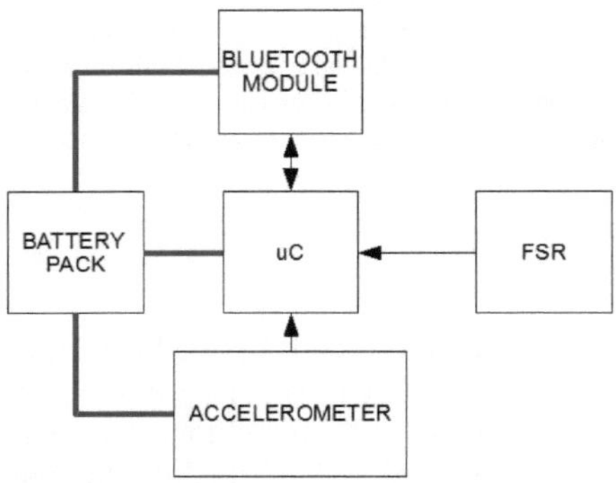

Figure 2. Wireless sensor's functional block diagram.

The ADXL321 is a small and thin, low power, dual-axis accelerometer with signal conditioned voltage outputs. It measures acceleration with a full-scale range of ±18 g. It can measure both dynamic acceleration (vibration) and static acceleration (gravity) [2]. For the presented project, only the vertical acceleration (i.e. along the tibia) has been considered.

The Force Sensing Resistors (FSR) is a polymer thick film (PTF) device which exhibits a decrease in resistance with an increase in the force applied to the active surface. FSRs are not a load cell or strain gauge, though they have similar properties [3]. For our purposes, the force sensors are used only as a switch, in order to detect foot-ground impacts.

The wireless sensors can be positioned around the ankles of a subject by means of elastic bands, while the FSR is inserted inside the shoes of the subject (a special insole has been made for this purpose).

Once turned on, the wireless sensors can be connected to a PC through a Bluetooth connection.

3. Software set-up

A LabVIEW software application has been developed to communicate with them. The user can manage the acquisition, visualizing signals on graphs, saving them in a text file, and doing off-line elaborations once the acquisition is finished. The software graphical user interface is very simple, consisting in large graphs, buttons and indicators that are strictly needed to send commands to the wireless sensor, and to see in real time the acquired signals and the battery status of the devices (Fig. 3).

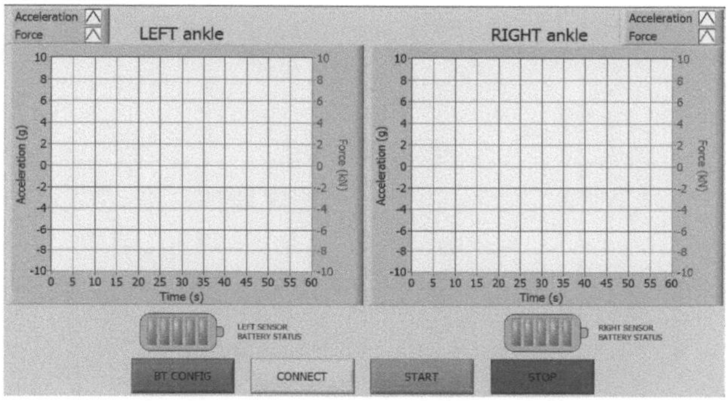

Figure 3. Graphical user interface of wireless sensors software.

4. Conclusions

As previously said, the system (hardware and software) has been successfully used for several studies in the field of sport biomechanics. In particular, it has been utilised to achieve information about ankles' vibrations of soccer players during standardized tasks performed both on natural and artificial pitches. The research has been supported by the Italian Soccer Federation, in order to understand which are the differences between different kind of surfaces. Also studies involving sport tracks and snowboards have been recently carried out (Fig. 4).

Figure 4. Three examples of applications involving the wireless sensors.

References

1. “Analog Devices JMicroConverter® 12-Bit ADCs and DACs with Embedded High Speed 62-kB Flash MCU ADuC841/ADuC842/ADuC843”, [Online document], Available at http://www.analog.com/static/imported-files/data_sheets/ADUC841_842_843.pdf
2. “Analog Devices Small and Thin ±18 g Accelerometer ADXL321”, [Online document], Available at http://www.analog.com/static/imported-files/data_sheets/ADXL321.pdf
3. “Force Sensing Resistor (FSR)”, [Online document], Available at http://www.interlinkelectronics.com/force_sensors/technologies/fsr.html

A HIGH-VOLTAGE DRIVER FOR A SCANNING MICROMIRROR

E. VOLPI[1], L. FANUCCI[1] AND F. D'ASCOLI[2]

[1]*Department of Information Engineer, University of Pisa, via Caruso, 16 I-56122, Pisa*
[2]*SensorDynamics AG, via Giuntini, 63, I-56023, Navacchio (Pisa), Italy*

Abstract. Modern micromachining technologies have made it possible to realize cost-effective and high-reliability Micro-Opto-Electro-Mechanical Systems (MOEMS). Among these devices, double axis micromirrors are widely used in optical telecommunication systems and they are becoming more and more popular for many applications towards the realization of high resolution projection displays. This paper presents a high-voltage driver designed for the actuation of a double axis scanning micromirror in a 0.18 μm Bipolar-CMOS-DMOS (BCD) technology.

1. Introduction

A micro mechanical scanning micromirror can be defined as an actuator designed for the continuous deflection of light and characterized by large scan angle and high scanning frequency [1]. There are different kinds of micromirrors depending on the physical principle exploited for the deflection of the micromirror itself: piezoelectrically, thermically, magnetically and electrostatically micromirrors. Electrostatically actuated micromirrors have several advantages compared to magnetically or piezoelectrically actuated micromirrors: they are much less sensible to the ambient temperature and they can be easily fabricated using micromachining techniques compatible with CMOS technologies. Moreover electrostatic actuators produce a high efficient electrostatic force in microscale dimensions allowing the use of lower driving voltages.

The electrostatic scanning micromirror used as reference for the design of the driver has been developed by FhG/ISiT (Fig. 1) [2]. It is a double axis scanning micromirror. In order to achieve high deflection angles with a lower actuation voltage, the micromirror must be actuated at its resonance frequency. Depending on the prototype, the "slow" axis frequency ranges from 300 Hz to 1 kHz, while the "fast" axis frequency ranges from 15 to 30 kHz. In a video projection system the two axes can be used to scan all the columns and all the rows of the screen, respectively. Both axes are actuated by electrostatic vertical comb drives. Each vertical comb drive consists of a set of moving mechanical electrodes and a set of rigid electrodes. Applying a voltage between the fixed fingers and the movable fingers, the movable fingers rotate around the torsional axis until the Electrostatic

P. Malcovati et al. (eds.), *Sensors and Microsystems: AISEM 2009 Proceedings*,
Lecture Notes in Electrical Engineering 54, DOI 10.1007/978-90-481-3606-3_83,

Torque (T_e) and the Mechanical restoring Torque (T_m) of the springs are equal. T_e and T_m can be expressed as:

$$T_e = \frac{1}{2} \cdot N_f \cdot V^2 \cdot \frac{dC}{d\theta} \qquad\qquad T_m = K \cdot \theta \tag{1}$$

where K is the torsional spring constant, θ is the deflection angle, V is the voltage applied to the fixed fingers, N_f is the number of the fingers of the comb drive and C is the capacitance between a fixed finger and a movable finger. If two sine waves of same amplitude and 180° out of phase (+/−v) summed to a constant voltage V_{cmfb} are applied to the fixed fingers of the mirror the total electrostatic torque T_{TOT} is proportional to:

$$T_{TOT}=[(V_{cmfb}+v)^2-(V_{cmfb}+v)^2] \approx 4 \cdot V_{cmfb} \cdot v \tag{2}$$

Equation (2) shows that the torque and the deflection angle have a linear relationship with the amplitude of the sinusoidal voltage v with a coefficient $4 \cdot V_{cmfb}$.

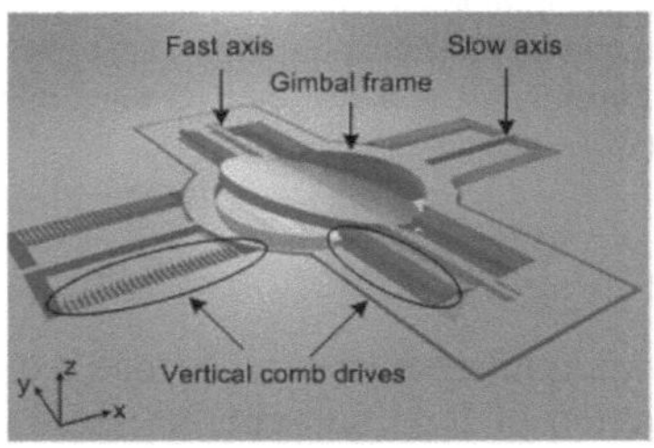

Figure 1. Simplified diagram of the micromirror.

2. Circuit description

The high-voltage driver is an inverting fully-differential amplifier (Fig. 2) [3–5]. The input stage of the operational amplifier (OPA1) is 1.8 V (V_{DD_low}) compliant. The output stage, connected to the two fixed fingers of one axis, must supply a sine wave with a peak-to-peak value that ranges from 15 V to 20 V. The ratio R_2/R_1 has been fixed to 16 so that the IPs coming before the high-voltage driver do not have to be rail-to-rail. The supply voltage (V_{DD_high}) of the output stage is 25 V in order to fulfill the high voltage requirements. This stage is divided in two sections (Fig. 3): the first one (M2/M2*) is the amplifying stage, while the second one (V_K, M3/3*, M4/4*) is a class AB output stage used as voltage buffer. The buffer is necessary to prevent stability issues due to the high equivalent load capacitance of the micromirror (up to 100 pF).

M2/2* are low-voltage devices and their V_{DS} voltage cannot exceed 1.8 V, so they must be correctly protected. M1/1* are NDMOS devices used for this purpose. They allow the drain of the 1.8 V devices M2/2* staying a V_{GS} (about 800 mV) down to V_{DD_low}. In this way the exceeding voltage is kept by M1/1* that are able to sustain high V_{DS} voltages.

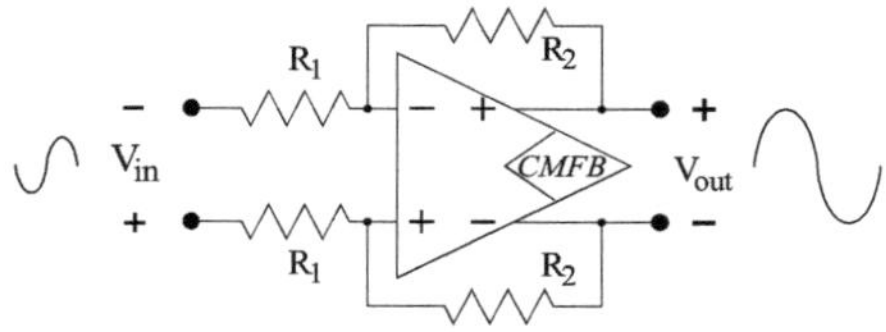

Figure 2. Simplified schematic of the high-voltage driver.

The output common mode voltage (V_{cmfb}) is fixed by a proper programmable common mode feedback circuitry (CMFB) composed by the OPA2, the R_p resistors and V_{ref} (a 1.2 V bandgap reference voltage). The loop of the CMFB is closed using the differential OPA2 that guarantees a high AC gain and performs the comparison between V_{ref} and V_{cmfb}. The output of OPA2 (V_{bias_ctrl}) is used to control the bias current of the gain stage inside OPA1 so that a variation of V_{cmfb} is compensated changing this current. Changing the value of R_{P3}, it is possible to choose four different value of V_{cmfb} between 5 and 12.6 V. The programmability of V_{cmfb} allows both driving different prototypes of the micromirror and obtaining different values of the electrostatic torque and the scan angle θ (Eq. (2)).

Another important parameter, which has been taken into account during the design of the high-voltage driver, is the Total Harmonic Distortion (THD). The THD, related to the slew rate factor, should be kept low (THD < 10^{-2}) to reduce the risk of the excitation of unwanted micromirror higher resonating modes.

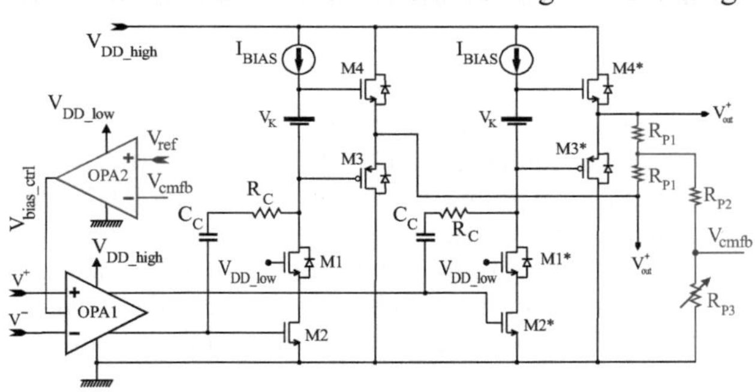

Figure 3. High-voltage driver schematic.

3. Simulation results

The circuit has been simulated with ELDO™ in the temperature range −40 ÷ 160°C and taking into account the process variations (corner). AC simulations were performed to check the stability of the high-voltage driver and of the CMFB control loop. In Fig. 4 is reported the high-voltage driver open loop gain (amplitude and phase) at 27°C versus process variations. Monte Carlo simulations have been performed to evaluate the THD (Fig. 5a) and the input offset voltage V_{io} (Fig. 5b). Several Monte Carlo simulations were performed in different conditions to verify that the system maintains a high linearity even in presence of devices mismatches. The extracted value of the third harmonic is about 80 dB below the fundamental value thus the driver has a good linearity. The THD mean value, as calculated from the Monte Carlo, is $4.74 \cdot 10^{-3}$.

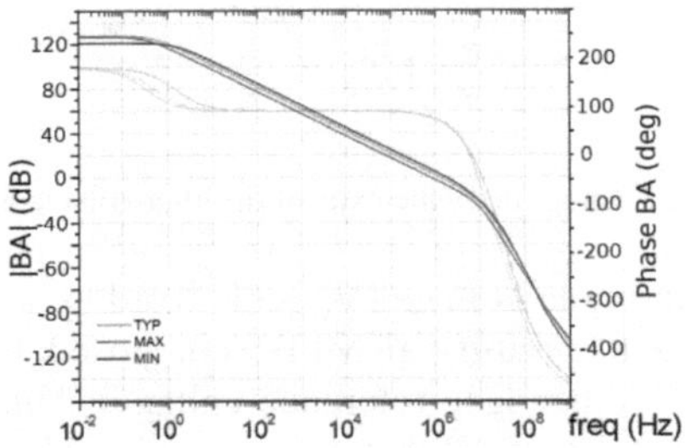

Figure 4. High-voltage driver open loop gain versus corner.

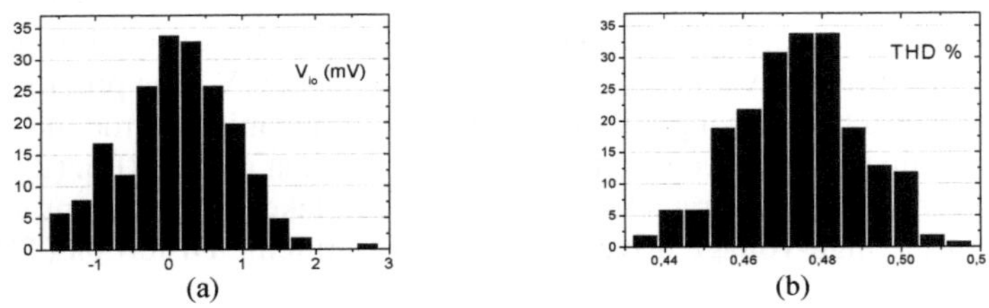

Figure 5. (**a**) THD statistical distribution; (**b**) V_{io} statistical distribution.

4. Conclusions

A high-voltage circuit has been designed for scanning micromirror driving purposes. The proposed circuit integrated in a 0.18 μm BCD technology guarantees good performances in all working conditions and a low THD fundamental to prevent the excitation of unwanted micromirror's higher resonating modes. The driver features programmability of the output common mode voltage. The circuit is also provided with a power-down circuitry that reduces the power consumption to zero (except for the leakage currents) when the block is not used.

References

1. H. Schenk, et al., "Large Deflection Micromechanical Scanning Mirrors for Linear Scans and Pattern Generation", *IEEE Journal of selected topics in quantum Electronics*, vol. 6, no. 5, Sep/Oct 2000.
2. U. Hofmann, M. Oldsen, H. Quenzer, et al., "Wafer-level vacuum packaged micro-scanning mirrors for compact laser projection displays", SPIE Photonics West 2008, San Jose, California USA, vol. 6887, pp. 100–114, 2008.
3. D. Mota, Term Paper "Fundamentals of Fully Differential Op-Amps and CMFB Circuit Design", University of Toronto Department of Electrical and Computer Engineering, Apr 9 2001.
4. G. Xu, S.H.K. Embabi, "A Systematic Approach in Constructing Fully Differential Amplifiers", *IEEE Transactions on circuits and systems—II. Analog and digital signal processing*, vol. 47, no. 12, Dec 2000.
5. L. Luh, et al., "A continuous-time common-mode feedback circuit (CMFB) for high-impedance current mode application", IEEE International Conference on Electronics, Circuits and Systems, vol. 3, pp. 347–350, 1998.

SYSTEM STUDY FOR A HEAD-UP DISPLAY BASED ON A FLEXIBLE SENSOR INTERFACE

E. VOLPI[1], F. SECHI[1], T. CECCHINI[1], F. BATTINI[1], L. BACCIARELLI[1], L. FANUCCI[1] AND M. DE MARINIS[2]

[1]*Department of Information Engineer, University of Pisa, via Caruso, 16 I-56122, Pisa*
[2]*SensorDynamics AG, via Giuntini, 63, I-56023, Navacchio (Pisa), Italy*

Abstract. This work presents the system study for an innovative and miniaturized head-up display based on a flexible sensor interface for automotive applications. After a brief introduction, in the second section the basic structure of a head-up display is described. Finally, in the third section our specific design is presented.

1. Introduction

Head-Up Displays (HUDs) are devices that allow the projection of images on semi-reflective surfaces allowing the user to receive valuable information without distracting themselves from the driving. These kinds of devices are widely used in military applications. HUDs are used in the aeronautical field to project a display in the pilot's line of sight as he looks out of the aircraft cockpit. Following the same philosophy, this concept was proposed for automotive field over 20 years ago [1] and a small number of experimental devices were developed and installed in motor cars for evaluation. Unfortunately, soon the project of an automotive HUD was stopped due to the high volume occupation of the overall system, the low performances of the sensors involved and the high manufacturing costs.

2. Head-up displays

A typical HUD structure is composed by an image source, a control electronics that handles the projection system and a semi-reflecting surface where the image is projected. The projection system is composed by a laser source and one or more mirrors; all these devices must be driven accordingly to the image from the image source.

Nowadays advances in technology have made it possible to realize cost-effective, high reliability and high performance sensors and systems such as the Micro-Opto-Electro-Mechanical Systems (MOEMS). MOEMS are employed to evaluate a new generation of low-cost HUD for automotive environment. In this way, using for example a single micromirror and three laser sources (Red, Green

P. Malcovati et al. (eds.), *Sensors and Microsystems: AISEM 2009 Proceedings*, Lecture Notes in Electrical Engineering 54, DOI 10.1007/978-90-481-3606-3_84,

and Blue), it is possible to realize a complete and high-quality image projection systems.

3. Proposed architecture

There is at the State-of-art a great variety of micromirrors with different designs, actuation methodologies and resonance frequencies. Our target is a two dimensional optical scanning micromirror, electrostatically actuated with staggered comb drive. Since this kind of micromirrors needs high actuation voltages in open air (up to 100 V), a vacuum package has been experimented [2]. This solution lowers significantly the actuation voltage (20–30 V) but introduces several reflection problems. Due to the absence of a definitive model, we decide to design a platform able to handle both mirrors with and without the vacuum package.

In order to achieve high deflection angles with lower actuation voltages, each axis of the micromirror must be actuated with a sinusoidal wave whose frequency is equal (or very close) to the resonance frequency of the micromirror axis. The resonance frequencies of the mirror are fixed by its design geometry. However, driving the mirror with sinusoidal waves has some drawbacks. First, these structures are usually characterized by high quality factors (Q) (in many cases up to tens of thousands), so the frequency of the driving signal must be locked with a proper control loop (closed loop driving of the micromirror).

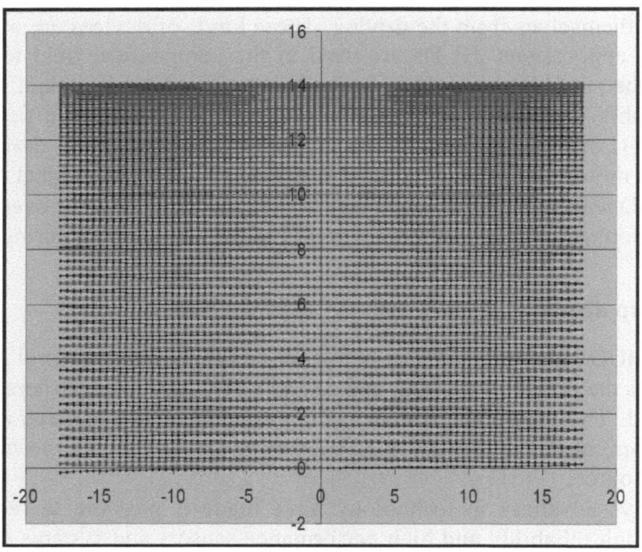

Figure 1. Although the beam is sweeping the screen in a sinusoidal way, it is possible to have the pixel correctly displayed in columns if a variable laser sampling frequency is used. x and y axes represent the position of the pixel on the screen.

Second, the laser beam does not sweep the screen with a raster scheme but follows a Lissajous pattern. One of the most evident consequences is the higher density of pixels in peripheral area where the punctual speed of the sinusoidal driving wave is lower. Therefore, while the usual projection systems use a fixed-time pixel projection, we need to realize a system able to handle a fixed-space pixel projection (Fig. 1). From this point of view, in order to calculate the laser sampling frequency, it becomes crucial to have the information about the position of the beam. The position of the beam can be determined exploiting a closed control loop that senses the position of the micromirror (e.g. evaluating its equivalent capacitance).

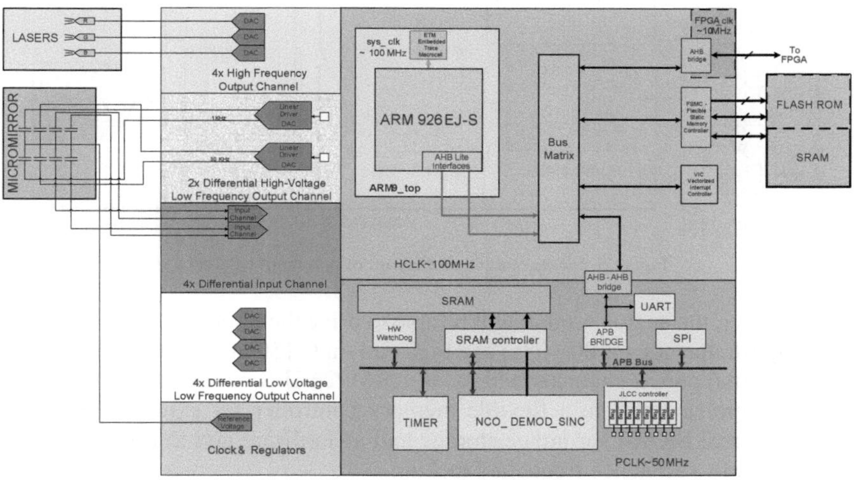

Figure 2. Block diagram of the system: the different inputs and driving channels are shown on the left; the digital section is shown on the right. The digital section is divided in an ARM and AHB section that work at higher frequency (~100 MHz) and the DSP section that works at lower frequency (25–50 MHz).

Another problem of a Lissajous projection scan is that to cover the screen completely with interlaced lines, only certain f_x and f_y ratios are allowed, with f_x and f_y as close as possible to the resonance frequencies of the micromirror.

The proposed architecture (Fig. 2) is designed following the Intelligent Sensor Interface paradigm proposed in [3]. The system aim is to be able to interface to a great number of MEMS sensors, with a specified focus over MOEMS sensors. So basically, it has to provide a great flexibility in the sensing channels and in the driving outputs. The system is a mixed signal ASIC, with a programmable analog section able to interface different kinds of MOEMS and a powerful digital section based on a ARM9 family microprocessor needed to implement the two closed control loops (one for each axis) and to manage the image projection algorithm (laser driving and evaluation of laser sampling frequency). The analog section will be able to sense and drive the micromirror by means of two high-voltage differential driving channels and four differential input channels in order to perform a closed loop driving. The simplified block diagram of the Differential Input Channel is reported in Fig. 3. Each channel is composed of different blocks. First an input charge amplifier is able to detect voltage, current or capacitance (thus covering the most of sensor typologies). After signal acquisition, differential amplifiers, low pass filters and level shifter provide proper analog conditioning with a high degree of configurability. Then the signal properly filtered and adjusted in gain and dynamic is converted by the Sigma Delta (or SAR) ADCs. Additional analog blocks provide voltage/current references, oscillators for clock generation and DACs converters.

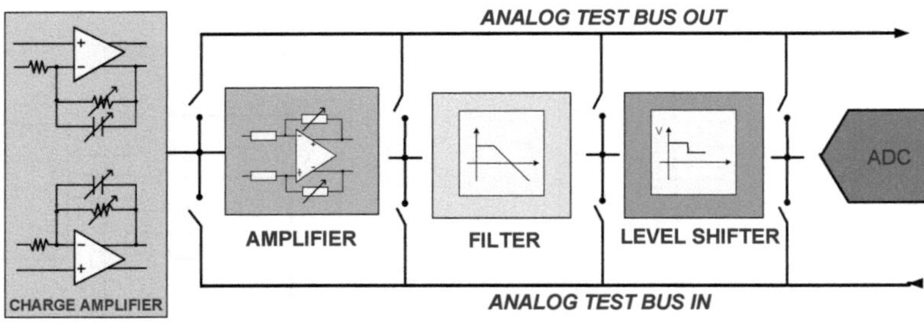

Figure 3. Simplified block diagram of one input channel.

Moreover, the analog section will be able to drive the laser sources by means of three high-frequency (~100 MHz) and high current (~150 mA) driving channels.

The digital part is characterized by an ARM926EJ-S on board, a native AHB Lite bus implementation, a flexible memory controller for external ROM and RAM, internal SRAM for image storage and a Hardware and Software DSP for mirror x and y resonance frequency locking.

4. Conclusions

In this paper we presented a cost-effective, high reliability and high performances MOEMS sensor interface that will be used to realize a projection system based on a single micromirror and three laser sources. The system is a mixed signal ASIC based on an ARM926EJ-S on-board processor used to manage the image projection algorithm and to control the analog section that includes two high-voltage differential channels for a closed loop driving of the micromirror and three high-current drivers for the laser source driving.

References

1. R. W. Evans, A. P. Ramsbottom, D. W. Sheel, "Head-up displays in motor cars", Second International Conference on Holographic Systems, Components and Applications, pp. 56-62, 11-13 Sep 1989.
2. U. Hofmann, M. Oldsen, H. Quenzer, J. Jane, M. Heller, M. Weiss, G. Fakas, L. Ratzmann, E. Marchetti, F. D'Ascoli, M. Melani, L. Bacciarelli, E. Volpi, F. Battini, L. Mostardini, F. Sechi, M. De Marinis, B. Wagner, "Wafer-level vacuum packaged micro-scanning mirrors for compact laser projection displays", SPIE Photonics West 2008, San Jose, CA, USA, vol. 6887, pp. 100-114, 2008.
3. M. De Marinis, L. Fanucci, A. Giambastiani, et al., "Sensor platform design for automotive applications", Euromicro Symposium on Digital System Design, 2003.

CAPACITIVE SENSOR SYSTEM FOR INVESTIGATION OF TWO-PHASE FLOW IN PIPES

M. DEMORI[1], V. FERRARI[1] AND D. STRAZZA[2]
[1]*Dipartimento di Elettronica per l'Automazione, Università degli Studi di Brescia, Via Branze 38, 25123 Brescia, Italy*
[2]*Dipartimento di Ingegneria Meccanica e Industriale, Università degli Studi di Brescia, Via Branze 38, 25123 Brescia, Italy*

Abstract. A sensor system for the study of a two-phase oil-water flow is proposed in particular for the estimation of the area fraction occupied by each fluid in a given section of the pipe. The system is based on a capacitive technique that takes advantage of the different electrical proprieties of the interested fluids. Capacitance measurements are made between two electrodes mounted on the external surface of a nonconductive section of the pipe. The system was designed and carefully dimensioned, also using electromagnetic simulation, to allow the application of the technique also in presence of conductive tap water used in a hydraulic plant. The system was experimentally tested with different flow patterns (geometrical arrangements of the fluids within the pipe cross section) and fluid fractions. Finally the results obtained in the fluid fraction estimation were compared to the values obtained with reference methods, demonstrating a good agreement.

1. Introduction

Liquid-liquid two-phase flows are characterized by the contemporary flux of two immiscible liquids in the same duct or pipe, which gives rise to different flow patterns [1]. This kind of flow is commonly found in many industrial applications, such as in oil industry or chemical plants.

In this work we propose the use of a capacitive technique for the investigation of liquid-liquid two-phase flows. The capacitive technique is based on the different electric characteristic of the two phases that, being located in various configurations and amounts inside the pipe, generate different dielectric permittivity distributions. These variations can be measured from a pair of electrodes placed on the external surface of a nonconductive section of the pipe to form a capacitor. The technique has the advantage to be nonintrusive hence it does not disturb the flow regime, therefore it is considered an optimal technique for the study and characterization of this kind of flows. Capacitive sensors have been developed with good results in gas-liquid and liquid-liquid flows for obtaining void fraction measurements or tomographic reconstruction [2]. Typically, the reported systems are applied to nonconductive liquids. In this work we present a system for the characterization of oil-water flows and, in particular, for estimating the relative fraction of the

P. Malcovati et al. (eds.), *Sensors and Microsystems: AISEM 2009 Proceedings*,
Lecture Notes in Electrical Engineering 54, DOI 10.1007/978-90-481-3606-3_85,

two phases in the pipe. Water and oil have a large difference in their dielectric constant, 80 versus 2.7, that should allow a good applicability of the capacitive technique. On the other hand, the high electrical conductivity of the tap water (~0.03 S/m), used in our plant, poses significant challenges and demands for innovative design solutions versus previous works [3].

2. System description

The developed measurement system is divided into three principal parts as shown in Fig. 1: the measurement head made by the capacitive sensor configuration, the interface and signal conditioning electronic circuit, and the measurement data acquisition and visualization system. Electrodes of the measurement head are directly placed facing each other on the external surface of the pipe. Each electrode is divided in guard sections and measurement section along the pipe. Guard sections are placed before and after the measurement section in order to minimize the perturbations on the measurement zone coupled from the outer portions of the pipeline due to the liquid conductivity. The interface electronic circuit excites the sensor capacitance *C(t)* with a sinusoidal voltage V_P at frequency f_P and measures the corresponding current by a transimpedance amplifier with gain determined by the reference capacitor C_F. The output voltage V_O of the transimpedance amplifier is a sinusoid at f_P with an amplitude modulated by *C(t)*. The information is carried back to DC by a demodulation stage where an operation of multiplication by V_P followed by low-pass filtering and amplification is performed. In this way the output signal V_{meas} is proportional to *C(t)*. This proportionality is controlled by design parameters and represents the system sensitivity for capacitance variations. Finally the signal V_{meas} is sampled and stored by means of a 16-bit data acquisition system connected to a personal computer. A dedicated program written in LabVIEW allows the visualization of the time behaviour of the measured capacitance in order to enable the estimation of relative fluid fractions and their fluctuations.

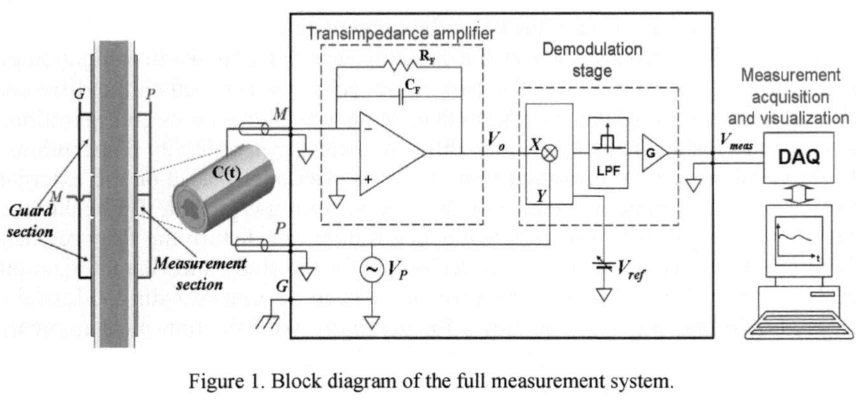

Figure 1. Block diagram of the full measurement system.

The developed measurement system has been mounted on an experimental plant that simulates an oil duct for the study of oil-water flows. The plant is realized with a Plexiglas tube with inner and outer diameters of 21 and 25 mm, respectively. Tap water has been used. The optimal system configuration has been determined by finite-element (FE) electromagnetic simulation and experimental tests, leading to: (1) an angle of 90° covered by each electrode for ensuring a good sensitivity to the phenomena in the centre of the tube; (2) a measurement section length of 10 cm as the best trade off between measurable capacitance variations and spatial resolution of the measurement; (3) a sensitivity of $2V/pF$ sufficiently high to counteract the attenuation imposed by the small capacitance of the pipe walls in series to the fluid capacitance inside the duct; (4) an operative excitation frequency f_P of 2 MHz sufficiently high to reduce the conductive influence of the water.

3. Experimental results

The performance of the system has been experimentally tested on various flow configurations realized in the duct. The measured capacitance behaviour *C(t)* is not evaluated as an absolute value but as a Relative Capacitance Difference (RCD) defined in Eq. (1). The dimensionless quantity RCD represents the relative variations of the measured value from the extreme conditions obtained with the pipe filled of either oil (C_{oil}, V_{oil}) or water (C_{water}, V_{water}).

$$RCD = \frac{C_{meas} - C_{oil}}{C_{water} - C_{oil}} \qquad RCD = 1 - \frac{V_{meas} - V_{water}}{V_{oil} - V_{water}} \tag{1}$$

To estimate the relative fraction of the two fluids, a capacitance-fluid fraction calibration relationship has been obtained by modellization and FE electromagnetic simulation with the fluids placed in the relevant configurations [4]. As an example, Fig. 2 shows the RCD expected trend versus oil fraction for the annular flow configuration, where water is located in a ring and oil is in the centre under the condition of centred cylindrical oil core.

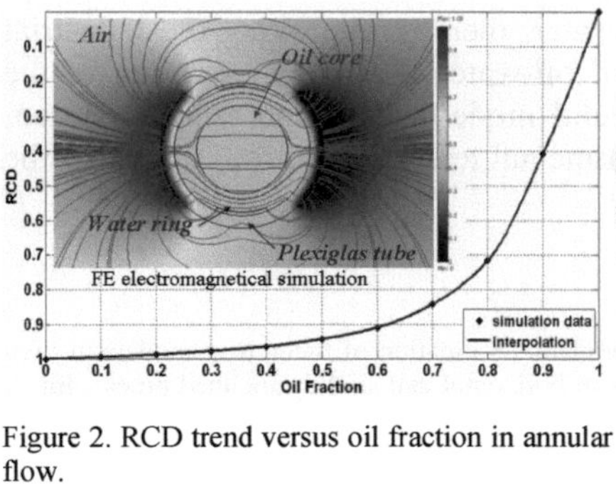

Figure 2. RCD trend versus oil fraction in annular flow.

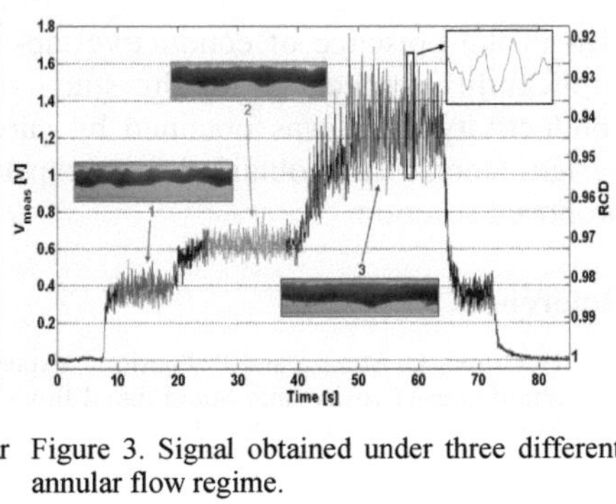

Figure 3. Signal obtained under three different annular flow regime.

The measured output of the system is illustrated in Fig. 3, which plots the signal V_{meas} measured during a sequence of three subsequent flows in annular configuration corresponding to three different increasing oil fractions. As expected, the system detects both the different oil quantities, by providing a different mean value of V_{meas}, and the flow fluctuations by oscillations around this value.

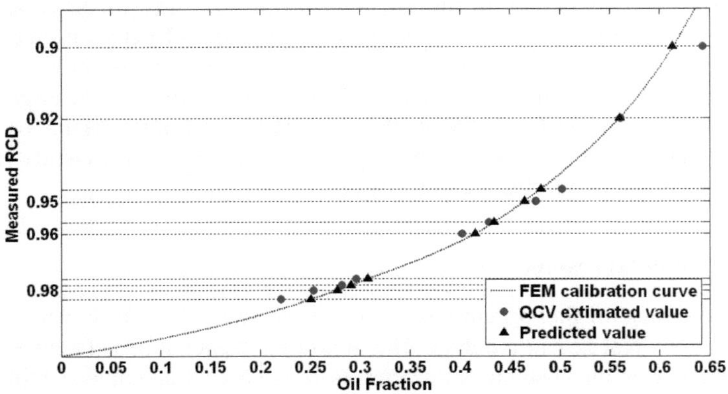

Figure 4. Comparison between sensor oil fraction prediction (by calibration curve) and QCV estimation technique.

The validity of the fluid fraction estimation was verified by a comparison of the results obtained from the system with the values obtained by an intrusive reference technique called QCV (Quick Closing Valve). The comparison results for the annular flow are shown in Fig. 4 where for each value of RCD, related to a given flux, the value estimated by the measurement system (black triangles) and the value obtained by the QCV technique (red dots) are reported.

4. Conclusions

The presented system for the analysis of two-phase flows addresses the operation also in the presence of conductive tap-water, thereby improving the industrial applicability, especially for the study of oil-water flows. The solution to the conductivity issue was obtained by careful modellization and tailored system design. Good results obtained from experimental tests confirm the validity of the proposed solution.

References

1. B. Grassi, D. Strazza and P. Poesio , "Experimental validation of theoretical models in two-phase high-viscosity ratio liquid-liquid flows in horizontal and slightly inclined pipes", Int. J. Multiphase Flow 34 (2008) 950–965.

2. N. Reinecke and D. Mewes, "Recent developments and industrial/research applications of capacitance tomography", Meas. Sci. Technol. 7 (1996) 233–246.
3. A. J. Jaworski and G. T. Bolton, "The design of an electrical capacitance tomography sensor for use with media of high dielectric permittivity", Meas. Sci. Technol. 11 (2000) 743–757.
4. J. J. M. Geraets and J. C. Borst, "A capacitance sensor for two-phase void fraction measurement and flow pattern identification", Int. J. Mutiphase Flow (1988) Vol. 14, No. 3, pp. 305–320.

SURFACE PLASMON RESONANCE IMAGING FOR AFFINITY-BASED BIOSENSORS

S. SCARANO, C. SCUFFI, M. MASCINI AND M. MINUNNI*

Dipartimento di Chimica, Università di Firenze, Polo Scientifico, Via della Lastruccia 3, 50019 Sesto Fiorentino (FI), e-mail: maria.minunni@unifi.it

Abstract. In the last 2 decades, Surface Plasmon Resonance (SPR) sensing has played a relevant role in many biochemical and biotechnological fields, and various kinds of biorecognition elements are currently employed in a wide range of fields such as food processing, environmental monitoring and clinical/diagnostics analyses. At present, SPR-based sensing represents one of the main optical biosensor technologies. The most attractive and powerful advancement of SPR-based optical detection is the SPR imaging (SPRi) technique (also termed "SPR microscopy"), which couples the sensitivity of scanning angle SPR measurements with the spatial capabilities of imaging. In recent years SPRi has gained attention mainly in affinity-biosensors research field and actually represents a promising sensing platform for the probing of biomolecules in array format, showing to be a highly versatile system. In this work we will report about the development of a multi-array SPRi affinity sensor. In particular the system has been studied using antibody-antigen interactions food and/or clinical controls. The analytical parameters of the systems will be discussed.

1. Introduction

SPR biosensors belong to the class of refractometric sensing devices and use evanescent waves as a valuable tool to investigate chemical and biological interactions. Since the output signal consists in a refractive index change directly induced by the binding of analyte molecule onto sensor surface, the labeling of biorecognition elements is not required. This feature joined the growing requirement of new non-invasive and continuous real time monitoring methods to be exploited in "-omics" research field, e.g. proteomic functional analyses [1]. Moreover, SPR-based biosensors offer several important features such as high sensitivity, insensitivity to electromagnetic interferences, and the requirement of none electric signal in the sensing area. Finally, SPR biosensors showed to be user friendly, versatile, and low-cost devices. The most attractive and powerful advancement of SPR-based optical detection is the SPR imaging (SPRi) technique (also termed "SPR microscopy"), which couples the sensitivity of scanning angle SPR measurements with the spatial capabilities of imaging. The two main breakthroughs offered by this innovative approach to SPR sensing are essentially the ability to visualize the entire biochip surface in real time and the chance to

P. Malcovati et al. (eds.), *Sensors and Microsystems: AISEM 2009 Proceedings*, Lecture Notes in Electrical Engineering 54, DOI 10.1007/978-90-481-3606-3_86,

design onto them two-dimensionally patterned microarrays of molecular probes. The first key feature, the real time monitoring of biochip surface, allows the continuous checking of size and shape of immobilized spot arrays, without labelling. This aspect is primarily crucial during the early phases of the measurement, in which is required to define position and size of the signals to be probed, also called regions of interest (ROIs). In fact, the "microscopic" vision of each ROI within the array allows an accurate and skilful choice of the best ROIs after that molecular probes have been deposited onto the biochip. Thus, this chance avoids the "blind" sampling of the signals and optimizes the results.

An additional outcome of the real time vision is the capability to monitor the unspecific binding all over the biochip surface. In fact, it is well known that non-specific adsorption represents a considerable disadvantage in SPR-based methods, since can lead to false signals and biosensor poisoning.

Shortly after its introduction [2], SPRi was first used in a biomolecular application for the imaging of phospholipid monolayer films [3]. Since these initial efforts, SPRi has also been used for surface morphological investigations of many surface systems, including self-assembled monolayer films [4], mono- and multilayer films prepared by Langmuir Blodgett techniques [5], and multilayer films built by alternate polyelectrolyte deposition [6]. At present, SPRi represents a promising optical technology for biosensor development in the field of biomolecules arrays on chips. In particular nucleic acid sequences (DNA, RNA and aptamers), proteins and short peptides, and carbohydrates have been tested for microarray development by SPRi technique. Here we report the development of a SPRi affinity-based sensor for the detection of anti-bovine Ig Gs in buffer and whole milk.

2. Experimental

2.1. *A SPRi immunosensor for anti-bovine IgGs detection in milk*

In order to anchor bovine IgGs to the gold layer of the SPRi chip, a dedicated procedure was developed. Figure 1a represents the main step of the procedure that first consists in the creation of a thin PDMS layer carrying 1-mm^2 holes in which the probes will be deposited. After that the mask is placed onto the gold surface, a self-assembled monolayer is created by adding a 1 mM aqueous solution of 11-mercaptoundecanoic acid in the wells. After treatment with NHS/EDAC solution, bovine Ig Gs were added to wells for the immobilization by amino-coupling chemistry. The last step of the procedure consists in the removal of the PDMS layer and the subsequent saturation of the gold surface with a mix of thiols. Figure 1b shows the final array of probes on the biochip.

The calibration curve for anti-bovine IgGs was carried out in buffer HEPES 10 mM pH 7.4 with 0.1% Tween20 (Fig. 2). The concentration range for the target was 0.05–10 ppm. To assess the sensor specificity, negative controls, i.e. anti-rabbit and anti-human IgGs, were tested. Moreover, the unspotted area (one well treated as others except for the probe addition) and the area surrounding the spots (saturated with thiols) were constantly monitored during the analyses to control that no unspecific binding was occurring. The main analytical parameters were determined.

The sensor was then tested in complex matrix, i.e. bovine whole milk. The results are showed in Fig. 3.

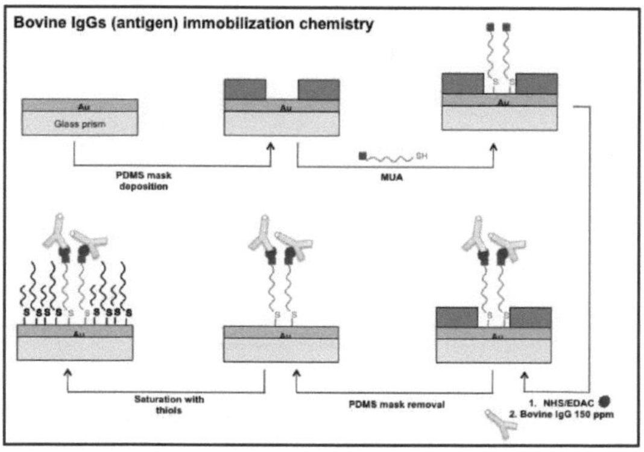

Figure 1. Immobilization scheme for bovine Ig Gs.

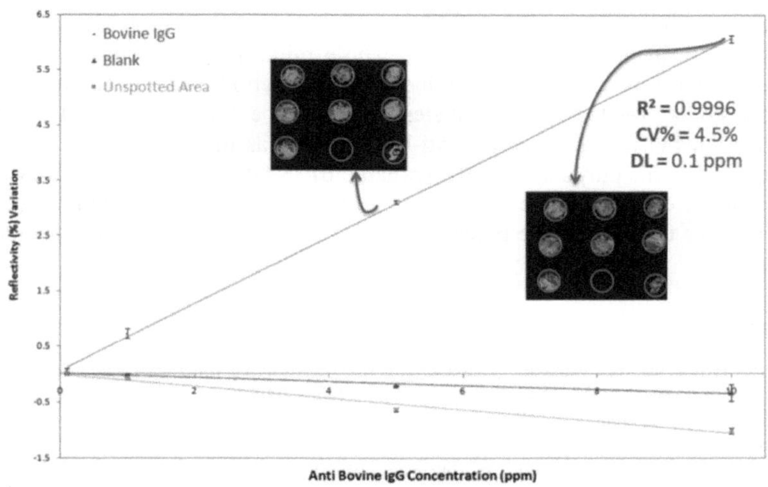

Figure 2. Calibration curve for anti-bovine IgGs in buffer. Images are relative to the biochip surface during measurements. Colored spots are relative to ROIs.

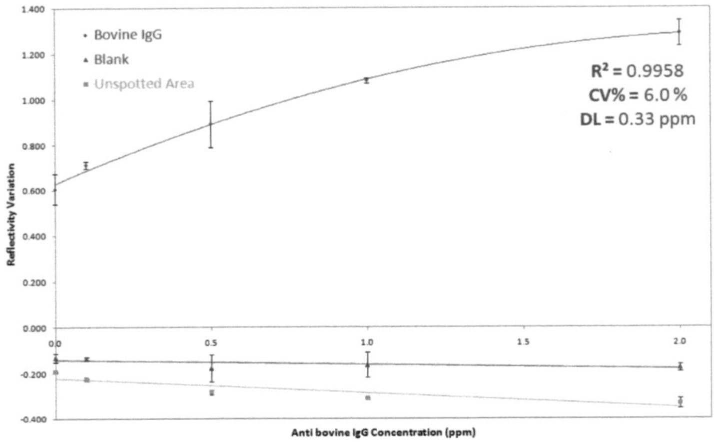

Figure 3. Calibration curve for anti-bovine IgGs in bovine milk.

3. Conclusions

These results demonstrate the applicability of SPRi to protein/protein interactions, even directly in raw matrices, such as whole milk. The detection of immuno-responses, such as specific anti-bovine IgGs production, has gained attention in clinical studies related to the diabetes mellitus type 1 (T1D) since it is under clarification the role of specific anti-bovine IgGs as markers of risk for T1D. Currently, for this purpose are only available ELISA-type tests. Further development will be directed to the analysis of suitable matrices such as human blood/ sera samples to evaluate the potential application of SPRi technique for clinical diagnostic.

References

1. Lee, H.J., Yan, Y., Marriott, G. & Corn, R.M., Quantitative functional analysis of protein complexes on surfaces. J Physiol, [vol. 563], pp. 61–71, 2005.
2. Yeatman, E. & Ash, E.A., Surface plasmon microscopy, Electronics Letters, [vol. 23], pp. 1091–1092, 1987.
3. Hickel, W., Kamp, D. & Knoll, W., Surface-plasmon microscopy, Nature, [vol. 339], pp. 186, 1989.
4. Evans, S.D. & Flynn, T.M., A self-assembled multilayer formation on predefined templates, Langmuir, [vol. 11], pp. 3811–3814, 1995.
5. Duschl, C., Liley, M., Lang, H., Ghandi, A. & Zakeeruddin, S.M., Sulphur-bearing lipids for the covalent attachment of supported lipid bilayers to gold surfaces: a detailed characterisation and analysis, Materials Science and Engineering: C, [vol. 4], pp. 7–18, 1996.
6. Nelson, B.P., Frutos, A.G., Brockman, J.M. & Corn, R.M., Near-infrared surface plasmon resonance measurements of ultrathin films. 1. Angle shift and SPR imaging experiments, Anal Chem, [vol. 71], pp. 3928–3934, 1999.

LASER BASED SCANNING SYSTEM FOR MONITORING ICE ACCRETION PHENOMENA ON HIGH VOLTAGE CONDUCTORS

E. GOLINELLI, U. PERINI, S. MUSAZZI AND G. PIROVANO
CESI RICERCA, Via Rubattino 54, 20134 Milano, Italy

Abstract. This paper will describe an innovative optical system properly designed for monitoring the ice accretion phenomena that, under critical weather conditions, may occur on the conductors of the high voltage aerial lines. The principle of operation of the instrument as well as preliminary experimental results obtained during laboratory tests will be presented.

1. Introduction

Ice accretion on overhead high voltage conductors is a critical phenomenon that sometimes occurs under peculiar weather conditions. Since the presence of a substantial amount of ice on the conductors may bring to dangerous situations (like e.g. broken lines and toppled towers), which consequently cause huge economic losses and operational difficulties in power industry, it would be of great importance to find a way to characterize this phenomenon and to identify typical weather conditions that make it happen. For this purpose we have developed a laser based scanning system capable of monitoring from a remote position both the ice accretion rate and the conductors sag.

2. Principle of operation

The principle of operation of this instrument relies on the detection of the intensity backscattered by the conductor under inspection once it is intercepted by a properly driven scanning laser beam (see Fig. 1).

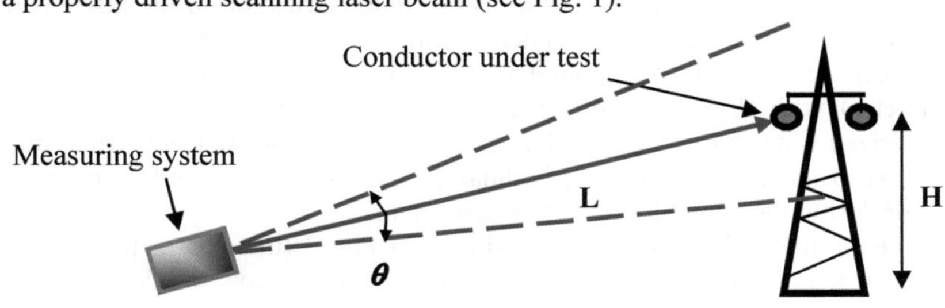

Figure 1. Schematic description of the laser based scanning system.

P. Malcovati et al. (eds.), *Sensors and Microsystems: AISEM 2009 Proceedings*,
Lecture Notes in Electrical Engineering 54, DOI 10.1007/978-90-481-3606-3_87,

Measurements are performed by positioning the measuring unit (shown in Fig. 2) on ground at a distance of approximately 20 m from the electric line. The unit is oriented in such a way the scanning beam oscillates in a plane perpendicular to the conductors. As reported in previous papers [1, 2] the wire diameter (and, in case of ice formation, the outer diameter of the ice cylinder) can be inferred from the time duration of the backscattered signal, while the conductor sag can be derived from the time delay between the backscattered signal and the scanner driving signal.

Figure 2. The measuring unit.

3. Experimental activity

The aim of the experimental activity is the characterization of the instrument in the presence of an ice accretion phenomenon. The ice accretion on a conductor has been simulated in our laboratory by filling a metallic pipe (30 mm outer diameter) with dry ice (sublimation temperature −78.5°C) and by spraying pulverized water on its surface. To get a radially symmetric ice distribution, the pipe was rotated around its longitudinal axis at a constant speed by means of a motor during the whole ice accretion procedure.

The simulated iced conductor is shown in Fig. 3 (the central portion of the accreted ice has been removed after measurements to show the thickness of the ice layer).

Figure 3. The experimental apparatus utilized to simulate the ice accretion phenomenon.

Preliminary measurements have been carried out by positioning the instrument at a distance of about 20 m from the simulated iced conductor. An example of these measurements is shown in Fig. 4. The two plots show the signals (amplitude vs. recording time) detected in the absence and in the presence of the ice layer, respectively (which correspondingly means 30 mm and ~60 mm outer diameter). The highest peaks represent the contribution of the light backscattered by the iced conductor, while secondary peaks are due to reflections from the background. As it can be noticed, the width of the highest peaks in the two plots is proportional to the conductor diameter (with and without ice, respectively). In particular, by taking into account that the width of the peak relevant to the iced conductor (lower plot) is 0.05 s, which corresponds to 2,000 samples of the recorded signal, it comes out that assuming a detection accuracy of about 1% (i.e. ± 20 samples) the ice accretion rate can be estimated with an accuracy of about 0.5 mm.

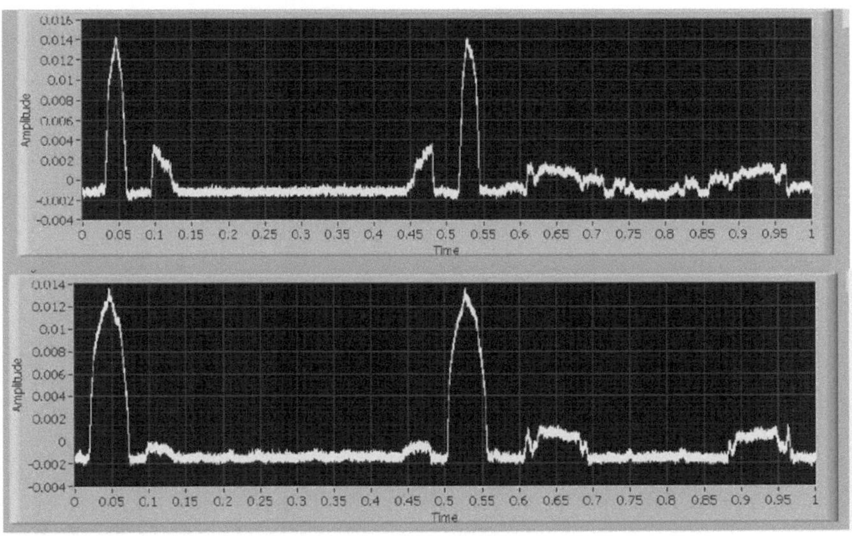

Figure 4. Backscattered signals generated by a 30 mm diameter conductor (upper plot) and by a 60 mm diameter iced conductor (lower plot) positioned 20 m far from the measuring system.

4. Conclusions

We have described a laser based scanning system properly developed for the detection of the ice layer that, under peculiar weather conditions, may growth on the conductors of the high voltage aerial lines. Tests carried out in our laboratory show that by positioning the measuring system at a distance of 20 m from the iced conductor, it is possible to estimate the ice accretion rate with an accuracy of about 0.5 mm.

Acknowledgment This work has been financed by the Research Fund for the Italian Electrical System under the Contract Agreement between CESI RICERCA and the Ministry of Economic Development – General Directorate for Energy and Mining Resources stipulated on June 21, 2007 in compliance with the Decree n.73 of June 18, 2007.

References

1. E. Golinelli, S. Musazzi, U. Perini, G. Pirovano: "Laser based scanning system for high voltage power lines conductors monitoring" Atti della XIII Conferenza Nazionale AISEM "Sensori e Microsistemi", Roma, 19–21 Febbraio 2008.
2. E. Golinelli, U. Perini, S. Musazzi, G. Pirovano: "Dispositivo laser per il monitoraggio della freccia e della formazione di manicotti di ghiaccio sulle linee AT" Atti del 10° Convegno Nazionale "Strumentazione e metodi di misura elettroottici", Milano, 10–12 Giugno 2008.

CAPACITIVE PROXIMITY SENSOR FOR CHAINSAW SAFETY

M. NORGIA AND C. SVELTO
Dipartimento di Elettronica e Informazione, Politecnico di Milano, Milano, Italy

Abstract. This work describes a customized capacitive sensor for the safety of a chainsaw. The sensor reveals the closeness of a dangerous object, in general a hand tool such as a chainsaw, to the operator body. The operator has to dress a garment with a wire cloth inserted, connected to a radio-frequency generator. The hand tool is provided with an electronic circuit which detects the radio frequency, and turn off the power in the case of proximity to the wire cloth.

1. Introduction

The aim of this work is the realization of a novel safety system, acting before the accidental contact between a dangerous part of a hand tool and the body of the operator. The security system has to detect the closeness of the blade with respect to a wire cloth, on purpose inserted in the wearable security garments. Several application areas are present, since in a number of working fields edge tools are used by human operators. Normally the operator has to dress safety garments (passive protective clothing) equipped with high mechanical resistance pads in the areas most exposed to danger. Moreover, when feasible, the tools themselves are provided with protective screens, in order to prevent the contact between the operator and the dangerous mechanical part. It is worth to note that these techniques operate in a passive way, indeed the protections are studied for limiting the physical harms of an accidental contact, but they are not able to control the danger situation. In other words, screens and protective clothing cannot stop or disable the lee shore, but only can limit the damages due to the contact.

2. Principle of operation

The proposed safety system has to detect the distance between the danger tool and the operator body, in order to stop the tool engine or to activate a rapid-stop system, in the case of dangerous proximity. The principle of operation for the proximity sensor is the measurement of an RF signal amplitude, as received by the danger tool, when the signal source is applied to a wire cloth inserted in the wearable security garments. This technique is a sort of capacitive measurement [1–3] of the distance between the wire cloth and the dangerous tool. The electric

P. Malcovati et al. (eds.), *Sensors and Microsystems: AISEM 2009 Proceedings*, Lecture Notes in Electrical Engineering 54, DOI 10.1007/978-90-481-3606-3_88,

scheme of the sensor is described in Fig. 1: the signal source is connected with the wire cloth (acting as the first capacitor wall) and the receiver is connected with the tool (second capacitor wall). The whole system operates in near field conditions, therefore there is no energy transmission and the electric fields are only reactive. In this condition the signal amplitude received by the tool depends on the geometry of the system, and especially on the distance *d* between the wire cloth and the tool itself.

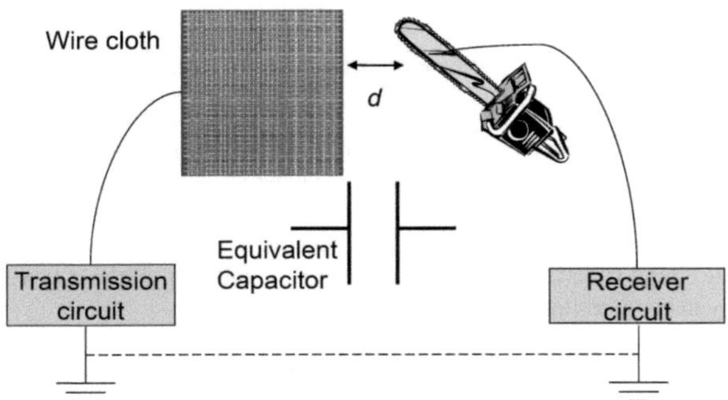

Figure 1. Experimental scheme of the proximity sensor.

The emergency signal switches on when the amplitude of the received RF signal overcomes a threshold level, whose value is defined depending on the particular application. For the case of the chainsaw the signal threshold value is fixed in correspondence to a few centimeters of minimum stop distance (usually $d \sim 10$ cm). The major requirements for this kind of security system are reliability and repeatability, but also the low cost of the sensor, with respect to the price of the chainsaw, is an important argument: the whole cost of the electronics has to be kept down to a few dollars.

3. Realized electronics

The transmission circuit, connected to the wire cloth, is a 80 kHz Wien bridge oscillator, made with a single low-power operational amplifier. The frequency of operation was chosen in order to maximize the system sensitivity, using low-cost and low-power electronics for the receiver. The receiver detection circuit has to measure the signal level, directly depending on the distance *d*, and it has to activate a stop system when the signal level is higher then a set threshold. The signal measurement is carried out directly in amplitude, by means of a rectified mean value detector, after ad hoc active filter stages useful for reducing disturbances. It is worth to note that in the near-field region (distances well below the electromagnetic wavelength $\lambda = c/f \cong 30$ km in our case) it is essential to close the circuit by means of a ground connection (dashed line in Fig. 1), in order to warrant the reliability requested by the system. This connection however could be capacitive too, and hence contactless, under the condition of exhibiting a capacitance value well above the value under measurement (equivalent capacitor between the wire cloth and the blade). In our prototype the ground connection is made between the metallic handle of the chainsaw and the operator glove, on purpose coated by a wire cloth. The ground of the receiver circuit is connected with the handle of the chainsaw and the ground of the transmitter is connected to the glove used by the human operator.

The overall gain of all the receiver filter stages is 10 dB at 80 kHz, and the quality factor of the band-pass resonance is 13. This value was chosen as a trade off between filtering action and need for allowing some frequency variation of the source signal (about 2% with respect to the nominal oscillator frequency).

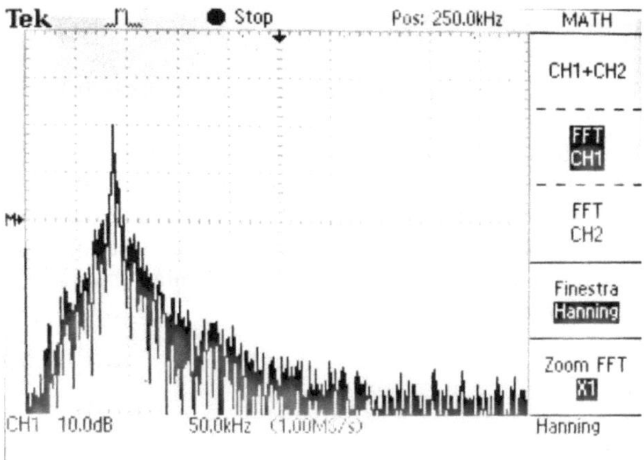

Figure 2. Measured spectrum of the signal received, after the filtering process.

4. Measurement results

Figure 2 shows the spectrum of the detected signal (0.5 V peak-to-peak, corresponding to $d \cong 10$ cm), acquired from the blade, when the electric chainsaw engine is turned on. The signal has been acquired by a digital oscilloscope with the FFT mathematical function, after the filtering stages of the receiver. The amplitude of this signal is measured by means of a rectified mean value detector, with rise time of about 10 ms, that represents also the response time of the whole system. This value of the response time is adequate for the application: considering a maximum blade speed equal to 2 m/s, in 10 ms time the distance covered is 2 cm, well below the 10 cm of blocking distance. In this way the electromotive force is certainly turned off when the blade eventually reaches the operator body. The output of the rectified mean value detector is fed to a comparator, with a threshold level set at a value of 300 mV, corresponding to a distance $d \cong 10$ cm. Finally, the output of the comparator drives a relay that disconnects the chainsaw power supply.

Figure 3 shows the Laboratory prototype of the sensorized chainsaw, at the instant of the motor-stopping for proximity to the helmet.

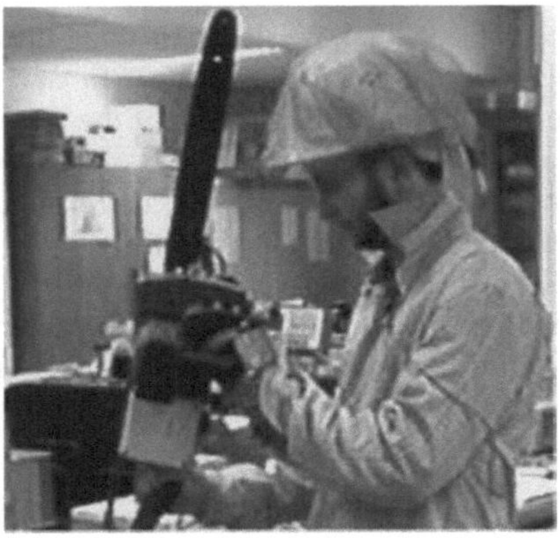

Figure 3. Photography of the sensorized chainsaw detecting a potentially harmful situation.

This safety system exhibits good repeatability and reliability: a number of trials for different positions of the chainsaw and movements of the operator have been carried out with the prototype. In all the cases the chainsaw has always stopped when the blade reached a minimum distance of about 10 cm. Moreover the circuit does not seem to interfere with the normal chainsaw operation (no false alarms were noticed).

5. Conclusions

A novel prototype of early warning system, applied to the chainsaw safety, has been designed and realized. This system switches off the chainsaw power supply when the distance between the blade and the operator body becomes less than 10 cm. This kind of RF-driven capacitive sensor can not measure distances with high accuracy, because the signal value depends also on the placement of the objects, not only on the distance, but it is really reliable as proximity sensor. The response time of the whole system has been set to 10 ms, low enough for tuning off the chainsaw before it can cause relevant damages to the operator. Recently the system has been realized also with a low-cost commercial charge-transfer sensor, that exhibits results similar to the ones obtained with our electronics. The working principle of the safety sensor has been patented [4] and an industrial evolution of the system is now under study for a final market product, probably available in the next year.

References

1. I. Sinclair, "Sensors and Transducers," Th. Ed., Newnes, Woburn, 2001.
2. J. Fraden, "Handbook of Modern Sensors," Sec. Ed., Springer-Verlag, New York, 2001.
3. J. G. Webster, "The Measurement, Instrumentation, and Sensors Handbook," CRC Press, Boca Raton, 1999.
4. M. Norgia, C. Svelto, G. A. Rossi, P. Cappellari, "Device for protection from accidents", patent n. WO/2007/060698, 2007.

AUTHOR INDEX